STRUCTURAL ANALYSIS SYSTEMS

Software — Hardware
Capability — Compatibility — Applications

Volume 1

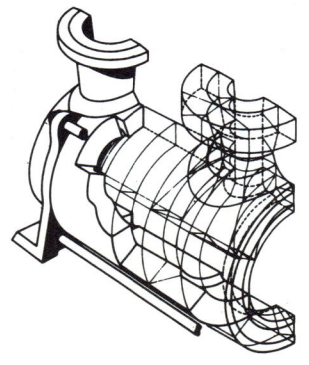

An international series of practical guidebooks
on structural analysis systems and their applications

Other Pergamon Titles of Interest

BATHE	Nonlinear Finite Element Analysis and ADINA 1983
COHN & MAIER	Engineering Plasticity by Mathematical Programming
COWAN	Predictive Methods for the Energy Conserving Design of Buildings
CROUCH	Matrix Methods Applied to Engineering Rigid Body Mechanics
GIBSON	Thin Shells
HARRISON	Structural Analysis and Design
HEARN	Mechanics of Materials, 2nd Edition
HOLLAND	Microcomputers and Their Interfacing
HORNE	Plastic Theory of Structures
JAMSHIDI & MALEK-ZAVAREI	Linear Control Systems
LEININGER	Computer Aided Design of Multivariable Technological Systems
LIVESLEY	Matrix Methods of Structural Analysis, 2nd Edition
NOOR & HOUSNER	Advances and Trends in Structural and Solid Mechanics
NOOR & McCOMB	Computational Methods in Nonlinear Structural and Solid Mechanics
PARKES	Braced Frameworks, 2nd Edition
ROZVANY	Optimal Design of Flexural Systems
SPILLERS	Automated Structural Analysis
WARBURTON	Dynamical Behaviour of Structures, 2nd Edition

Pergamon Related Journals *(Free Sample Copy Gladly Sent on Request)*

BUILDING AND ENVIRONMENT

CEMENT AND CONTRETE RESEARCH

CIVIL ENGINEERING FOR PRACTICING AND DESIGN ENGINEERS

COMPUTERS AND GRAPHICS

COMPUTERS AND INDUSTRIAL ENGINEERING

COMPUTERS AND STRUCTURES

FATIGUE AND FRACTURE OF ENGINEERING MATERIALS AND STRUCTURES

INTERNATIONAL JOURNAL OF APPLIED ENGINEERING EDUCATION

INTERNATIONAL JOURNAL OF SOLIDS AND STRUCTURES

JOURNAL OF ENGINEERING AND APPLIED SCIENCES

MATHEMATICAL MODELLING

STRUCTURAL ANALYSIS SYSTEMS

Software — Hardware
Capability — Compatibility — Applications

A. NIKU-LARI
Director, Institute for Industrial Technology Transfer
24 Rue des Mimosas, Gournay s/Marne
F93460 France

Volume 1

PERGAMON PRESS
OXFORD · NEW YORK · TORONTO · SYDNEY · FRANKFURT

U.K.	Pergamon Press Ltd., Headington Hill Hall, Oxford OX3 0BW, England
U.S.A.	Pergamon Press Inc., Maxwell House, Fairview Park, Elmsford, New York 10523, U.S.A.
CANADA	Pergamon Press Canada Ltd., Suite 104, 150 Consumers Road, Willowdale, Ontario M2J 1P9, Canada
AUSTRALIA	Pergamon Press (Aust.) Pty. Ltd., P.O. Box 544, Potts Point, N.S.W. 2011, Australia
FEDERAL REPUBLIC OF GERMANY	Pergamon Press GmbH, Hammerweg 6, D-6242 Kronberg, Federal Republic of Germany
JAPAN	Pergamon Press Ltd., 8th Floor, Matsuoka Central Building, 1-7-1 Nishishinjuku, Shinjuku-ku, Tokyo 160, Japan
BRAZIL	Pergamon Editora Ltda., Rua Eça de Queiros, 346, CEP 04011, São Paulo, Brazil
PEOPLE'S REPUBLIC OF CHINA	Pergamon Press, Qianmen Hotel, Beijing, People's Republic of China

Copyright © 1986 Pergamon Press Ltd.

All Rights Reserved. No part of this publication may be reproduced, stored in a retrieval system or transmitted in any form or by any means: electronic, electrostatic, magnetic tape, mechanical, photocopying, recording or otherwise, without permission in writing from the publishers.

First edition 1986

Library of Congress Cataloging in Publication Data
Structural analysis systems.
Includes indexes.
1. Structures, Theory of—Data processing—
Addresses, essays, lectures. I. Niku-Lari, A.
TA647.S77 1985   624 1'7'0285   85-9419

British Library Cataloguing in Publication Data
Niku-Lari, A.
Structural analysis systems: software, hardware, capability, compatibility, applications.
Vol. 1.
1. Structures, Theory of—Data processing
I. Title
624.1'71'02854   TA647
ISBN 0-08-032577-7

Cover drawing: Centrifugal pump casing.
Manufacturer: C.C.M.-Sulzer, France.
Software used: CA.ST.OR

Printed in Great Britain by A. Wheaton & Co. Ltd., Exeter

INTERNATIONAL EDITORIAL ADVISORY COMMITTEE

Dr. T. ANDERSSON, *Sweden*
Prof. J. H. ARGYRIS, *Federal Republic of Germany*
Prof. K. J. BATHE, *USA*
Prof. T. BELYTSCHKO, *USA*
Dr. M. BERNADOU, *France*
Prof. B. A. BILBY, *UK*
Dr. A. CHAUDOUET, *France*
Prof. R. D. COOK, *USA*
Dr. T. FUTAGAMI, *Japan*
Prof. GUO YOUZHONG, *People's Republic of China*
Dr. L. IMRE, *Hungary*
Prof. H. LIEBOWITZ, *USA*
Mr. J. MACKERLE, *Sweden*
Mr. W. M. MAIR, *UK*
Dr. G. A. MILIAN, *Mexico*
Dr. D. NARDINI, *Yugoslavia*
Dr. I. PACZELT, *Hungary*
Prof. G. SANDER, *Belgium*
Prof. R. P. SHAW, *USA*
Prof. M. TANAKA, *Japan*
Prof. W. N. WENDLAND, *Federal Republic of Germany*

PREFACE

Recent years have seen a rapid increase in the number of structural analysis software existing in the world market.

Most of the current software journals are based on theoretical background. They provide academics and software developers with a very useful tool. People from industry however, who are non-specialists in finite or boundary element methods, have often difficulty in finding industry-oriented documents to help them select the software and hardware most suited to their needs.

This guidebook series aims to provide the engineer with up-to-date information about structural analysis systems existing around the world.

Each paper gives detailed information about a specific software, its capability, its limitations and several practical examples from industry with computer and user cost. It also gives to the user the necessary information about postprocessor capabilities, computer-aided design connection and software compatibility with the most common computers.

Most papers published in this volume follow the same logical structure to allow interactive comparison.

Our concern is to promote international co-operation on this important subject and to contribute to a better understanding between research and industry.

I would like to thank the distinguished members of the editorial committee for their scientific and technical help which made the publication of the present volume possible.

A Niku-Lari
Editor

EDITORIAL

Structural analysis aims to construct numerical models which represent the best behaviour of the actual engineering material and component. These models are used in research for better understanding of experimental results. In industry the structural analysis models allow both the optimization of design and the prediction of failure.

The structural analysis (SA) is therefore a multidisciplinary problem which demands knowledge of several scientific and industrial disciplines such as, engineering sciences, mechanical or civil engineering, informatics, applied mathematics, computer sciences, etc.

International competition gives to industry the necessary impulse to optimise the design of parts and structures.

The engineer should save material and energy and use new and lighter materials such as composites. No longer is one allowed to over-design parts for "security reasons", and new international criteria have to be considered.

Industry needs to design sophisticated parts working in very special environments, in space, in the human body, in the sea, etc.

The compatibility of the structural analysis systems with modern micro-computers allow small and medium size companies to make use of these new technologies. New super computers help to find rapid solutions to complex industrial design problems.

The evolution of interactive graphics allow the full integration of structural analysis programs in a computer aided design and manufacturing environment. Expert systems, application of artificial intelligence and computer-aided decision making bring new developments in this field.

Structural analysis systems existing in the world market are powerful bridges between research and industry. They bring theory in direct physical contact with the industrial application.

The SA technology is in a rapid evolution. More and more new computers and powerful software appear in the market and the industry faces a new problem – that of selecting the optimum structural analysis system.

The choice of a structural analysis software is an important decision which can often exercise a significant influence over the successful development of a research, manufacture, or design project. Depending on the engineering problem and hardware available, a good choice of the computer program can be both cost and time effective. This new international guidebooks series aims to provide the engineer with the most up-to-date information about structural analysis systems currently available in the world market, and their capabilities.

Editorial

Published under the guidance of a distinguished scientific committee whose members are internationally recognised specialists of finite or boundary element methods, the series should be considered as an essential practical reference tool for the modern engineer involved in such areas as structural, mechanical, civil, nuclear, aeronautical and design engineering, computer science and software development.

Each volume gives detailed information about a wide range of selected software packages describing their purpose, capabilities and limitations and provides several practical examples of industrial applications, often supported by case studies. It also gives to the user the necessary information about postprocessor capabilities, computer-aided design integration and software compatibility with the most commonly used computers.

The guidebooks are industry-oriented and should prove indispensible in helping potential users to select the soft and hardware most suited to their needs. Each volume commences with a program description in tabular form, rapidly directing readers to the program most likely to solve their industrial problems, and concludes with a case-study index.

Main areas covered in the series:

- Finite and boundary element programs
- Finite difference and other methods
- Computer graphics
- Artificial intelligence and expert systems
- Computer-aided decision making in engineering
- Computer-aided design and manufacturing (CAD/CAM)
- Integration of structural analysis and expert systems in engineering CAD/CAM environment
- Hard and software selection
- Micro-computer applications in engineering
- New development in structural analysis software and interactive graphics
- Industrial case study
  . Mechanical engineering
  . Aeronautics and nuclear
  . Biomechanics
  . New materials (composites, plastics, etc)
  . Civil engineering (offshore, seismic, earthquake, etc).

Authors wishing to submit a paper under one of the above headings for possible publication in future volumes are invited to submit their manuscript for editorial consideration of the international scientific committee to the address below.

Dr A Niku-Lari, Director
Institute for Industrial Technology Transfer
I.I.T.T.-international
24 Rue des Mimosas
93460 Gournay-sur-Marne
FRANCE
TEL: (1) 43.05.17.19

CONTENTS

Program Description Tables	xiii
The ADINA System K.J. Bathe and G. Larsson	1
AIT: Analysis Interpretive Treatise W. Kanok-Nukulchai	25
ANSYS: Engineering Software with the Design and Analysis Answers C. Ketelaar	31
AXISYMMETRIC: Microcomputer Programs for Axisymmetric Problems in Structural and Continuum Mechanics C. T. F. Ross	41
BEASY: A Boundary Element System of Structural Analysis C. A. Brebbia	49
CA.ST.OR: Finite Elements and Boundary Elements Analysis System M. Afzali, A. Turbat, J. F. Billaud and J. Peigney	61
DAPST: A Finite Element Package for the Dynamical Analysis of Structures H. Sol, M. Van Overmeire and W. P. de Wilde	89
DEFOR: Program for Statical Analysis of Structures Composed of One-dimensional Elements V. Kolář and I. Němec	97
FLASH: An Analysis and Design Tool for Engineers D. Pfaffinger and U. Walder	103
IBA: Interactive Building Analysis F. Braga, M. Dolce, C. Fabrizi and D. Liberatore	111
KYOKAI: A User-friendly BEM.FEM Solver Y. Ohura, K. Obata and K. Onishi	121

Micro STRESS: Structural Analysis Program   133
    D. Nardini

NE-XX: A Finite Element Program System   141
    V. Kolář and I. Němec

PAFEC: The PAFEC Finite Element Analysis System   151
    P. M. Wheeler

PAID: Piping Analysis and Interactive Design with the Stand-alone Graphics Package   161
    Zs. Révész

PANDA: Interactive Program for Minimum Weight Design of Composite and Elastic-Plastic Stiffened Cylindrical Panels and Shells   171
    D. Bushnell

S and CM: Microcomputer Programs for Structural and Continuum Mechanics   203
    C. T. F. Ross

SESAM '80: A General Purpose Structural Analysis System   211
    A. Berdal

THERMAL: Microcomputer Programs for Thermal Stress Analysis   225
    C. T. F. Ross

TITUS: A General Finite Element System   231
    D. Halbronn

UCIN-GEAR: A Finite Element Computer Program for Determining Stresses in Spur Gears   249
    S-H. Chang and R. L. Huston

Survey of General Purpose Finite Element and Boundary Element Computer Programs for Structural and Solid Mechanics Applications   257
    J. Mackerle

Case Study Index   299

PROGRAM
DESCRIPTION
TABLES

Program Description Tables

	METHOD				ELEMENT LIBRARY								GEOMETRY			SEE VOL
	Finite element	Boundary element	Finite difference	Other	Truss/beams	2D membranes	Plates	Shells	Axisymmetric	3D solids	Boundary elements	Special elements	2D analysis	3D analysis	Axisymmetric	
ADINA	x				x	x	x	x	x	x			x	x	x	1
AFAG	x			x	x									x		2
AIT	x				x	x	x	x	x				x	x	x	1
ALSA	x				x	x	x	x					x	x		3
ANSYS	x				x	x	x	x	x	x	x	x	x	x	x	1
AQUADYN		x		x						x				x		2
ASE	x				x	x	x	x	x	x	x	x	x	x	x	2
AXISYMMETRIC	x							x	x			x			x	1
BEASY		x									x		x	x	x	1
BEFE	x	x			x	x	x	x		x	x	x	x	x		3
BEMFFT		x		x							x		x			2
BEWAVE		x									x	x	x	x		3
BOSOR4	x		x			x	x						x		x	2
BOSOR5	x		x			x	x						x		x	3
CASTEM	x				x	x	x	x	x	x		x	x	x	x	3
CASTOR	x	x			x	x	x	x	x	x	x	x	x	x	x	1
DAPST	x				x	x	x	x	x	x			x		x	1
DEFOR	x				x									x		1
ELASTODYNAMICS (2D)		x									x	x	x			2
ESA	x				x	x	x	x	x			x	x	x	x	2
FEMFAM	x				x	x	x	x	x	x		x	x	x	x	3
FEMPAC	x				x	x	x	x	x	x			x	x	x	3
FENRIS	x				x	x	x	x	x	x		x	x	x	x	3
FIESTA	x									x				x		3
FLASH	x				x		x	x				x		x	x	1
FLEXAN	x											x		x		3
HYBRID	x					x						x	x			2
IBA	x				x	x	x	x	x				x	x	x	1
INFESA	x					x							x			2

TABLE 1. Modelization and Type of Discretization.

Program Description Tables

	METHOD				ELEMENT LIBRARY								GEOMETRY			SEE VOL
	Finite element	Boundary element	Finite difference	Other	Truss/beams	2D membranes	Plates	Shells	Axisymmetric	3D solids	Boundary elements	Special elements	2D analysis	3D analysis	Axisymmetric	
KYOKAI	x	x				x			x	x	x	x	x	x	x	1
LASSAQ	x				x		x	x					x			3
MEF/MOSAIC	x				x	x	x	x	x	x		x	x	x	x	2
MICRO STRESS	x				x								x	x		1
MODULEF	x				x	x	x	x	x	x		x	x	x	x	3
MSRC-RB	x									x				x		3
NE XX	x				x					x				x		1
OSTIN		x								x			x		x	3
PAFEC	x	x			x	x	x	x	x	x	x	x	x	x	x	1
PAID	x				x							x		x		1
PANDA				x									x			1
PDA/PATRAN	x			x	x	x	x	x	x	x			x	x	x	2
RAPS	x				x	x	x	x	x	x		x	x	x	x	3
RCAFAG	x			x	x								x			2
REST	x				x	x	x	x	x	x			x	x	x	2
ROBOT	x						x	x	x				x		x	3
S AND CM	x				x	x		x				x	x	x		1
SAMKE	x				x	x	x	x	x	x		x	x	x	x	3
SESAM '80	x				x	x	x	x	x	x		x	x	x	x	1
SIMP	x		x	x	x	x	x	x					x	x		2
STAR 2	x				x						x	x		x		2
STDYNL	x	x			x	x	x	x			x		x			3
STRUGEN				x	x	x	x	x		x			x	x		2
SURFOPT	x					x			x				x		x	3
THERMAL	x				x	x							x	x		1
TITUS	x				x	x	x	x	x	x		x	x	x	x	1
UCIN GEAR	x					x							x			1
Y12M																2
ZERO-4	x					x		x	x	x	x	x	x	x	x	3

TABLE 1. Modelization and Type of Discretization (continued).

Program Description Tables

	MATERIAL								CAPABILITIES							SEE VOL
	Linear elastic isotropic	Linear elastic anisotropic	Elasto-plastic	Nonlinear elastic	Viscoelastic/creep	Composites	Soil	Concrete	Static analysis	Dynamic analysis	Geometric nonlinear	Buckling/postbuckling	Heat transfer	Fracture mechanics	Fluid/structure inter	
ADINA	x	x	x	x	x	x	x	x	x	x	x	x	x	x	x	1
AFAG	x								x							2
AIT	x								x		x				x	1
ALSA	x	x							x	x						3
ANSYS	x	x	x	x	x	x		x	x	x	x	x	x	x	x	1
AQUADYN										x					x	2
ASE		x							x	x						2
AXISYMMETRIC	x								x	x	x	x	x		x	1
BEASY	x		x						x				x	x		1
BEFE	x		x			x			x							3
BEMFFT	x								x							2
BEWAVE					x					x				x		3
BOSOR4	x	x			x				x	x	x	x				2
BOSOR5	x	x	x		x	x			x		x	x				3
CASTEM	x	x	x	x	x	x	x	x	x	x	x	x	x	x	x	3
CASTOR	x	x			x				x	x		x	x	x	x	1
DAPST	x	x				x			x	x						1
DEFOR	x	x							x		x					1
ELASTODYNAMICS (2D)	x									x						2
ESA	x	x		x					x	x			x			2
FEMFAM	x	x							x	x	x	x	x		x	3
FEMPAC	x	x							x	x			x			3
FENRIS	x		x	x					x	x	x	x			x	3
FIESTA	x	x							x				x	x		3
FLASH	x	x							x		x					1
FLEXAN		x							x	x					x	3
HYBRID	x								x							2
IBA	x	x			x	x			x	x			x			1
INFESA	x				x				x							2

TABLE 2. Materials and Analysis Capabilities.

Program Description Tables

	MATERIAL							CAPABILITIES							SEE VOL
	Linear elastic isotropic	Linear elastic anisotropic	Elasto-plastic	Nonlinear elastic	Viscoelastic/creep	Composites	Soil	Concrete	Static analysis	Dynamic analysis	Geometric nonlinear	Buckling/postbuckling	Heat transfer	Fracture mechanics	Fluid/structure inter	
KYOKAI	x	x				x	x		x				x			1
LASSAQ		x				x			x							3
MEF/MOSAIC	x	x		x		x			x	x	x	x	x	x		2
MICRO STRESS	x								x							1
MODULEF	x	x	x	x		x			x	x	x		x			3
MSRC-RB		x	x						x							3
NE XX	x								x							1
OSTIN				x		x			x							3
PAFEC	x	x	x	x	x	x	x		x	x	x	x	x	x	x	1
PAID	x								x	x			x		x	1
PANDA	x	x	x			x			x			x				1
PDA/PATRAN	x	x							x							2
RAPS									x	x			x	x		3
RCAFAG	x			x	x	x			x							2
REST	x									x						2
ROBOT	x	x							x							3
S AND CM	x								x	x			x			1
SAMKE	x	x				x			x			x		x		3
SESAM '80	x	x	x				x		x	x				x	x	1
SIMP	x								x	x						2
STAR 2				x		x		x	x							2
STDYNL	x			x			x		x	x		x				3
STRUGEN																2
SURFOPT	x								x							3
THERMAL	x								x							1
TITUS	x	x	x	x	x		x		x	x	x	x	x	x	x	1
UCIN GEAR	x								x							1
Y12M																2
ZERO-4	x								x	x					x	3

TABLE 2. Materials and Analysis Capabilities (continued).

Program Description Tables

	LOADING								PRE/POSTPROCESSING							SEE VOL
	Nodal/line	Pressure	Selfweight	Centrifugal	Thermal	Heat flux	Prescribed displacement	Other	Free format input	Mesh generation	Plot routines	Automatic node number	Combinations of load cs	Interactive graph	CAD interfaces	
ADINA	x	x	x	x	x	x	x	x	x	x	x	x	x	x	x	1
AFAG	x	x	x		x		x	x	x	x	x	x	x	x		2
AIT	x	x	x		x		x		x							1
ALSA	x	x	x				x	x	x	x	x	x				3
ANSYS	x	x	x	x	x	x	x	x	x	x	x	x	x	x	x	1
AQUADYN								x	x	x	x	x		x		2
ASE	x	x	x		x		x		x	x	x	x	x			2
AXISYMMETRIC	x	x					x					x				1
BEASY	x	x	x	x	x	x	x	x	x	x	x	x	x	x	x	1
BEFE	x	x	x	x	x		x	x		x	x		x	x		3
BEMFFT							x			x						2
BEWAVE		x												x	x	3
BOSOR4	x	x		x	x		x	x	x	x	x	x				2
BOSOR5		x		x			x	x	x	x	x	x				3
CASTEM	x	x	x	x	x	x	x	x	x	x	x	x	x	x		3
CASTOR	x	x	x	x	x	x	x	x	x	x	x	x	x	x	x	1
DAPST	x	x	x				x		x		x	x				1
DEFOR	x		x		x		x	x	x		x		x	x		1
ELASTODYNAMICS (2D)	x						x	x			x					2
ESA	x	x	x	x	x	x	x	x	x	x	x	x	x	x	x	2
FEMFAM	x	x	x	x	x	x	x	x	x	x	x	x	x	x		3
FEMPAC	x	x	x	x	x	x	x	x	x	x	x	x	x	x	x	3
FENRIS	x	x	x		x		x	x	x	x	x	x	x	x		3
FIESTA	x	x	x	x	x	x	x		x	x	x	x				3
FLASH	x	x	x	x			x	x	x	x	x	x	x	x	x	1
FLEXAN	x		x		x		x	x	x	x	x			x		3
HYBRID	x						x		x	x	x	x				2
IBA	x	x	x		x		x	x	x	x	x	x	x	x		1
INFESA	x	x					x		x	x	x	x		x		2

TABLE 3. Loadings and User Comfort.

Program Description Tables

	LOADING							PRE/POSTPROCESSING							SEE VOL
	Nodal/line	Pressure	Selfweight	Centrifugal	Thermal	Heat flux	Prescribed displacement	Other	Free format input	Mesh generation	Plot routines	Automatic node number	Combinations of load cs	Interactive graph	CAD interfaces	
KYOKAI	x	x			x	x	x	x	x	x	x	x		x		1
LASSAQ	x															3
MEF/MOSAIC		x	x	x	x	x	x	x	x	x	x	x		x		2
MICRO STRESS	x	x	x		x		x		x		x					1
MODULEF	x	x	x	x	x	x	x	x	x	x	x	x		x		3
MSRC-RB	x	x	x				x	x	x				x			3
NE XX	x		x				x	x	x			x	x	x		1
OSTIN								x		x			x			3
PAFEC	x	x	x	x	x	x	x	x	x	x	x	x	x	x	x	1
PAID	x	x	x			x	x	x			x		x	x		1
PANDA								x	x				x			1
PDA/PATRAN	x	x	x		x		x	x	x		x	x	x	x	x	2
RAPS									x		x		x	x		3
RCAFAG	x	x	x		x		x	x	x	x	x	x		x		2
REST								x		x						2
ROBOT	x	x	x	x	x		x	x				x	x			3
S AND CM	x	x											x	x		1
SAMKE	x	x	x		x		x	x	x	x	x	x	x	x	x	3
SESAM '80	x	x	x	x	x		x	x	x	x	x	x	x	x	x	1
SIMP	x			x		x	x	x	x	x	x	x	x	x	x	2
STAR 2	x	x	x		x		x	x	x	x	x	x				2
STDYNL	x	x	x		x		x	x	x	x		x	x			3
STRUGEN									x	x	x	x		x		2
SURFOPT	x						x		x	x	x	x	x			3
THERMAL	x	x			x							x	x			1
TITUS	x	x	x	x	x	x	x	x	x	x	x	x	x	x	x	1
UCIN GEAR		x							x	x	x	x				1
Y1 2M																2
ZERO-4	x	x	x		x		x			x	x	x		x		3

TABLE 3. Loadings and User Comfort (continued).

Program Description Tables

	HARDWARE													SEE VOL
	CDC	IBM	Univac	Cray	Amdahl	Honeywell	Data General	Prime	VAX, DEC	HP	Apollo	Microcomputers	Other mainframes	Other minicomputers	
ADINA	x	x	x	x		x	x	x	x			x	x	x	1
AFAG										x					2
AIT	x	x										x			1
ALSA		x			x								x		3
ANSYS	x	x	x	x	x		x	x	x	x	x		x	x	1
AQUADYN			x						x						2
ASE	x	x										x			2
AXISYMMETRIC												x			1
BEASY	x	x	x	x				x		x			x	x	1
BEFE		x						x	x						3
BEMFFT									x						2
BEWAVE	x								x					x	3
BOSOR4	x	x	x						x						2
BOSOR5	x	x	x						x						3
CASTEM	x	x	x	x				x	x					x	3
CASTOR	x	x	x	x					x						1
DAPST	x							x							1
DEFOR		x							x	x				x	1
ELASTODYNAMICS (2D)		x											x		2
ESA												x		x	2
FEMFAM										x					3
FEMPAC	x	x	x		x	x	x	x	x	x	x	x	x	x	3
FENRIS		x		x					x		x			x	3
FIESTA	x								x						3
FLASH	x	x	x				x	x	x		x		x	x	1
FLEXAN	x	x		x					x						3
HYBRID													x		2
IBA		x							x	x				x	1
INFESA									x					x	2

TABLE 4. Hardware Compatibilities.

Program Description Tables

	CDC	IBM	Univac	Cray	Amdahl	Honeywell	Data General	Prime	VAX, DEC	HP	Apollo	Microcomputers	Other mainframes	Other minicomputers	SEE VOL
KYOKAI								x	x						1
LASSAQ									x						3
MEF/MOSAIC	x	x	x			x	x	x	x						2
MICRO STRESS												x			1
MODULEF	x	x	x	x		x			x		x		x		3
MSRC-RB													x		3
NE XX										x					1
OSTIN								x	x						3
PAFEC	x	x	x	x	x	x	x	x	x	x	x		x	x	1
PAID	x							x							1
PANDA	x	x							x						1
PDA/PATRAN	x			x			x	x	x		x		x	x	2
RAPS		x						x	x		x				3
RCAFAG										x					2
REST	x														2
ROBOT	x												x		3
S AND CM												x			1
SAMKE	x	x	x						x						3
SESAM '80	x	x						x	x					x	1
SIMP	x		x		x			x							2
STAR 2	x	x										x			2
STDYNL	x	x												x	3
STRUGEN	x	x							x				x		2
SURFOPT			x												3
THERMAL												x			1
TITUS	x	x	x	x					x		x				1
UCIN GEAR		x			x										1
Y1 2M	x	x	x												2
ZERO-4		x												x	3

TABLE 4. Hardware Compatibilities (continued).

THE ADINA SYSTEM

Klaus-Jürgen Bathe* and Gunnar Larsson**

*Department of Mechanical Engineering, Massachusetts Institute of Technology,
Cambridge, MA 02139, USA
**ADINA Engineering AB, Munkgatan 20D, S-722 12 Västeros, Sweden

ABSTRACT

The ADINA 84 system is briefly described. The finite elements, material models and solution capabilities are summarized and a few applications are briefly presented.

INTRODUCTION

The finite element analysis of a proposed structure is now in many cases an important ingredient of the design process. For the analysis a suitable computer program must be employed.

The objective in our development of the ADINA system is to provide one program system that can be employed very effectively in the analysis of a wide variety of structures and continua. The specific aims are to have one program system that (1) can be employed effectively for linear, nonlinear, static and dynamic analysis, (2) contains few effective well-chosen elements which can be employed to analyze a large variety of problems, (3) displays optimum reliability, efficiency and accuracy in the solution procedures.

In this short paper we briefly describe the ADINA system.[1,2] We first summarize the current program capabilities, and then present some applications.

PRESENT STATE OF THE ADINA SYSTEM

The complete ADINA system consists of the programs ADINA for displacement and stress analysis, ADINAT for analysis of heat transfer and field problems, ADINA-IN for preparation and display of the input data and ADINA-PLOT for display of the calculated results.

Incremental Equilibrium Equations

The basic equations that ADINA is operating on are[3]:

in static analysis or implicit time integration (Newmark or Wilson θ methods):

$$M \, {}^{t+\Delta t}\ddot{U}^{(i)} + C \, {}^{t+\Delta t}\dot{U}^{(i)} + {}^{t}K \, \Delta U^{(i)} = {}^{t+\Delta t}R - {}^{t+\Delta t}F^{(i-1)} \tag{1}$$

$${}^{t+\Delta t}U^{(i)} = {}^{t+\Delta t}U^{(i-1)} + \Delta U^{(i)}$$

in explicit time integration (central difference method):

$$M \, {}^{t}\ddot{U} + C \, {}^{t}\dot{U} = {}^{t}R - {}^{t}F \tag{2}$$

where

M = constant mass matrix, lumped or consistent,

C = constant damping matrix, nodal point concentrated damping or Rayleigh damping,

${}^{t}K$ = tangent stiffness matrix at time t,

${}^{t}R, {}^{t+\Delta t}R$ = external load vector applied at time t, $t+\Delta t$,

${}^{t}F, {}^{t+\Delta t}F^{(i-1)}$ = nodal point force vector equivalent to the element stresses at time t, time $t+\Delta t$ and iteration (i-1),

${}^{t+\Delta t}U^{(i-1)}$ = vector of nodal point displacements at time $t+\Delta t$ and iteration (i-1),

and a superimposed dot denotes time derivative, e.g., ${}^{t}\dot{U}$ = nodal point velocities at time t.

It should be noted that time is a dummy variable in static analyses when the material properties are time-independent. However, the use of the time variable to define load intensities when time is a dummy variable enables the analyst to continue with a dynamic analysis (after a static analysis) with a minimum amount of input changes.

Linear Analysis. In linear analysis the tangent stiffness matrix in Eq.(1) is constant. In this case, no geometric or material nonlinearities are included. It should be noted that ADINA is very effective in linear analysis.

Time integration. An implicit time integration method would generally be employed for analysis of structural vibration problems, i.e., when the system is primarily excited in a few vibration modes. An explicit time integration solution would primarily be performed to predict wave propagation phenomena. Since no stiffness matrix is assembled when operating on Eq. (2), relatively large-order systems can be analyzed in explicit time integration using relatively small high-speed core storage.

Solution of equations. In static analysis, implicit time integration, frequency calculations (say, for mode superposition analysis), the total stiffness matrix and (consistent) mass matrix are assembled in blocks depending on the high-speed storage available in the computer, and only the elements below the skyline are stored. A skyline out-of-core column solver is employed to obtain the solution of the equations, see Figs. 1 and 2.[3]

Of much concern is the stability and accuracy of the incremental solution of the nonlinear equations. In ADINA the solution of Eq. (1) can be obtained using the modified or full Newton-Raphson methods with or without line searches,[3,4,5] the BFGS method,[3,4] or an automatic load-stepping scheme for collapse and post-collapse response calculations.[6]

Substructuring. To take advantage of some repetitive geometric and material conditions in the system under consideration, substructures of linear elements

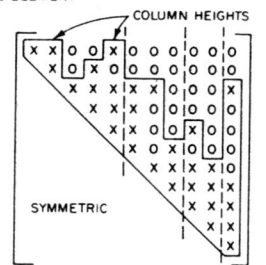

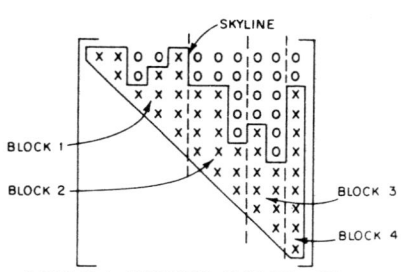

Fig. 1. Storage in solution of equations

can be defined. Each substructure can be employed as a "superelement" a number of times in the complete element assemblage. This option can be effective in some linear analyses to solve large systems. However, the option has also been incorporated into ADINA for nonlinear analyses in which only local nonlinearities are encountered. In these analyses it can be efficient to statically condense out the major part of the linear degrees of freedom prior to the incremental solution of the nonlinear equations. The substructuring capability is demonstrated in Fig. 3 and can be employed in static and dynamic lumped mass analysis (without any constraints on the lumped mass matrix to be used).[7]

Constraint equations. In some analyses it is necessary to prescribe displacements at some nodal points, and/or impose constraints between some nodal displacement components. In ADINA the following relationships can be specified

$$U_i = f_\ell(t) \tag{3}$$

and

$$U_k = \sum_j \beta_j U_j \tag{4}$$

where f_ℓ is a general time function (input to ADINA) that prescribes the nodal point displacement U_i. The constraints in Eq. (4) are frequently simply specified as rigid links, in which case the program establishes the constraint equations automatically. The prescribed displacement and displacement constraint options can be employed in static and dynamic analysis.

Solution of frequencies and mode shapes. Desired frequencies and corresponding vibration mode shapes can be calculated using the determinant search and an improved subspace iteration algorithm.[8]

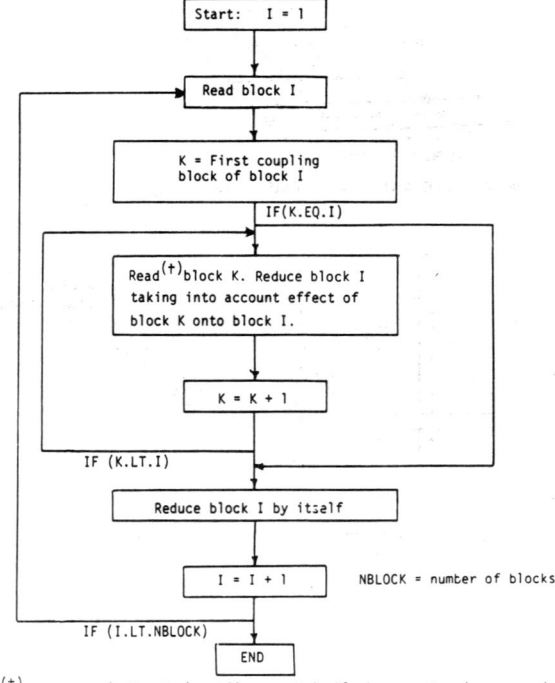

Fig. 2. Block operations in solution of equations

Mode superposition. The mode superposition method can be employed to calculate the time history response of linear systems, or to perform a response spectrum analysis. The mode superposition technique can also be used effectively for nonlinear analysis if the system contains only local nonlinearities.[7,8]

Load definition. Prescribed concentrated, pressure and temperature loading can be defined; mass proportional loading and centrifugal loading can be applied, and (as given in Eq. (3) already) nodal displacements can be imposed. The pressure loading can be deformation dependent. Also, an interface is provided through which the user can program an own load definition using the current displacements, velocities and so on (e.g., to define hydrodynamic loading).

Linearized buckling analysis. A linearized buckling analysis by an eigensolution can be performed at any load level. The calculated buckling mode shapes can be used to define geometric imperfections on the structural model to simulate nonperfect conditions in the collapse analysis.

Fracture mechanics. Stress intensity factors can be calculated using the algorithm of reference 10. The procedure can be employed in plane stress, plane strain, axisymmetric analyses and in 3-D analysis (using the 3-D solid elements).

Solution of contact problems. A very useful, new Lagrange multiplier/segment algorithm is available in ADINA to solve contact problems, in which the contact area is initially unknown and varies during the response. The bodies in contact may be flexible or rigid with Coulomb frictional conditions, and the bodies may

The ADINA System

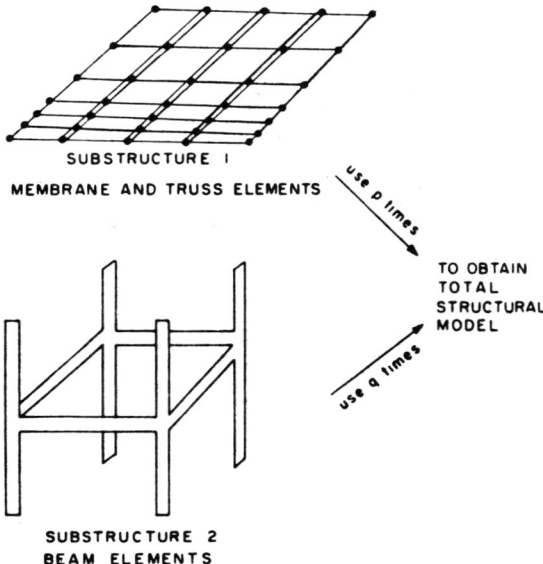

Fig. 3. Example of substructuring

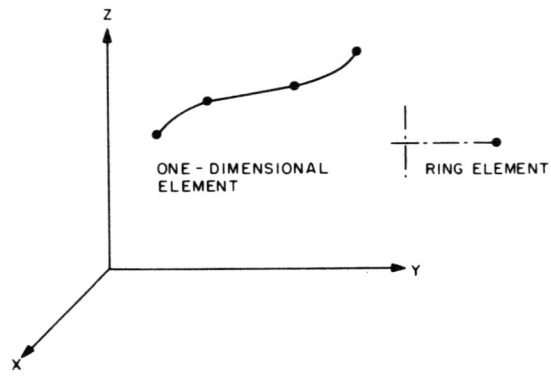

AVAILABLE NONLINEAR FORMULATIONS	AVAILABLE MATERIAL MODELS
a. LINEAR ANALYSIS	a. LINEAR ELASTIC
b. MATERIALLY NONLINEAR ONLY	b. NONLINEAR ELASTIC
c. UPDATED LAGRANGIAN WITH LARGE DISPLACEMENTS BUT SMALL STRAINS	c. THERMO-ELASTIC
	d. ELASTIC-PLASTIC
	e. THERMO-ELASTIC-PLASTIC and CREEP

Fig. 4. Truss and cable element (2, 3 or 4 nodes)

undergo large deformations in sliding.

ELEMENT LIBRARY AND MATERIAL MODELS

The program ADINA contains a few most effective elements which can be used to model a large variety of problems. The elements are depicted in Figs. 4 to 11. These figures also give the kinematic formulations and the material models with which the elements can currently be employed. The 2-D and 3-D elements of Figs. 5 and 6 can also be used as (acoustic) fluid elements for analysis of fluid-structure interactions.

The finite elements used in ADINA are described in detail in [3,13-28] and additional applications are given in ref. 2. In the following we only briefly summarize the various analysis options.

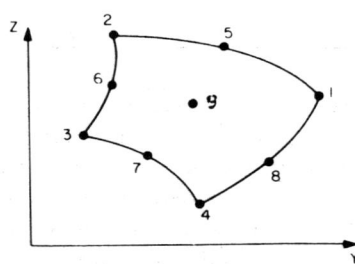

AVAILABLE NONLINEAR FORMULATIONS	AVAILABLE MATERIAL MODELS
a. LINEAR ANALYSIS	a. ISOTROPIC LINEAR ELASTIC
b. MATERIALLY NONLINEAR ONLY	b. ORTHOTROPIC LINEAR ELASTIC
c. UPDATED LAGRANGIAN	c. ISOTROPIC THERMO-ELASTIC
d. TOTAL LAGRANGIAN	d. CURVE DESCRIPTION NONLINEAR MODEL FOR ANALYSIS OF GEOLOGICAL MATERIALS
	e. CONCRETE MODEL
	f. ISOTHERMAL PLASTICITY MODELS, VON MISES YIELD CONDITION OR DRUCKER-PRAGER YIELD CONDITION WITH CAP
	g. THERMO-ELASTIC-PLASTIC and CREEP MODELS, VON MISES YIELD CONDITION
	h. ISOTROPIC NONLINEAR ELASTIC, INCOMPRESSIBLE (MOONEY-RIVLIN MATERIAL) (plane stress only)

Fig. 5. 2-D solid (plane stress/strain and axisymmetric) element (variable number of nodes)

Three different analysis procedures can be considered for an element.

Linear Elastic Analysis. The displacements of the element are assumed to be negligibly small and the strains infinitesimal. The material is isotropic or orthotropic linear elastic.

Materially-Nonlinear-Only Analysis. The displacements of the element are

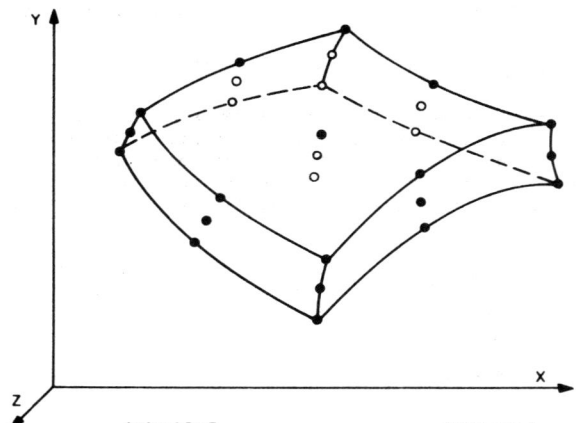

Fig. 6. 3-D solid element (variable number of nodes)

negligibly small, and the strains are infinitesimal. The material stress-strain description is nonlinear.

Large Deformation Analysis. The element may experience large displacements and large strains (described by an updated or total Lagrangian formulation[3]). The material stress-strain relationship is linear or nonlinear.

The linear elastic analysis does not allow for any nonlinearities, whereas the materially-nonlinear-only analysis includes material nonlinearities, but no geometric nonlinearities. The large deformation solution using the total Lagrangian and updated Lagrangian formulations may include all nonlinearities, and which formulation should be employed depends essentially on the definition of the material model used. These material models are briefly described below.

Isotropic and orthotropic linear elastic material. The stress-strain relationships are defined by means of constant Young's moduli and Poisson's ratios. In orthotropic analysis, different axes of orthotropy can be used for each element.

Nonlinear elastic material. The nonlinear elastic material behaviour is defined by specifying the stress as a piece-wise linear function of the current strain.

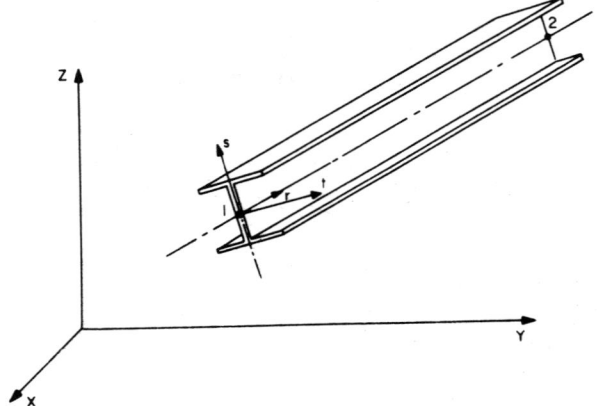

AVAILABLE
NONLINEAR FORMULATIONS

a. LINEAR ANALYSIS
b. MATERIALLY NONLINEAR ONLY
c. UPDATED LAGRANGIAN WITH LARGE DISPLACEMENTS BUT SMALL STRAINS

AVAILABLE
MATERIAL MODELS

a. ISOTROPIC LINEAR ELASTIC
b. ELASTIC-PLASTIC
   VON MISES YIELD CONDITION

Fig. 7. Two-node beam element

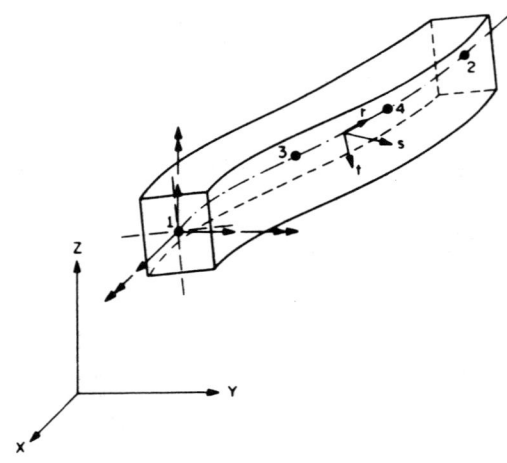

AVAILABLE
NONLINEAR FORMULATIONS

a. LINEAR ANALYSIS
b. MATERIALLY NONLINEAR ONLY
c. TOTAL LAGRANGIAN WITH LARGE DISPLACEMENTS BUT SMALL STRAINS

AVAILABLE
MATERIAL MODELS

a. ISOTROPIC LINEAR ELASTIC
b. ISOTHERMAL PLASTICITY,
   VON MISES YIELD CONDITION

Fig. 8. Isoparametric beam element (2, 3 or 4 nodes)

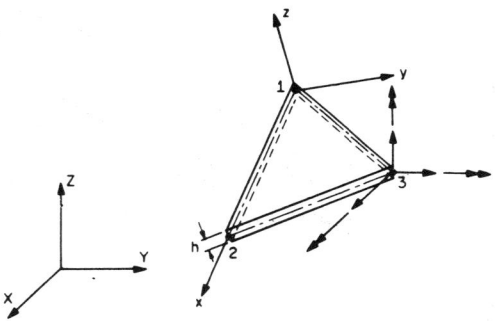

AVAILABLE NONLINEAR FORMULATIONS	AVAILABLE MATERIAL MODELS
a. LINEAR ANALYSIS	a. ISOTROPIC LINEAR ELASTIC
b. MATERIALLY NONLINEAR ONLY	b. ISOTHERMAL PLASTICITY, ILYUSHIN YIELD CONDITION
c. UPDATED LAGRANGIAN WITH LARGE DISPLACEMENTS BUT SMALL STRAINS	

Fig. 9. Three-node plate/shell element

Isotropic thermo-elastic model. The Young's modulus, Poisson's ratio and the mean coefficient of thermal expansion vary as a function of temperature.

Curve description model. The instantaneous bulk and shear moduli are defined as a function of the volumetric strain. The material tensile failure is modeled using tension cut-off conditions.

Concrete model. The multi-axial stress-strain relation is defined using for the principal stress directions a uniaxial stress-strain relation that is modified for biaxial and triaxial stress conditions. The material tensile failure, compression crushing, and post-failure behaviour including strain-softening conditions are modeled[29].

Cap model. Flow theory of plasticity is used with a yield surface, a cap and a tension cut-off criterion. The model can be reduced to the Drucker-Prager model[30].

Von Mises isothermal plasticity. The plasticity relations are based on the Prandtl-Reuss equations. Multilinear isotropic and kinematic hardening, small strain or large strain conditions can be analyzed[30,31].

Thermo-elasto-plasticity and creep. The material is described using the Prandtl-Reuss equations for plasticity. The yield stress, Young's modulus, Poisson's ratio and coefficient of thermal expansion are temperature-dependent. Linear isotropic or kinematic hardening is assumed. Creep strain accumulation in cyclic loading conditions is taken into account using auxiliary hardening rules[32-34].

Mooney-Rivlin model. The stress-strain relationship of this hyperelastic rubber model is defined using the Mooney-Rivlin constants.

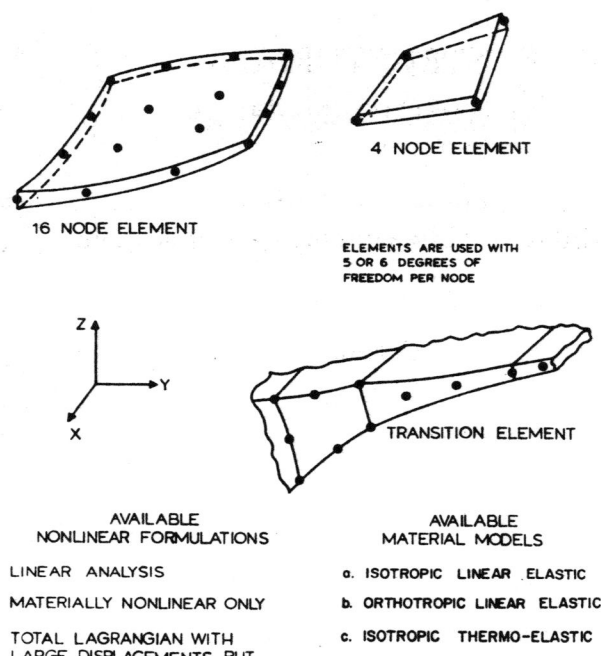

Fig. 10. Thin shell elements

Element birth and death options. All elements in ADINA (and ADINAT) can be employed with the element birth and element death options. Using the element birth option, the element is not present in the finite element assemblage until its time of birth, whereas in the element death mode the element is initially present but then vanishes at the time of death. These options are very useful in modeling for example, the construction or repair of a structure, or the excavation of a tunnel[2].

User-supplied element. To enable the use of special finite elements, the user-supplied element option reads and assembles general element stiffness, mass and damping matrices provided by the user. This option is only available for linear elements (the stiffness coefficients are constant).

User-supplied material model. The ADINA program provides also an option whereby the user can program through an interface an own material model, which may include time and temperature effects. The interface is designed so that the user can program directly complex linear or nonlinear material models with relatively little knowledge of ADINA.

HEAT TRANSFER ANALYSIS AND SOLUTION OF FIELD PROBLEMS

For heat transfer analysis the ADINAT program can be employed. The temperatures calculated using ADINAT can directly be used for a thermal stress analysis with ADINA.

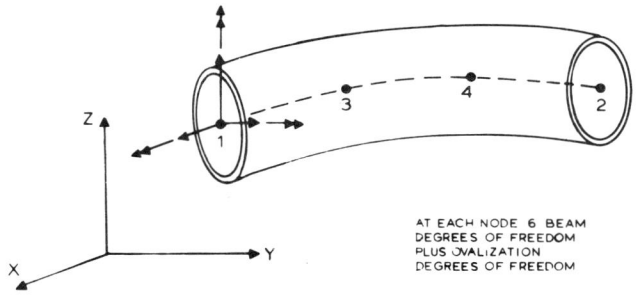

Fig. 11. Pipe element with ovalization

ADINAT can also be employed for the solution of other field problems, such as seepage, electrostatic field analysis, torsion, electric conduction and so on[2,3].

PRE- AND POST-PROCESSING

For the efficient practical use of a finite element program effective pre- and post-processing capabilities can be most important. The programs ADINA-IN and ADINA-PLOT have been developed for the pre- and post-processing of the ADINA input data and solution results[1]. The finite element model can be generated and plotted to verify the input, and the calculated results (displacements, velocities, accelerations, mode shapes and stresses, nodal forces, and so on) can be displayed in an effective manner. ADINA-IN and ADINA-PLOT have been written in free-format command language for batch or interactive usage.

SUPPORTING LITERATURE ON THE ADINA SYSTEM

Precise and comprehensive documentation of a computer program is most important. For the ADINA system the following documentation is available.

Users manuals. Reference 1 gives the users manuals of the ADINA, ADINAT, ADINA-IN and ADINA-PLOT programs.

Theory documentation. A large number of papers and reports (e.g., references 4 to 38) and the book by Bathe[3] describes the theory used in the ADINA and ADINAT computer programs. The theory is also presented in the video-course of reference 39. In addition, the proceedings of the ADINA Conferences held biyearly at M.I.T. provide valuable information on experiences using the ADINA system[2].

Modeling guide and verification manual. Since a large number of analysis

options are available in the ADINA system, a user may well be initially
overwhelmed with the different analysis choices and the theoretical basis of the
computer programs. Also, the literature of references 3 to 38 is very
comprehensive and frequently provides more detail than the user needs to consult.
The purpose of the ADINA System Theory and Modeling Guide is to provide a bridge
between the practical application of the ADINA system and the detailed
documentations on the theory used[40]. The report provides a document that
summarizes in a compact manner the methods and assumptions used and gives in-
detail references for further study. The report is meant to "guide" the user
through the process of finite element modeling and study.

The ADINA System Verification Manual is also meant to help the user in studies of
the capabilities of the ADINA system, while at the same time it is used to verify
the installation of the program on a specific computer[41]. The description,
solution and data input for a large number of problems is given in the manual.
These problems can directly be executed with the programs.

INSTALLATION OF THE ADINA SYSTEM

The computer programs have been written in standard FORTRAN and can be obtained
for IBM, CDC, VAX, UNIVAC, MASSCOMP and Cray computers. The programs have also
been installed on PRIME, BURROUGHS, DATA GENERAL and other machines.

SOME APPLICATON AREAS

Apart from being employed abundantly in almost routine linear and nonlinear
analyses, the ADINA program has been applied in some interesting complex
application areas[2], for example, in the areas of,

> nonlinear analysis of reinforced and prestressed concrete structures;
> analysis of tunnels and earth structures;
> analysis of metal forming problems;
> thermo-elastic-plastic and creep analysis of heat-treatment processes;
> nonlinear dynamic analysis of large structural systems;
> solution of frequencies and mode shapes of very large systems;
> analysis of fluid-structure systems;
> buckling and collapse analysis of complex shell structures.

A large number of industrial applications using ADINA are found in the
proceedings of the ADINA Conferences, see reference 2. To very briefly illustrate
some capabilities, we refer to Figs. 12 to 18 and give a short explanation. The
details on these analyses are found in the references.*

Figure 12 illustrates the analysis of a contact problem with friction. The
problem considered arises when a wire in a continuous wiring around a prestressed
cylinder snaps. The analysis was conducted to predict the increase in stress in
the intact wiring[11]. Figure 13 shows the results of a collapse analysis of a
concrete slab supported at its four corners[29]. Figure 14 gives the results
obtained in the large displacement elasto-plastic collapse analysis of a shell[21].
This solution was obtained using the automatic load incrementation algorithm[6].
Figure 15 illustrates the large displacement solution of the motion of a cable
that is subjected to a horizontal displacement at its upper support point (node
11)[4]. Figure 16 gives the results obtained in the dynamic analysis of a water-
filled pipe subjected to a pulse[9,27], and Fig. 17 summarizes the results
calculated in the analysis of the quenching process of a cylinder[34]. Finally,

---

*Since some of these results have been obtained in the development of the program,
slightly different results may be obtained with the newest version of the
program, ADINA 84.

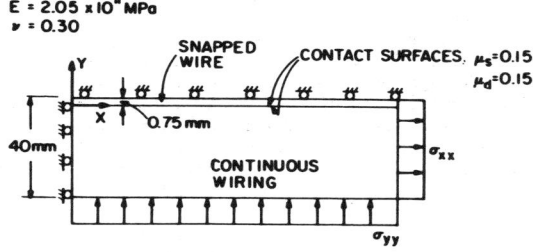

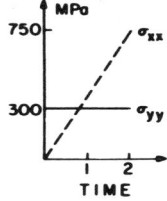

(a) Model used

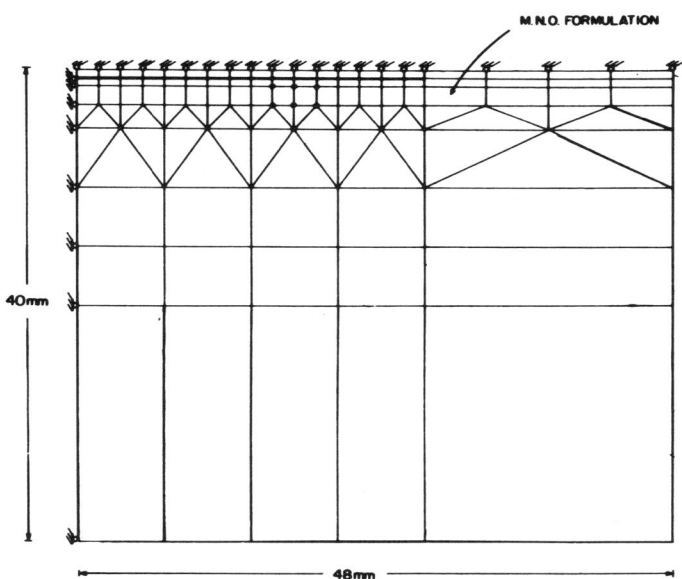

(b) Finite element mesh

Fig. 12. Analysis of wiring around cylinder when one wire snaps

Fig. 18 shows the results obtained in a pipe bend test[26]. The pipe bend is once modeled with the special pipe element available in ADINA (see Fig. 11) and once with the higher-order 16-node shell element.

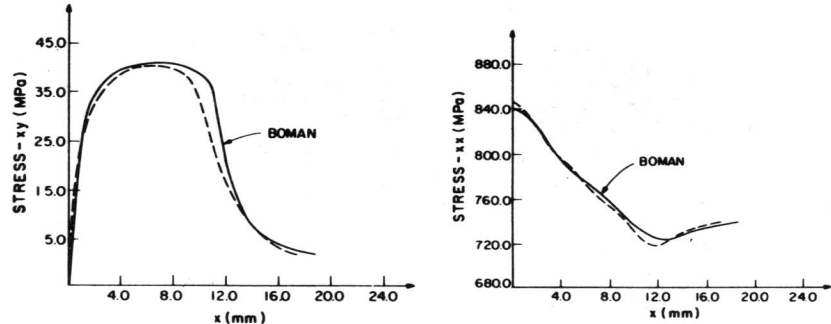

(c) Predicted stresses just below snapped wire

Fig. 12 (continued)

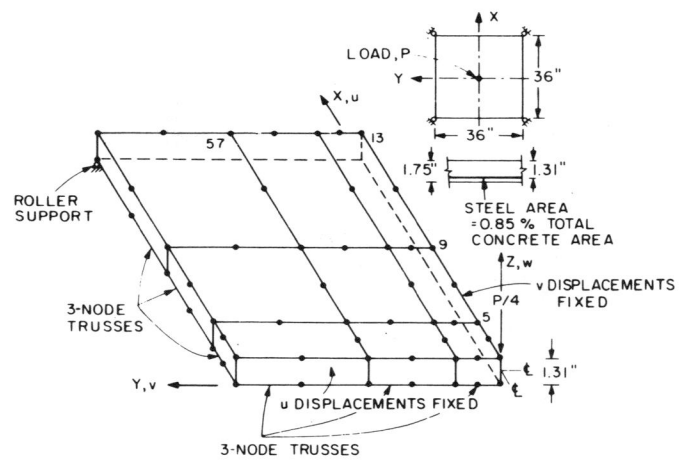

MATERIAL PROPERTIES:

STEEL:
$A_{st} = 0.536$ in^2
$E_{st} = 29000$ ksi
$\sigma_y = 50$ ksi
$E_T = 29$ ksi

CONCRETE:
$\tilde{E}_0 = 4150$ ksi
$\nu = 0.15$
$\tilde{\sigma}_t = 0.77$ ksi
$\tilde{\sigma}_c = -5.5$ ksi
$\tilde{\sigma}_u = -4.7$ ksi
$\tilde{e}_c = -.002$ in/in
$\tilde{e}_u = -.003$ in/in

ANALYSIS PARAMETERS:
$\gamma = 1.$; $\kappa = 0.7$; $\eta_n = 0.0001$; $\eta_s = 0.5$

9 SIXTEEN-NODE ISOPARAMETRIC ELEMENTS WITH
3 x 3 x 3 INTEGRATION
24 THREE-NODE TRUSS ELEMENTS

(a) Model used

Fig. 13. Analysis of corner supported concrete slab

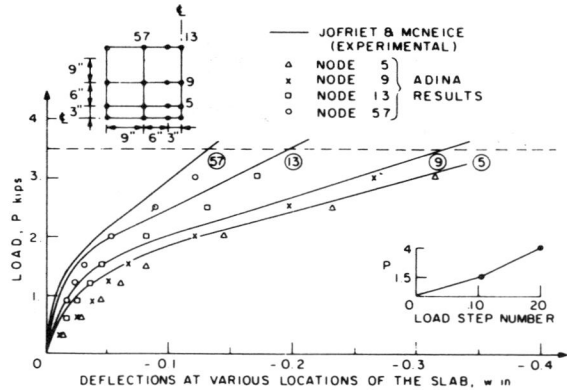

(b) Predicted displacements

Fig. 13 (continued)

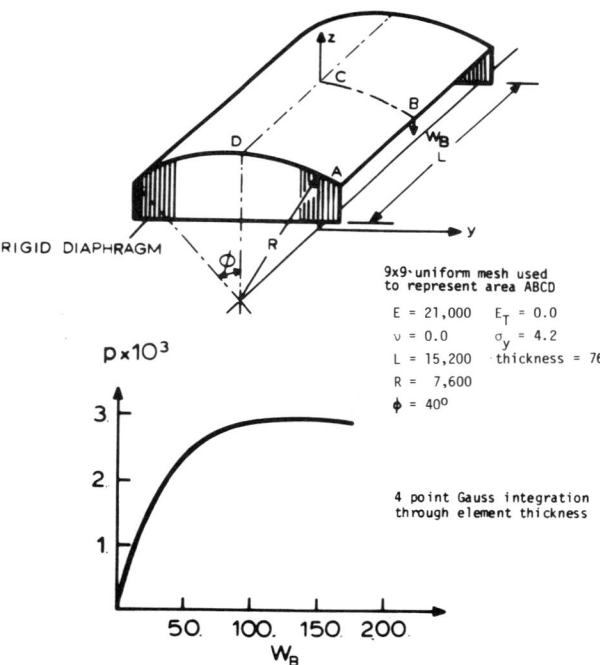

Fig. 14. Collapse analysis of shell using 4-node shell elements, p=load/unit hor. area

CONCLUDING REMARKS

The objective in this paper was to briefly present the ADINA system. For details on the capabilities and experiences with the system, it is necessary to study the references.

We actively continue with the development of the ADINA system, both to increase the applicability of ADINA to engineering problems and to render the program more easily usable in the engineering profession[42-46]

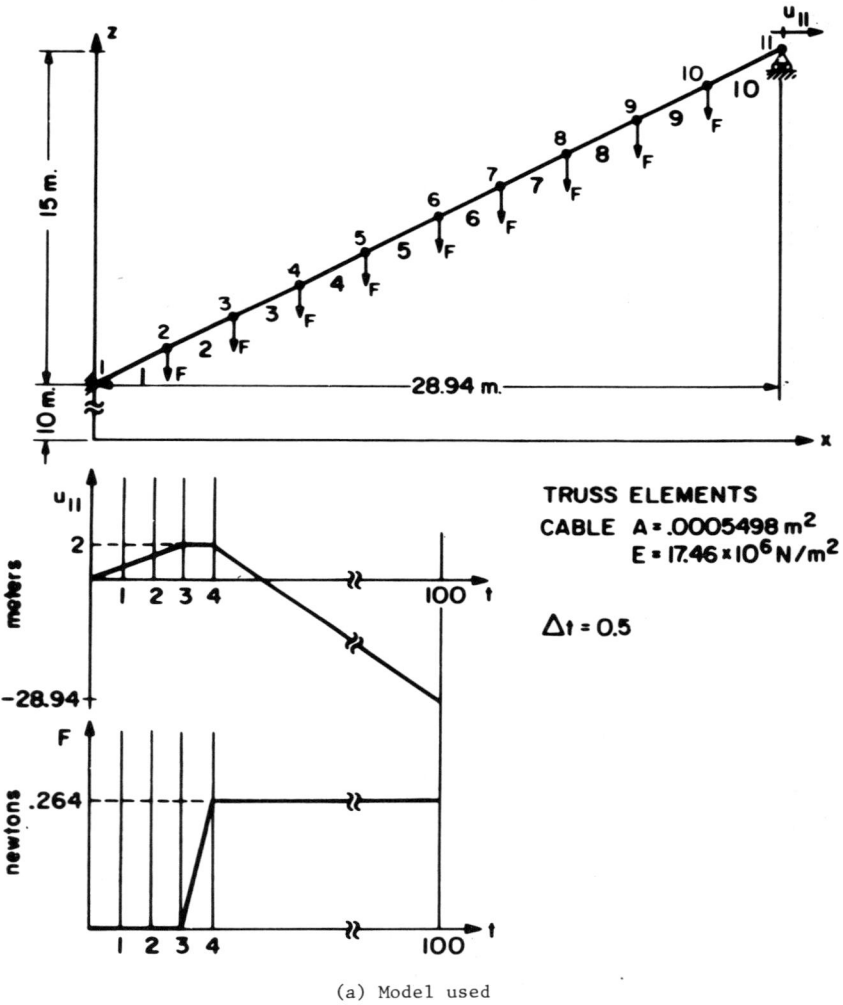

(a) Model used

Fig. 15. Large displacement analysis of cable motion

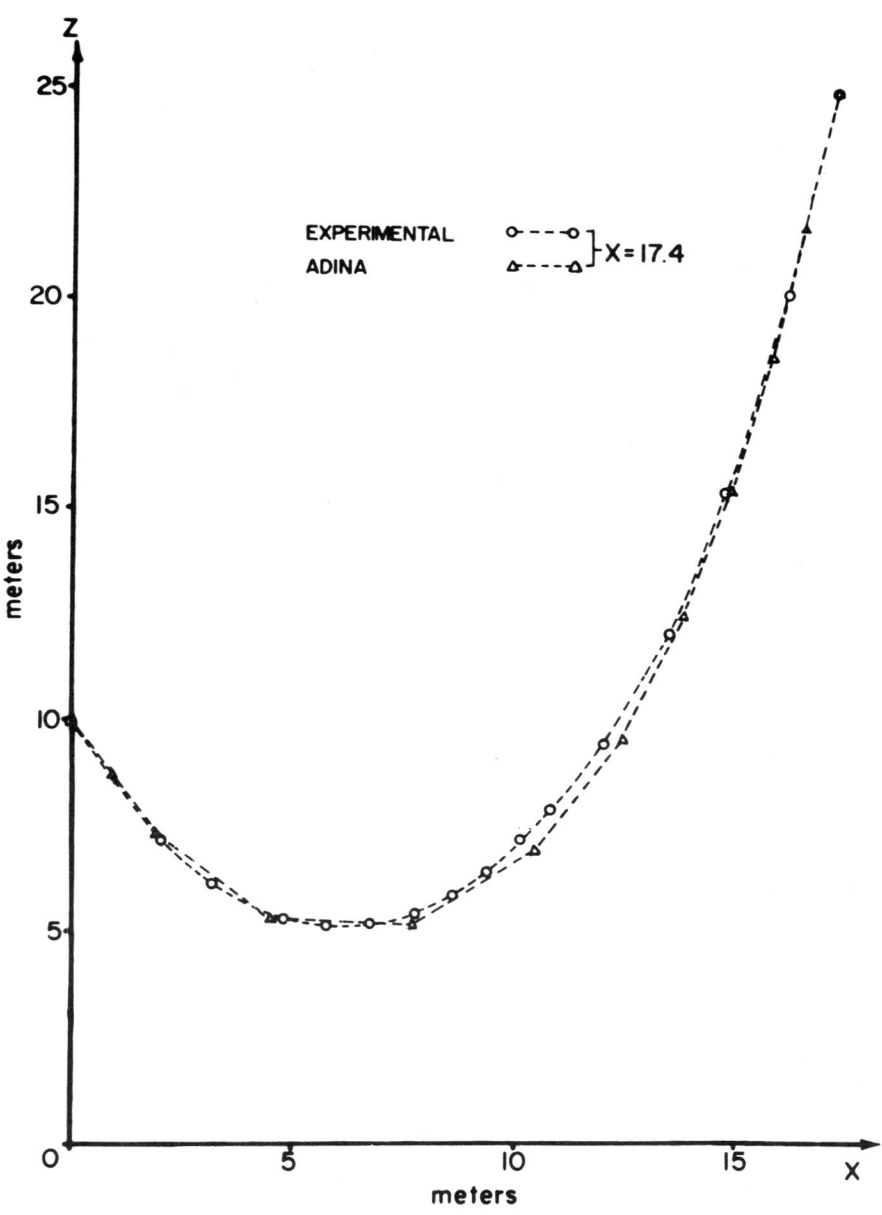

(b) Deflected shapes of cable

Fig. 15 (continued)

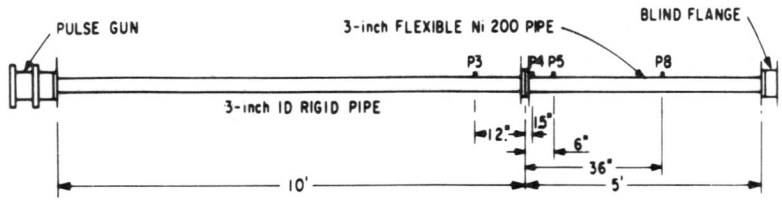
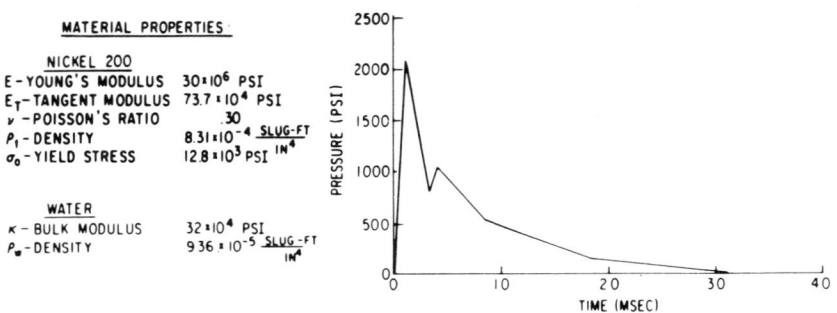
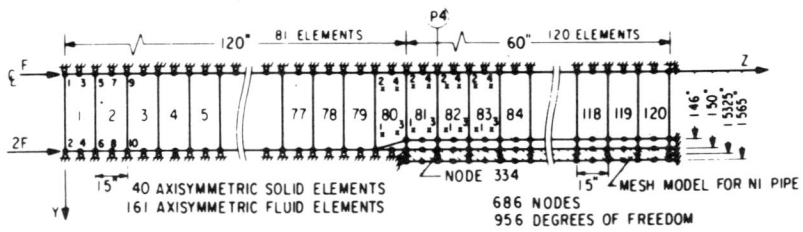

(a) Structure, model and loading

Fig. 16. Analysis of (water-filled) pipe

The ADINA System 19

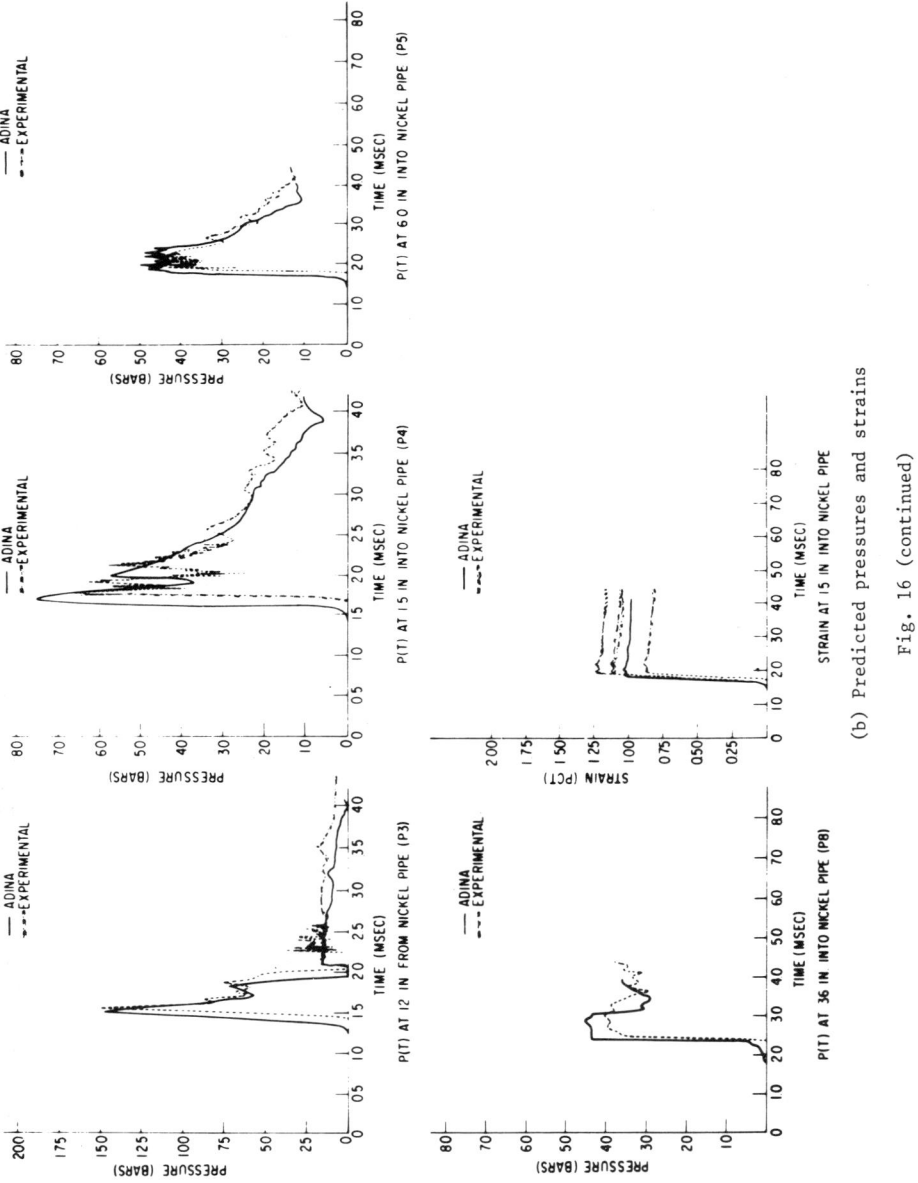

(b) Predicted pressures and strains

Fig. 16 (continued)

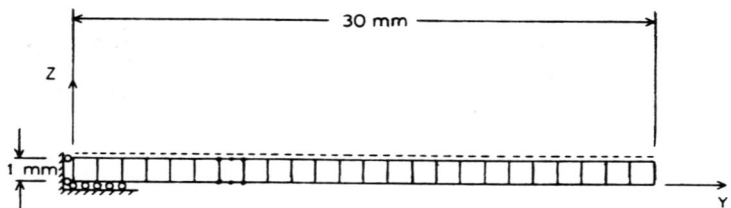

(a) Model used for temperature and stress analyses

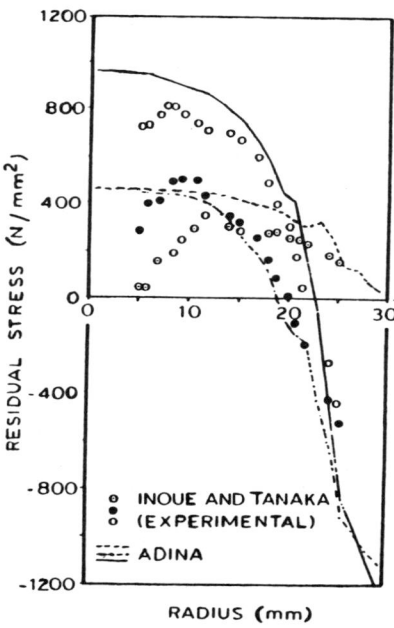

(b) Predicted residual stresses

Fig. 17. Analysis of quenching process of a cylinder

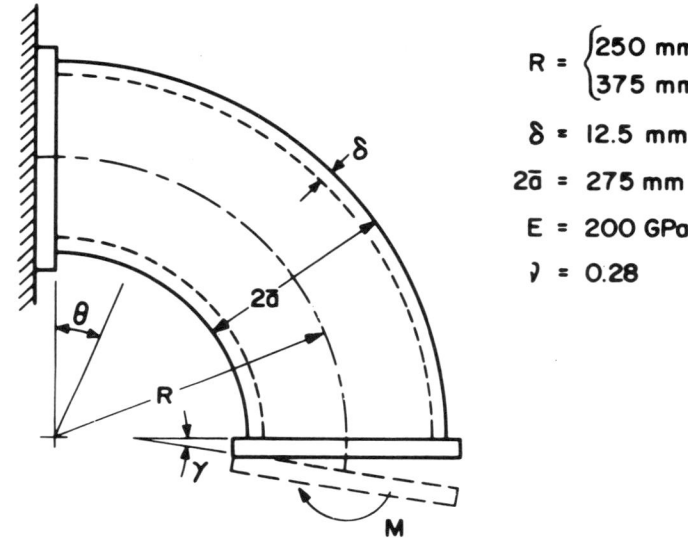

(a) Pipe bend considered.

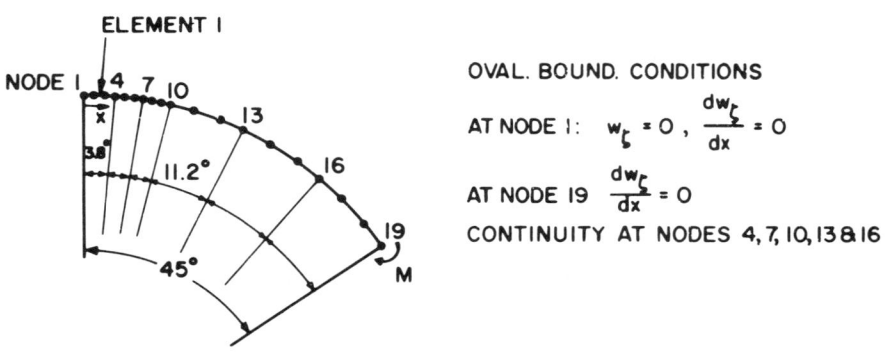

(b) Model of six elbow elements

Fig. 18. Collapse analysis of flanged pipe bend

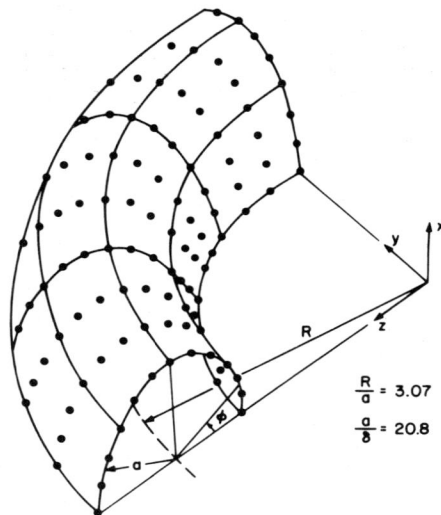

(c) Model of twelve 16-node shell elements

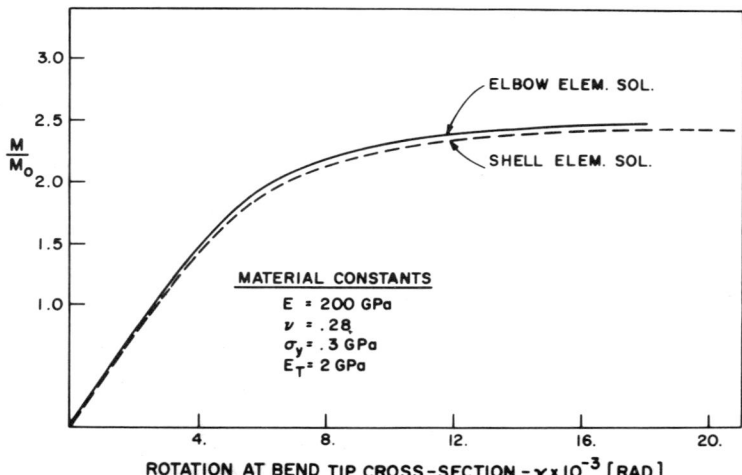

(d) Predicted moment response; M_0 is the moment at first yield

Fig. 18 (continued)

REFERENCES

1. ADINA, ADINAT, ADINA-IN and ADINA-PLOT Users Manuals, ADINA Engineering AB, Munkgatan 20D, S-722 12 Västerås, Sweden, and ADINA Engineering Inc., 71 Elton Ave., Watertown, MA 02172, U.S.A.
2. Proceedings of the ADINA Conferences, (K. J. Bathe, ed.)

- Aug. 1977, M.I.T. Report AVL 82448-6, Mech. Eng. Dept.
- Aug. 1979, M.I.T. Report AVL 82448-9, Mech. Eng. Dept.
- June 1981, *J. Computers and Structures*, Vol. 13, No. 5-6, 1981.
- June 1983, *J. Computers and Structures*, Vol. 17, No. 5-6, 1983.
- June 1985, *J. Computers and Structures*, Vol. 21, No. 1-2, 1985.
3. Bathe, K. J. *Finite Element Procedures in Engineering Analysis*, Prentice-Hall, Inc. 1982.
4. Bathe, K. J. and Cimento, A. P. Some Practical Procedures for the Solution of Nonlinear Finite Element Equations, *J. Comp. Meth. in Appl. Mech. and Eng.*, 22, 59-85, 1980.
5. Bathe, K. J., Chaudhary, A., Dvorkin, E. and Kojić, M. On the Solution of Nonlinear Finite Element Equations, Proceedings Int. Conf. on Computer-Aided Analysis and Design of Concrete Structures, Split, Yugoslavia, Sept. 1984.
6. Bathe, K. J. and Dvorkin, E. On the Automatic Solution of Nonlinear Finite Element Equations, *J. Computers and Structures*, 17, No. 5-6, pp. 871-879, 1983.
7. Bathe, K. J. and Gracewski, S. On Nonlinear Dynamic Analysis Using Substructuring and Mode Superposition, *J. Computers and Structures*, 13, 699-707, 1981
8. Bathe, K. J. and Ramaswamy, S. An Accelerated Subspace Iteration Method, *J. Comp. Meth. in Appl. Mech. and Eng.*, 23, 313-331, 1980.
9. Bathe, K. J., Finite Element Formulation, Modeling and Solution of Nonlinear Dynamic Problems, Chapter in *Numerical Methods for Partial Differential Equations* (S. V. Parter, ed.), Academic Press, 1979.
10. Sussman, T. and Bathe, K. J. The Gradient of the Finite Element Variational Indicator with Respect to Nodal Point Coordinates: An Explicit Calculation and Applications in Fracture Mechanics and Mesh Optimization, *Int. J. Num. Meth. in Eng.*, 21, 763-774, 1985.
11. Bathe, K. J. and Chaudhary, A. A Solution Method for Planar and Axisymmetric Contact Problems, *Int. J. Num. Meth. in Eng.*, 21, 65-88, 1985.
12. Bathe, K. J. and Chaudhary, A. On Finite Element Analysis of Large Deformation Frictional Contact Problems, in *Unification of Finite Element Methods*, (H. Kardestuncer, ed.), North-Holland Publ. Co. 1984.
13. Bathe, K. J. and Bolourchi, S. Large Displacement Analysis of Three-Dimensional Beam Structures, *Int. J. Num. Meth. in Eng.*, 14, 961-986, 1979.
14. Bathe, K. J. and Bolourchi, S. A Geometric and Material Nonlinear Plate and Shell Element, *J. Computers and Structures*, 11, 23-48, 1980.
15. K. J. Bathe and P. M. Wiener. On Elastic-Plastic Analysis of I-Beams in Bending and Torsion, *J. Computers and Structures*, 17, 711-718, 1983.
16. Bathe, K. J. and Chaudhary, A. On the Displacement Formulation of Torsion of Shafts with Rectangular Cross-Sections, *Int. J. Num. Meth. in Eng.*, 18, 1565-1568, 1982.
17. Batoz, J. L., Bathe, K. J. and Ho, L. W. A Study of Three-Node Triangular Plate Bending Elements, *Int. J. Num. Meth. in Eng.*, 15, 1771-1812, 1980.
18. Bathe, K. J. and Ho, L. W. Some Results in the Analysis of Thin Shell Structures, in *Nonlinear Finite Element Analysis in Structural Mechanics*, (Wunderlich, W., *et al.* eds.) Springer-Verlag, 1981.
19. Bathe, K. J. and Ho, L. W. A Simple and Effective Element for Analysis of General Shell Structures, *J. Computers and Structures*, 13, 673-681, 1981.
20. Dvorkin, E. and Bathe, K. J., A Continuum Mechanics Based Four-Node Shell Element for General Nonlinear Analysis, *Engineering Computations*, 1, 77-88, 1984.
21. Bathe, K. J. and Dvorkin, E. A Four-Node Plate Bending Element Based on Mindlin/Reissner Plate Theory and a Mixed Interpolation, *Int. J. Num. Meth. in Eng.*, 21, 367-383, 1985.
22. Bathe, K. J., Dvorkin, E., and Ho, L. W. Our Discrete Kirchhoff and Isoparametric Shell Elements for Nonlinear Analysis — An Assessment, *J. Computers and Structures*, 16, 89-98, 1983.
23. Bathe, K. J. and Almeida, C. A. A Simple and Effective Pipe Elbow Element — Linear Analysis, *J. Appl. Mech.*, 47, 93-100, 1980.
24. Bathe, K. J. and Almeida, C. A. A Simple and Effective Pipe Elbow Element —

Interaction Effects, *J. Appl. Mech.*, 49, 165-171, 1982.
25. Bathe, K. J. and Almeida, C. A. A Simple and Effective Pipe Elbow Element — Pressure Stiffening Effects, *J. Appl. Mech.*, 49, 914-916, 1982.
26. Bathe, K. J., Almeida, C. A. and Ho, L. W. A Simple and Effective Pipe Elbow Element — Some Nonlinear Capabilities, *J. Computers and Structures*, 17, No. 5-6, 659-667, 1983.
27. Bathe, K. J. and Hahn, W. On Transient Analysis of Fluid-Structure Systems, *J. Computers and Structures*, 10, 383-391, 1978.
28. Olson, L. and Bathe, K. J. A Study of Displacement-Based Fluid Finite Elements for Calculating Frequencies of Fluid and Fluid-Structure Systems, *J. Nuclear Eng. and Design*, 76, 137-151, 1983.
29. Bathe, K. J. and Ramaswamy, S. On Three-Dimensional Nonlinear Analysis of Concrete Structures, *J. Nucl. Eng. and Design*, 52, 385-409, 1979.
30. Bathe, K. J., Snyder, M. D., Cimento, A. P. and Rolph III, W. D. On Some Current Procedures and Difficulties in Finite Element Analysis of Elastic-Plastic Response, *J. Computers and Structures*, 12, 607-624, 1980.
31. Ishizaki, T. and Bathe, K. J. On Finite Element Large Displacement and Elastic-Plastic Dynamic Analysis of Shell Structures, *J. Computers and Structures*, 12, 309-318, 1980.
32. Snyder, M. D. and Bathe, K. J. Finite Element Analysis of Thermo-Elastic-Plastic and Creep Response, Report AVL 82448-10, Dept. of Mech. Eng., M.I.T., Dec. 1980.
33. Snyder, M. D. and Bathe, K. J. A Solution Procedure for Thermo-Elastic-Plastic and Creep Problems, *J. Nucl. Eng. and Design*, 64, 49-80, 1981.
34. Cesar, F. and Bathe, K. J. A Finite Element Analysis of Quenching Processes, in *Numerical Methods for Non-Linear Problems*, (Taylor, C. et al., eds), Pineridge Press, 1984.
35. Bathe, K. J. and Khoshgoftaar, M. R. Finite Element Formulation and Solution of Nonlinear Heat Transfer, *J. Nuclear Eng. and Design*, 51, 389-401, 1979.
36. Rolph III, W. D. and Bathe, K. J. An Efficient Algorithm for Analysis of Nonlinear Heat Transfer with Phase Changes, *Int. J. Num. Meth. in Eng.*, 18, 119-134, 1982.
37. McTaggart, P. E. and Bathe, K. J. On Finite Element Analysis of Fluid Flow in Ducts with Boundary Layer Correction, *J. Computers and Structures*, 21, 105-111, 1985.
38. Bathe, K. J. and Khoshgoftaar, M. Finite Element Free Surface Seepage Analysis Without Mesh Iteration, *Int. J. Num. and Anal. Meth. in Geomech.*, 3, 13-22, 1979.
39. Bathe, K. J. *Finite Element Methods in Engineering Mechanics*, Center for Advanced Engineering Study, Massachusetts Institute of Technology, Cambridge, MA 02139, video-course and study guide.
40. ADINA System Theory and Modeling Guide, Report AE 83-4, ADINA Engineering.
41. ADINA System Verification Manual, Report AE 83-5, ADINA Engineering.
42. Bathe, K. J., Sonnad, V. and Domigan, P. Some Experiences Using Finite Element Methods for Fluid Flow Problems, Proceedings 4th Int. Conf. on Finite Element Methods in Water Resources, Hannover, W. Germany, June 1982.
43. Bathe, K. J. and Sussman, T. D. An Algorithm for the Construction of Optimal Finite Element Meshes in Linear Elasticity, in *Computer Methods for Nonlinear Solids and Structural Mechanics*, A.S.M.E., AMD-54, 15-36, 1983.
44. Bathe, K. J., Slavković, R. and Kojić, M. On Large Strain Elasto-Plastic and Creep Analysis, in Finite Element Methods for Nonlinear Problems, (Bergan, P. G. et al., eds), Springer-Verlag, 1985.
45. Olson, L. G. and Bathe, K. J. An Infinite Element for Analysis of Transient Fluid-Structure Interactions, Engineering Computations, in press.
46. Bathe, K. J. and Dvorkin, E. N. A Formulation of General Shell Elements — the Use of Mixed Interpolation of Tensorial Components, Int. J. Num. Methods in Eng., in press.

AIT: ANALYSIS INTERPRETIVE TREATISE

W. Kanok-Nukulchai

*Division of Structural Engineering and Construction, Asian Institute of Technology,
P.O. Box 2754, Bangkok 10501, Thailand*

ABSTRACT

Analysis interpretive treatise (AIT) is a powerful symbolic manipulation program. It was written to serve as a comprehensive "computing tool" for students who need to exercise real-world structural problems in the classes of modern structural analysis and elementary finite element methods. The program is organized in modules and all data units are stored in the centralized data bank. Each individual "processor" module has one specific purpose for data processing. Current program processors include those for matrix operations, and for generation and operations of finite element characteristics. The program can also handle interactions of multiple domains as well as substructuring. Its efficient database concept provides an easy program maintenance and allows future expansion.

THEORETICAL BACKGROUND

The main objective of the analysis interpretive treatise (AIT) program is to enable graduate students to understand the computational procedure while solving large practical structural problems, using modern methods of structural analysis. Most commercial programs (black-box type) are not appropriate for this purpose, as students would not be able to learn basic principles from its usage. Three groups of array processors are available in the program:

(1) *Matrix Manipulations and operations.*

(2) *Generation of characteristic arrays* for a given finite element mesh. A built-in finite element library can be used to generate element arrays of ten standard element types. In addition, the user can include in this library his own element type(s) with little effort.

(3) *General finite element operations* such as assembly of element characteristics, static condensation and substructuring, linear system solution by active-column profile solver, as well as eigen solution by the Jacobi diagonalization method (only in the mainframe version of AIT).

FIELD OF APPLICATION

The program is applicable to matrix and finite element solutions of linear structural problems, subjected to an algorithm proposed by the user through a sequence of "macro" commands. Currently, the program has the following capabilities:

1. General matrix operations and manipulations.

2. Finite element mesh generation. Several meshes (domains) can be handled in the same job for multidomain interactions. Each mesh and all its finite element characteristics will be identified by a user's assigned "mesh name".

3. Formulation of characteristic arrays of finite element mesh(es).

4. Substructuring capability.

5. Multilevel looping capability.

6. Restart capability.

7. Option for in-core or secondary storage for larger arrays.

8. Option for array storage form, either in rectangular form or profile (compacted) form for symmetric, diagonally dense square matrices.

9. An element library which can be easily expanded to include user's own element type(s) in addition to the ten standard element types.

10. In the microcomputer version, the program also allows graphic presentation of 2D finite element mesh and results.

PROGRAM DESCRIPTION

Method. Matrix and finite element methods.

Standard element types. Standard element types currently available in the library are (1) plane stress/strain element, (2) axisymmetric solid element, (3) plane truss element, (4) space truss element, (5) plane beam element, (6) space beam element, (7) St. Venant torsion element, (8) Hertzian contact element, (9) Mindlin plate element and (10) four-node bilinear degenerated shell (BDS) element.

Program structure. AIT is a modularized program with a central databank accessible and amendable to all processor modules. Individual data units are stored in arrays (as compacted form) and can physically reside in core or in a secondary storage. A directory is used to keep track of information about these data units, such as the user's assigned name, array form, length, and location of storage.

HARDWARE COMPATIBILITIES

1. *Mainframe AIT program*

    1. Mainframe computer (IBM, CDC).
    2. Fortran compiler.
    3. The program source (5140 lines) is available in 8 inch standard IBM floppy disk or a magnetic tape.

AIT: Analysis Interpretive Treatise

2. *Microcomputer MICRO-AIT program*
   1. Apple II series microcomputer with two disk drives, and a printer. Future conversion for IBM-PC is planned.
   2. Apple disk operating system (DOS) and Applesoft BASIC language.
   3. The program source (copyable but modification protected to preserve its reliability) is available in one $5\frac{1}{4}$ inch floppy disk.

   EXAMPLES OF APPLICATION

Three examples, which also serve as exercises for students in a course work, appear in the user manual with details of input and the results. The three examples are:

1. *General matrix operations.* This example includes various matrix operations. Students can use this example for practice at the start to gain familiarity with the program as well as its input convention.

2. *A plane strain finite element problem.* This example demonstrates the program finite element capabilities.

3. *Analysis of a five-story building system* by substructuring technique. The building model, depicted in Fig. 1, employs the commonly-used rigid floor slab assumption and thus treats plane frames and shear walls as 2D substructures

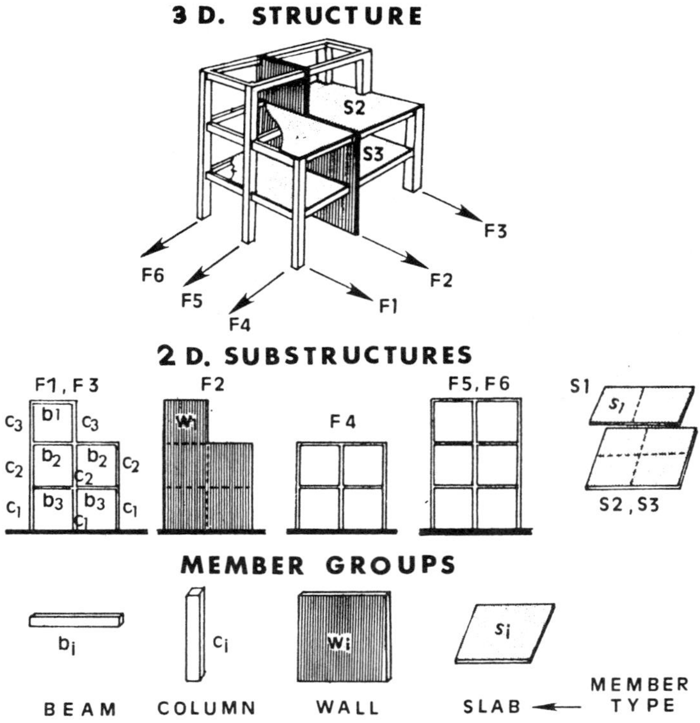

Fig. 1. An example of AIT capabilities - 3D building substructure analysis.

interconnected by floor slabs. This example allows the user to specify an
algorithm for the complete solution of a building system, using substructure
technique such as the flow chart shown in Fig. 2. In this example, students need
to understand the complete procedure to be able to specify a sequence of
operations necessary for the final solution. The following sequence of symbolic
macro-commands is a sample input for establishing an algorithm to form the lateral
stiffness of a frame (substructure) of the five-story building.

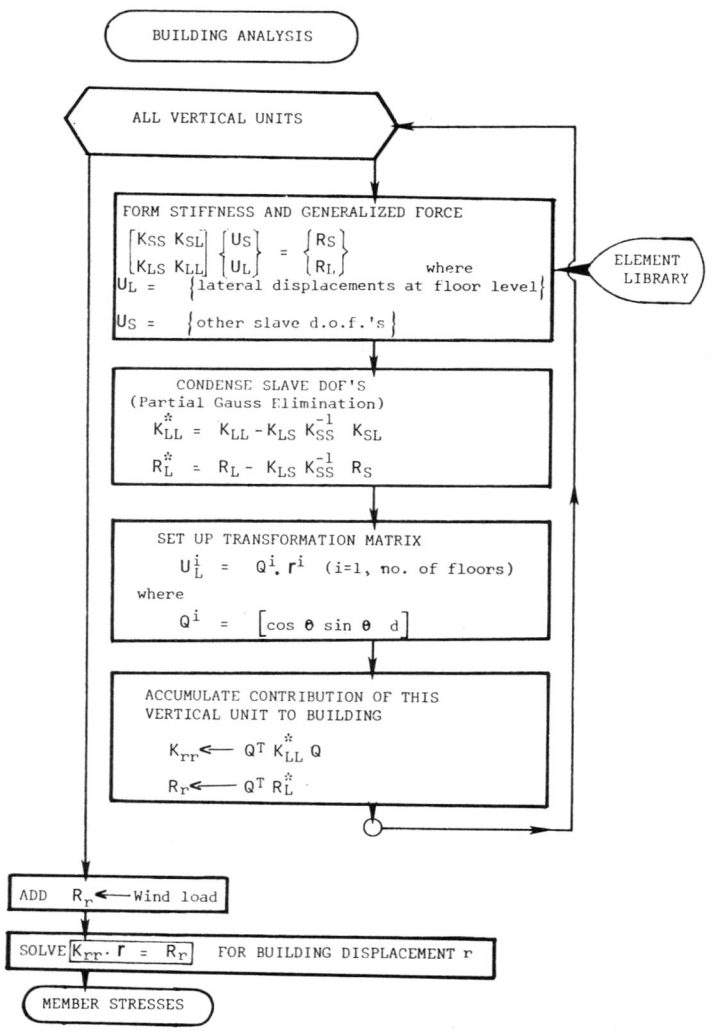

Fig. 2. Algorithm for lateral stiffness of 3D building with rigid floors.

Macro-commands (parameters)	Explanations
MESH(VERT1) - Nodal coordinates - Element connectivity - Boundary conditions - Material properties - Nodal forces/settlements	VERT1 = User's assigned mesh (substructure) name. Data units to be requested for VERT1.
STIFF(VERT1,K1,R1)	Form stiffness matrix K1 and internal force vector R1 for mesh VERT1.
ORDER(K1,R1,ID)	Order K1 and R1 according to vector ID (previously defined) for proper order of partial Gauss elimination.
GAUSS(K1,R1,KL1,RL1,45,40)	Condense the first forty equations and store the remaining in KL1 and RL1.
LOAD(Q,5,15)	Load a transformation matrix Q (5*15).
TRANS(Q,QT)	Transpose Q and store the result in QT.
MULT2(QT,KL1,Q,KS1)	Perform KS1=QT*KL1*Q.
MULT(QT,RL1,RS1)	Perform RS1=QT*RL1.
ASSEMB(KRR,KS1)	Assemble KS1 into KRR.
ASSEMB(RR,RS1)	Assemble RS1 into RR.

AVAILABILITY

The mainframe version of the AIT program is useful and suitable for supplementing graduate students taking courses in modern structural analysis, such as matrix methods of structural analysis, as well as the elementary finite element methods.

The microcomputer version (MICRO-AIT) will be helpful for individuals who hope to learn and master computational methods of structural and general finite element analyses by themselves. With some experience in its usage, practising engineers can also use this program for general computations as well as analysis of medium-size structural problems.

The AIT program is available from Division of Structural Engineering & Construction, Asian Institute of Technology, P.O. Box 2754, Bangkok 10501, Thailand. The mainframe version of AIT is available only to academic institutes free of charge (with nominal handling costs). The microcomputer (Apple II family) version, MICRO-AIT, is available to general users at US$100 including user manual and a demo disk.

ANSYS®: ENGINEERING SOFTWARE WITH THE DESIGN AND ANALYSIS ANSWERS

C. Ketelaar

Swanson Analysis Systems Inc., Johnson Road, P.O. Box 65, Houston, PA 15342-0065, U.S.A.

ABSTRACT

ANSYS is a general purpose, finite element computer program for engineering analysis which is developed, marketed, and supported by Swanson Analysis Systems, Inc. in Houston, PA. ANSYS has the ability to solve a wide range of structural, electromagnetic and heat transfer problems and is used by the design engineer to determine displacements, forces, stresses, strains, temperatures and magnetic fields. Graphics, preprocessing, solution and postprocessing are all integrated in this complete package. These extensive analytic capabilities, in addition to quality customer support and unmatched ease of use, have attracted ANSYS users from many industries including nuclear, aerospace, transportation, medical, petrochemical, steel, electronics, farm equipment and civil construction.

ANSYS is an integral part of the overall CAD environment. It can provide information about physical structures - information which is essential to proper design decisions. Engineers may select optimum materials and construction designs which are indicated by the analysis results and incorporate these modifications early in the design cycle. ANSYS users can simulate two- and three-dimensional models including surfaces, shells, springs, beams and others. These models can be subjected to proposed loading and the resulting stress effects are then available for detailed study.

THEORETICAL BACKGROUND

The matrix displacement method of analysis based upon finite element idealization is employed in the ANSYS program. The structure being analysed is mathematically modelled as an assembly of discrete structural regions (called elements) connected at a finite number of points (called nodal points). If the force-displacement relationship for each of these discrete structural elements is known (the element "stiffness" matrix), then the force-displacement relationship for the entire structure can be assembled using standard matrix methods.[1]

Thermal analyses are performed analogously.

---

*Superscript numbers refer to References at the end of the article.

Plasticity is based on the von Mises yield function and the Prandtl-Reuss flow equations. The material is assumed to be isotropic and the basic technique is the initial stress method. The basic reference used for the development of the plastic load vector is *Plasticity: Theory and Applications* by A. Mendelson.[2]

Each element in the ANSYS program is discussed in section 2 of the *ANSYS Theoretical Manual*. The assumed shape function and the integration points used to generate the element matrices and load vectors are given, as well as further assumptions and restrictions. Certain other aspects of the elements are also discussed in the chapters on the nonlinear capabilities of ANSYS.[3]

The large deflection procedure used in ANSYS is based on an updated stiffness matrix and a changing load vector. Large deflections in ANSYS are defined as including both large displacements and large rotations. A large displacement solution is required whenever the displacements are large enough such that the structure stiffness matrix based on the geometry at the beginning of the iteration does not characterize the deformed structure at the end of the iteration.[4] To accomplish large deflection analyses in ANSYS, an updating of the nodal point locations is made. These updated locations are used in determining the latest element matrices and vectors. The basic working variable for the large displacement option is the total displacement vector, rather than an incremental displacement vector. The updating is done after the element stresses are computed but before the calculation of reactions.[5]

Large rotations are commonly defined as those cases where the sine function of the angle starts to be significantly different than the angle itself. A general rotation is described by a unique axis of rotation (oriented by direction cosines) and by a magnitude of rotation (measured in radians).

FIELD OF APPLICATION

A powerful mesh generation feature in ANSYS gives the user many options for describing the system geometry. The ANSYS element library gives the user the ability to analyze two- and three-dimensional frame structures, piping systems, two-dimensional plane and axisymmetric solids, three-dimensional solids, flat plates, axisymmetric and three-dimensional shells and nonlinear problems including interfaces and cables.

Material behaviours can be simulated and studies in isotropic, orthotropic and anisotropic modes. All elements in ANSYS include isotropic material behaviour. Orthotropic material properties may be used with all plane and solid elastic structural elements and with all heat transfer elements. An elastic structural solid element allows user input of a complete material matrix for general anisotropic behaviour. Also, layered shell elements are available with each layer having its own material and orientation.

The following analysis types are available in ANSYS:

Static - used to solve for displacements, strains, stresses and forces in structures under applied loads. Elastic, plastic, creep and swelling material behaviours are available. Stress stiffening and large deflection effects may be included. Bilinear elements such as interfaces (with or without friction) and cables can be used.

Eigenvalue buckling - used to calculate critical loads and buckling mode shapes for linear bifurcation buckling based on the stress state from a previous static analysis.

Mode frequency - used to solve for natural frequencies and mode shapes of a structure. Stresses and displacements may be obtained by using the displacement, velocity, acceleration, or force spectrum analysis or modal PSD options. A

ANSYS: Engineering Software   33

spectrum may represent a seismic loading.

Nonlinear transient dynamic - used to determine the time history solution of the response of a structure to a known force, pressure, and/or displacement forcing function. Stiffness, mass and damping matrices vary with time and may be functions of the displacements. Friction, plasticity, large deflection and other nonlinearities may be included. An automatic time step procedure is available.

Linear transient dynamic - used to determine the time history solution of the response of a linear elastic structure to a known forcing function. A quasi-linear option includes interfaces (gaps) within the structure or to ground.

Harmonic response - used to determine the steady-state response of a linear elastic structure to a set of harmonic loads of known frequency and amplitude. Complex displacements or amplitudes and phase angles can be input and calculated for output. Stresses may be calculated at specified frequencies and phase angles.

Heat transfer - used to solve for the steady-state or transient temperature distribution in a body. Conduction, convection, radiation and internal heat generation may be included. The calculated temperature distribution may then be used as input to a structural analysis. Other options include phase change, nonlinear flow in porous media, thermal-electric and thermal-fluid flow. The heat transfer analysis can also be used to solve many classes of analogous field equation problems such as torsion of shafts, ideal fluid circulation, pressurized membranes and electrostatics.

Magnetics - used to determine steady-state electromagnetic fields and magnetic forces with either the 2-D vector potential or the 3-D scalar potential method. Linear permanent magnets can be modelled and coil, bar, and arc current sources can be specified. Electric-magnetic field coupling is available so that other current-carrying members of arbitrary geometry can be modelled. Nonlinear materials can be modelled.

Substructures - used to assemble a group of linear elements into one "element" (a superelement) to be used in another ANSYS analysis. It is advantageous to use substructures to isolate the linear portion of a structure within an iterative solution.

Loading inputs for structural analyses include nodal forces and moments, body forces, displacements, pressures and temperatures. Structural constraints may be applied in a user-defined coordinate system. In dynamic analyses, these loads may be random, sinusoidal or arbitrary functions of time or frequency. Loading inputs for heat transfer analyses include fluid convective heating, internal heat generation, radiation and known temperatures or heat flows.

PROGRAM DESCRIPTION

ANSYS employs the finite element analysis (FEA) method for the solutions. The FEA method is a mathematical technique for constructing approximate solutions to boundary value problems. The method involves dividing the solution domain into a finite number of subdomains, or elements, and constructing an approximate solution over the collection of elements. The elements connect at points called nodes, at which continuity of the approximating functions between elements is maintained. FEA programs are general purpose in that they apply the technique to a wide variety of scientific and engineering boundary value problems. Most major civil and mechanical engineering design firms use FEA programs as a design tool since they are very accurate and cost-effective. An engineer can analyze an object and make predictions about its performance before the first model is manufactured.

The ANSYS element library contains over seventy distinct elements. Each element may simulate several different theories as special element options.

Structural

* 2- and 3-D beams and pipes
* axisymmetric with axisymmetric loading
* axisymmetric with nonaxisymmetric loading
* shells and plates
* 2- and 3-D solids
* interface/gaps
* immersed pipe with wave loading
* 3-D reinforced composite/concrete
* crack tip
* plastic hinge
* nonlinear spring

Heat transfer

* compatible structural elements for thermal-stress analyses
* conducting bars, areas, and volumes
* convection links
* radiation links
* surface effect element

Electromagnetic

* 3-D scalar potential with nonlinear BH curves
* 2-D vector potential with nonlinear B-µ curves

Fluid

* contained fluid, 2-D and 3-D
* incompressible fluid flow

Because ANSYS is one integrated package, users have access to easy-to-use pre and postprocessors, as well as many analytic capabilities. Routines are available for facilitating data input, including model generation. A complete set of graphics provides geometry and loading verification in addition to results interpretation. ANSYS can be used in interactive and batch modes of operation. These features, many of which result in cost savings, are discussed in further detail in the following paragraphs.

The input data for the ANSYS program has been designed to make it as easy as possible to define the analysis to the computer. A preprocessor contains powerful mesh generation capability as well as being able to define all other analysis data (real constants, material properties, constraints, loads, etc.). Geometry plotting is available for all elements in the ANSYS library, including isometric, perspective, section, edge and hidden-line plots of three-dimensional structures. ANSYS also generates substructures (or superelements). These substructures may be stored in a library file for use in other analyses. Substructuring portions of a model can result in considerable computer-time savings for nonlinear analyses.[6]

ANSYS data may include parametric input. By varying some data items, while holding others constant, the user can conveniently input a series of repeating commands. This may be used to create models for design studies. Anticipated variations of the geometry are defined as parameters. Each solution then requires changing only one or two items instead of dozens.

ANSYS uses the wave-front (or "frontal") direct solution method for the system of simultaneous linear equations developed by the matrix displacement method, and gives results of high accuracy in a minimum of computer time. The program has the ability to solve large structures. There is no limit on the number of elements used in an analysis. There is no "band width" limitation in the analysis definition; however, there is a "wave-front" restriction. The "wave-front" restriction depends on the amount of memory available for a given problem. Up to 2000 degrees of freedom on the wave-front can be handled in a large-core.[7] With virtual memory machines and a virtual equation solver in ANSYS, very large analyses can be solved.

Postprocessing routines are available for algebraic modification, differentiation, and integration of calculated results. Root-sum-square operations may be performed on seismic modal results. Response spectra may be generated from dynamic analysis results. Results from various loading modes may be combined for harmonically loaded axisymmetric structures. Post routines also plot distorted geometries, stress contours, safety factor contours, temperature contours, mode shapes, time history graphs and stress-strain curves.

Graphics capabilities provide many options for verification of model geometry and loads. Windowing on a model can be done by limiting included nodes, included elements, or included geometric distances. The geometry may be limited in any defined coordinate system. Surfaces and defined coordinate systems may be plotted. All boundary conditions (displacements, forces, moments, pressures and master degrees of freedom) may be displayed on element or node plots. Shrinking elements helps the user to verify that no elements are missing. Element, node, material, type or member property numbering can be shown on plots of the model. Section views through three-dimensional structures, plots of model edges and hidden line plots are all available for further checking of model geometry and for presentation in reports. ANSYS graphics also provide many plot display options such as multiple windows on one screen, choice of focus point, varying distance from object, perspective and zoom. Colour graphics is available in pre and postprocessing. These colour plots help interpret both model geometry and results

All portions of ANSYS can be operated in either an interactive or a batch mode. When running interactively, the user can take advantage of on-line documentation which provides an explanation for each command at the terminal as well as immediate graphic displays to verify the model and correct input errors. Analyses can be done efficiently by creating the model interactively, obtaining the solutions in a batch mode and interpreting the results interactively.

ANSYS's position in the CAD/CAM market is that of a computer-aided analysis (CAA) tool. ANSYS can be used to efficiently generate the finite element model, but it is also able to evaluate a CAD-created design and modify it, if necessary. The ability of ANSYS to interface with many CAD systems makes it a powerful product in this cycle.

ANSYS is written in FORTRAN and although the source program is not available, SASI (Swanson Analysis Systems, Inc.) does work with hardware vendors to recode key routines in assembly languages or microcode when appropriate. Since ANSYS is a complete program with preprocessing, solution, postprocessing, and graphics, it is made up of many subroutines. The program is modular; that is, logical functions, such as matrix operations, a command processor or graphics, are done in one group of subroutines. An updated version of ANSYS will be released in the spring of 1985. Some new developments include allowing the user to create and input his/her own element routine, as well as giving the user options for entering his/her own creep law or preprocessing command.

HARDWARE COMPATIBILITIES

The minimal hardware configuration typically includes:

floating point in hardware
2 MB main memory (more is desirable)
150 MB disc (more is desirable)
tape drive (or distribution media)
printer
graphic terminal

ANSYS currently supports over fifteen types of computer hardware, including mainframes, minis and workstations. In the near future, ANSYS will support additional computers - some of which will be micros. A wide variety of plotters and graphics devices are also supported. A 1-2 page FORTRAN source program to link with plotters is available. ANSYS is distributed on either floppy disk or magnetic tape, depending on the computer hardware.

EXAMPLES OF APPLICATIONS

ANSYS has been used to determine bolt preload in the analysis of bolted assemblies as reported by F. S. Kelley in *Finite Element Stress Analysis of Bolted Connections*. Bolted assemblies (Fig. 1) occur in many machines. They contain two or more nearly axisymmetric parts connected by a large number of bolts. They may be loaded by spin, torque, axial force or temperature gradient. The analysis of a bolted connection is sometimes performed as though the structure were axisymmetric. The analyst supplies orthotropic properties in the area of the bolt holes and assumes that nothing varies significantly in the circumferential direction. A healthy stress concentration factor is expected in the results of such an analysis.

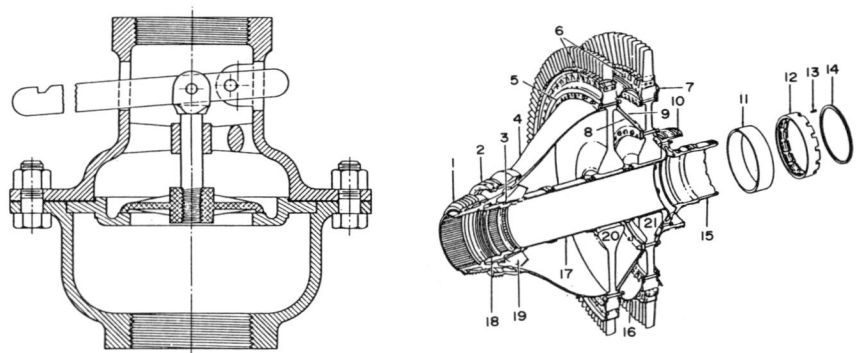

Fig. 1. Typical bolted assemblies.

Whether the analysis is two- or three-dimensional, nonlinear interface (gap) elements must be placed in all areas of potential contact between the components. Interface elements transmit load only where the surfaces are in contact. This means that the assembly components must be modelled so that all potential contact areas have matching node locations. The interface elements make an iterative solution necessary. The solution time can be reduced dramatically by substructuring. The total cost of the analysis with substructuring may be less than the cost of two iterations without substructuring.[9]

In *Use of Finite Element Analysis for Improving Disk Drive Servo Performance*. Dr. Hartem R. Radwan, Jai V. Chokshi, and L. C. Goss summarize the results of a dynamic FE analysis using ANSYS to investigate the cause of a servo loop instability in a disk drive feedback system. The feedback system is used to force the disk drive actuator mechanism (Fig. 2) to remain "on-track" during data transfer to and from the disk drive, or to move from one track to another.

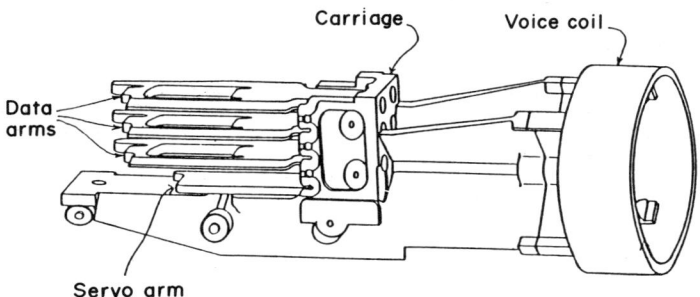

Fig. 2. Actuator carriage with data and servo arms.

Table 1 shows the number of elements and element types used for each component in the analysis. Thus, both the load arm and gimbal are represented by thin shell/plate elements, while the slider and data arm are modelled by 3-D isoparametric solid elements. The air bearings supporting the slider are represented by one-dimensional axial spring elements.[10]

TABLE 1 Finite Element Types Used in Analysis

Modelled region	Element type	ANSYS element	No. of elements	Option keys
Load Arm	Thin Shell/Plate	STIF63	284	000000
Gimbal	Thin Shell/Plate	STIF63	72	000000
Slider	3-D Isopar. Solid	STIF45	8	110000
Air Bearing	Axial Spring	STIF14	4	000000
Data Arm	3-D Isopar. Solid	STIF45	212	000000

A mode/frequency analysis of the load arm assembly alone was performed, as well as the data arm assembly which includes four load arms, modelled as substructures. The model was then used to perform a frequency response analysis to estimate slider vibration sensitivity to carriage motions in order to determine the cause of the servo system instability at 2.8 kHz. The analysis showed the flexibility of the gimbal caused excessive displacements near the resonant frequency. As a relatively simple remedy, a flat gimbal was considered. An analysis of this modified gimbal showed the excessive displacements could be reduced, but displacements at other points could inadvertantly be increased if caution was not used. Hence, the behaviour of the design modification, as well as the original design, was predicted.

ANSYS has also been used to develop a thermal analysis design tool for X-ray vacuum tube anode assemblies as reported by Jeffery W. Herrick and Leonard J. Vetrone in their paper *Finite Element Thermal Analysis of an X-ray Vacuum Tube*. The finite element technique was employed to thermally represent the X-ray tube anode. This was accomplished by using heat transfer elements to model conduction and radiation with assumed boundary conditions. These elements and boundary conditions were varied to investigate possible improvements in heat transfer. The obvious benefit is the ability to optimize designs and design revisions prior to hardware fabrication and test.

The study focused on the rotating anode of the X-ray tube shown in Fig. 3. Heat input and many components of the anode assembly are axisymmetric. Thus, much of the conductive heat transfer mechanism was modelled using an axisymmetric ring element with two-dimensional conduction capability. However, several components of the anode assembly are non-axisymmetric. The bearings conduct heat through a finite number of balls; the sliding rear bearing retainer makes only line contact with the anode stem and heat is transferred by conduction through four bolts connecting the rotor to the bearing hub. An axisymmetric heat conducting bar was used to approximate heat transfer through these components. Radiation to the bell jar surroundings and between individual anode components was modelled using radiation link elements.

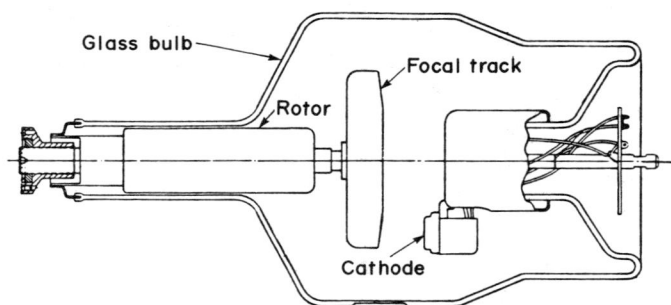

Fig. 3. Schematic diagram of an X-ray tube with rotating anode.

A bell jar experiment was performed to verify the finite element results. The anode was mounted on a water-cooled room temperature base inside an evacuated glass bell jar. Current was passed through a toroidal filament coil to produce thermionic emission of electrons. The electrons were accelerated towards the anode by an electric potential. A suitable heating effect with negligible X-ray generation was obtained by this process. Thermocouples were located at numerous points of interest to measure temperatures. Figures 4 and 5 show plots of these measurements together with temperatures obtained by the finite element model. These results confirm that both transient and steady state temperature distributions can be predicted with a high degree of accuracy.[11]

REFERENCES

1. Kohnke, Peter C. (1983). *ANSYS Theoretical Manual*. Houston, PA: Swanson Analysis Systems, Inc.
2. Kohnke, p. 4.1.8.
3. Kohnke, p. 2.0.1.
4. DeSalvo, Gabriel J. and John A. Swanson.(1983). *ANSYS User's Manual*. Houston, PA: Swanson Analysis Systems, Inc.
5. Kohnke, p. 3.2.2.
6. DeSalvo and Swanson, p. a.1.
7. DeSalvo and Swanson, p. a.1.
8. DeSalvo and Swanson, p. a.1.
9. Kelley, F. S. (1983). *Finite Element Stress Analysis of Bolted Connections*. Houston, PA: Swanson Analysis Systems, Inc.
10. Radwin, Dr. Hatem R., Jai V. Chokshi and L. C. Goss (1983). Use of finite element analysis for improving disk drive servo performance. *ANSYS Conference Proceedings*. Ed. D. E. Dietrich. Houston, PA: Swanson Analysis Systems Inc.

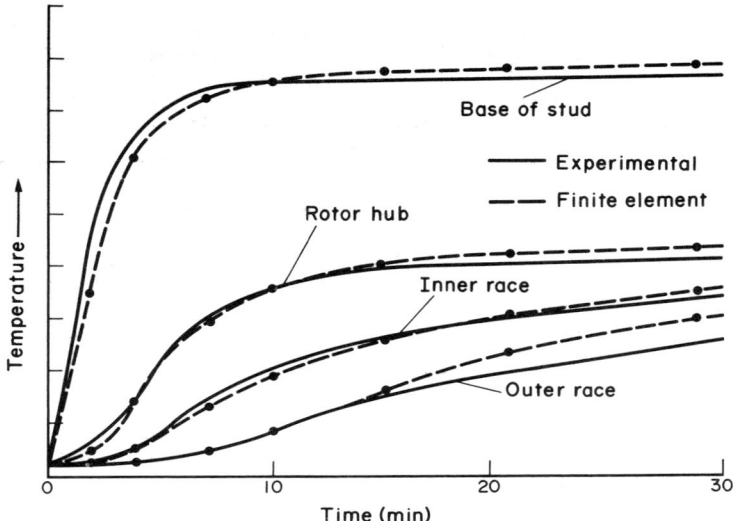

Fig. 4. Comparison of experimental (bell jar) and analytical (finite element) temperatures.

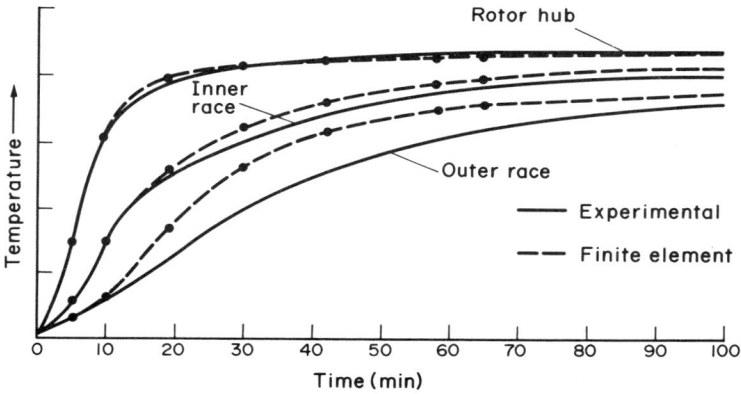

Fig. 5. Comparison of experimental (bell jar) and analytical (finite element) temperatures.

11. Herrick, Jeffery W. and Leonard J. Vetrone (1983). Finite element thermal analysis of an X-ray vacuum tube. *ANSYS Conference Proceedings*. Ed. D. E. Dietrich. Houston, PA: Swanson Analysis Systems, Inc.

AXISYMMETRIC: MICROCOMPUTER PROGRAMS FOR AXISYMMETRIC PROBLEMS IN STRUCTURAL AND CONTINUUM MECHANICS

C. T. F. Ross

Department of Mechanical Engineering, Portsmouth Polytechnic, U.K.

ABSTRACT

This package contains thirteen computer programs, which have been written in BASIC for microcomputers The programs include the static analysis of plates, ring-stiffened cylinders, domes and plate/cylinder combinations, and the dynamic analysis of cones and domes in-vacuo, and plates submerged in a liquid.

In addition to this, the package contains two programs on the static analysis of thick-walled cones and domes, two programs on the elastic instability of thin-walled cones and domes, and two programs on axisymmetric field problems.

THEORETICAL BACKGROUND

For *static analysis*, the stiffness matrices $[K]$ were of banded form for all cases.

To take advantage of the sparsity of $[K]$, the upper half of the band was rotated in a clockwise direction, as shown by equation (1), and the solution of the simultaneous equations was carried out through triangulation. This process considerably reduced the computational time and space, so that the problem could be tackled on a 64k microcomputer. The process also had the advantage of increasing the numerical precision of the solution.

$$[K] = \begin{bmatrix} \text{Non-zero elements} \\ 0 \end{bmatrix} \qquad (1)$$

For *vibration analysis*, without damping, the equation of motion is given by:

$$\left| [K] - \omega^2 [M] \right| = 0 \qquad (2)$$

where $[M]$ = mass matrix of the entire structure,

ω = radian frequency.

Unfortunately, however, equation (2) is not very suitable for a microcomputer, as satisfactory solution of many problems would have required the inverse of matrices larger than of order 80. To overcome this problem, Irons's[1]* eigenvalue economizer technique was adopted, where all displacements, except for one, were eliminated at all the nodes, with the exception of the last node.

This process had the effect of reducing the size of the problem by a factor of nearly four, so that solution on a microcomputer became feasible. Thus, the equation of motion took the form of equation (3):

$$\left| [K_r] - \omega^2 [M_r] \right| = 0, \quad (3)$$

where $[K_r]$ = the reduced stiffness matrix,

$[M_r]$ = the reduced mass matrix.

A similar process was adopted for *elastic instability*, where the matrix equation took the form of equation (4):

$$\left| [K_r] - \lambda [K_{Gr}] \right| = 0, \quad (4)$$

where $|K_{Gr}|$ = the reduced geometrical stiffness matrix,

λ = eigenvalue (proportional to the buckling load).

For the vibration of *submerged plates*, the added virtual mass was determined by the finite element method.[2]

FIELDS OF APPLICATION

CIRCPLATE. This program can calculate nodal displacements and bending stresses in tapered circular plates under lateral pressure and lateral concentrated ring loads. The solution is based on small deflection elastic theory.

CYLPLATE. This program can calculate nodal displacements and stresses in thin-walled cylinders, which can be blocked off at one end by a flat circular plate. Step variation is allowed for in wall and plate thickness and the vessels can be subjected to internal or external pressure.

RINGCYL. This program can calculate nodal displacements and stresses in thin-walled cylinders under uniform or hydrostatic pressure. Step variation is allowed for in wall thickness and the stiffener can be stiffened by rings of unequal size, spaced at unequal intervals. The stiffening rings can be both internal and external.

ACMC. This program can calculate nodal displacements and stresses in thin-walled axisymmetric shells with constant meridional curvature. The pressure loading can be internal or external.

THICKCONE. This program can calculate nodal displacements and stresses in thick-walled cones under lateral pressure. The wall thickness of the element is of uniform taper.[3]

THICKAXI. This program is a more sophisticated version of THICKCONE, which allows for a parabolic variation in the wall thickness and in the meridional curvature of the element.

---

*Superscript numbers refer to References at the end of the article.

AXISYMMETRIC: Microcomputer Programs for Axisymmetric Problems

VIBCONE. This program can calculate the natural frequencies and eigenmodes of thin-walled cylinders and cones and cone-cylinder combinations.

CONEBUCKLE. This program can calculate the elastic instability buckling pressures of cylinders, cones and cylinder-cone combinations under uniform external pressure.

VIBDOME. This program can calculate the natural frequencies and eigenmodes of thin-walled domes and cylinder-cone-dome combinations.

DOMEBUCKLE. This program can calculate the elastic instability buckling pressures of thin-walled domes and cylinder-cone-dome combinations under uniform external pressure.

2DAXIFIELD. This program solves the axisymmetric two-dimensional Poisson's equation with complex boundary conditions.

3DAXIFIELD. This program solves the axisymmetric two-dimensional Poisson's equation with complex boundary conditions.

VIBSUBPLATE. This program calculates the natural frequencies and eigenmodes of an axisymmetric plate submerged in a liquid. The liquid is assumed incompressible and can be on one or both sides of the plate.

PROGRAM DESCRIPTION

The method adopted in twelve of the thirteen programs was the finite element method, but for the program RINGCYL, the solution was based on a slope-deflection method.

All elements were axisymmetric, but the deformation pattern for VIBCONE, CONEBUCKLE, VIBDOME and DOMEBUCKLE was of sinusoidal form in the circumferential direction. For all the other cases, the function was assumed to be axisymmetric.

All the programs were written in interactive BASIC, which allowed the capabilities of introducing new lines of subroutines with little effort.

HARDWARE

The minimal configuration is a microcomputer with a TV/monitor, a home tape recorder and a printer, although it is possible to do without the printer by altering the program.

The programs have been written for CBM PET/64 and for Apple II +/e, and it is a simple matter to modify the programs so that they are suitable for other micros, minis and mainframes.

The programs are available on cassette tapes and on $5\frac{1}{4}$ inch floppy disks.

EXAMPLES OF APPLICATION

CIRCPLATE. This program can be used in circular plates, as shown in Fig. 1.

The boundary conditions can be simply-supported or clamped, and can be on the periphery or at the centre of somewhere in between.

CYLPLATE. This program can be used in thin-walled circular cylinders and plate-cylinders combinations, as shown in Fig. 2.

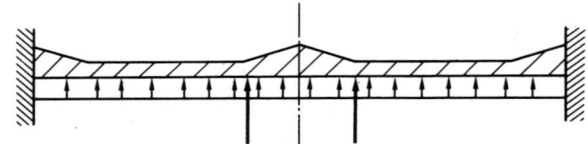

Fig. 1. Tapered disc under lateral loading.

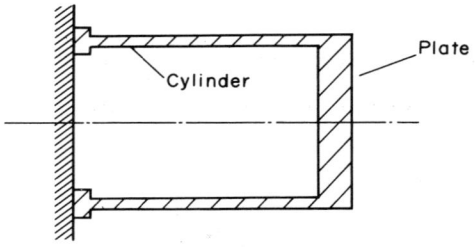

Fig. 2. Cylinder blocked off by a disc.

RINGCYL. This program can be used in circular stiffened cylinders, as shown in Fig. 3.

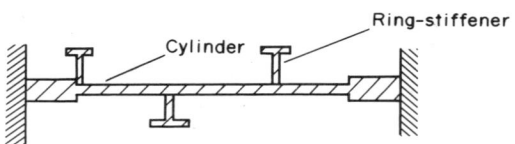

Fig. 3. Ring-stiffened cylinder.

The lateral pressure can be hydrostatic or uniform and it can be either internal or external.

ACMC. This program can be used in thin-walled axisymmetric shells, with or without meridional curvature. Apart from analysing domes, the program can analyse dome-cylinder-cone combinations and hour-glass figures, as shown in Fig. 4.

Fig. 4. Dome-cylinder-cone combination.

THICKCONE. This program can be used in thick-walled axisymmetric structures. This work is of much importance in civil and ocean engineering, where for the latter the program can be applied to submersibles, as shown in Fig. 5.

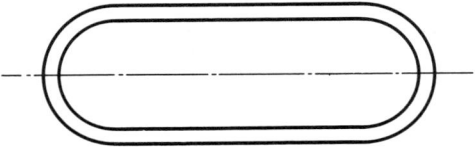

Fig. 5. Thick-walled submersible.

THICKAXI. This program is a more sophisticated version of THICKCONE, as it allows for a parabolic variation in wall thickness and meridional curvature. In addition to solving problems of the type shown in Fig. 5, it can be analyse domes, as shown in Fig. 6, and also thick-walled dome-cylinder-cone combinations.

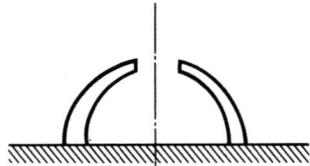

Fig. 6. Thick-walled dome.

VIBCONE. The program can be applied to structures of the type shown in Fig. 4 and also to cooling towers, as shown in Fig. 7.

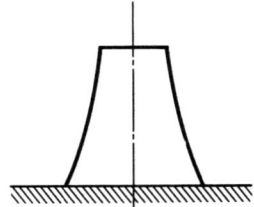

Fig. 7. Cooling tower.

CONEBUCKLE. The pressure must be external and a typical buckling mode, which is of lobar form, as shown in Fig. 8.

VIBDOME. The program can be applied to structures as in Fig. 4, with positive meridional curvature or to problems with negative meridional curvature, such as cooling towers.

DOMEBUCKLE. This program can calculate the elastic buckling pressures of domes under external pressure. If the domes are oblate, they tend to buckle axisymmetrically, as shown in Fig. 9, but if the dones are hemispherical or prolate, they tend to buckle in a lobar manner, as shown in Fig. 10.

2DAXIFIELD. The program is applicable to:

(1) heat conduction;

Fig. 8. Shell instability of circular cylinders.

Fig. 9. Buckling pattern for oblate domes.

(2) fluid flow;
(3) seepage.

3DAXIFIELD. The program can be applied to a large number of problems including those mentioned under the previous sub-heading.

VIBSUBPLATE. This package is applicable to ocean, aerospace, mechanical, civil and nuclear engineering and to pressure vessel and boiler technology.

The computation time varies for each program and for the problem solved, and in

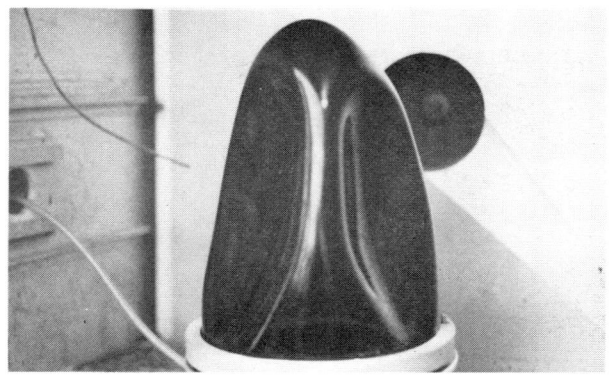

Fig. 10. Buckling pattern for hemispherical or prolate domes.

the case of VIBCYL, CONEBUCKLE, VIBDOME and DOMEBUCKLE, the computation time on a CBM PET for some problems amounted to several hours.

However, by using compiled BASIC, it was possible to considerably decrease the computation time. For example, by using the PETSPEED compiler, the computation time on a CBM PET was decreased by a factor of about four. Two problems that occurred when using PETSPEED were:

(1) Dynamic arrays could not be used.
(2) The priority of unary minus was reversed to its BASIC equivalent.

REFERENCES

1. Irons, B. (1965). Structural eigenvalue problems: Elimination of unwanted variables. *J.A.I.A.A.* **3**, 961.

2. Zienkiewicz, O. C. (1971). *The Finite Element Method in Engineering Science*. McGraw-Hill.

3. Ahmad, S., B. M. Irons and O. C. Zienkiewicz (1968). Curved thick shell and membrane elements with particular reference to axisymmetric problems. *2nd Conf. Matrix Meth. Struct. Mech.* Wright-Patterson A. F. Base, Ohio.

BEASY: A BOUNDARY ELEMENT SYSTEM FOR STRUCTURAL ANALYSIS

C. A. Brebbia

Computational Mechanics Institute, Ashurst Lodge, Southampton, U.K.

ABSTRACT

This paper describes the structural analysis capabilities of the boundary element analysis system (BEASY) developed at the Computational Mechanics Centre, Southampton, England. The code is based on the boundary element technique and can be used to solve two-dimensional, axisymmetric and three-dimensional potential and elastostatics problems. Only the boundary needs to be discretized, thus reducing by one dimension the data required to run a problem. The code has excellent pre- and postprocessing facilities and can be interfaced to a range of well-known data generating and postprocessing systems.

THEORETICAL BACKGROUND

The finite element method (FEM) has become well established as a valuable tool for the solution of a wide variety of engineering problems. The FEM requires the discretization of the domain into a number of elements over which the solution is approximated by some function which satisfies in an average form the governing equations. A set of equations is then set up from which the solution at various points in the domain is obtained, to the best approximation allowed by the approximating functions and the boundary conditions.

An alternative approach to FEM is to use functions which satisfy the differential equation in the domain but not the boundary conditions. The boundary is then divided into elements and the boundary values assumed to vary in some manner within these elements. A set of equations may then be formulated in temrs of nodal values, with the nodes this time only on the boundaries. The attractions of this type of approach, known as boundary solutions, are obvious. Only the boundary needs to be discretized, thus reducing by one dimension the nodal coordinates and connectivity tables which makes the finite element method so tedious. The resulting equations are fewer in number. This idea is the basis of the boundary element method.[1,2]*

The mathematical formulation of the boundary element method is more complicated than for the FEM. It is usually based in applying fundamental solution - i.e. Green functions in infinite domains - as influence functions, which identically

---

*Superscript numbers refer to References at the end of the article.

satisfy the governing equations of the problem. An added advantage of this
approach is that it intrinsically allows for the solution of problems with
infinite domains.

Although the mathematical formulation of BEM has been successfully applied to
nonlinear and time-dependent problems[2] as well as to linear static analysis, the
latest version of the BEASY code is concerned with the solution of linear static
cases. Further options and new versions will deal with elastoplastic, time-
dependent potential, elastodynamics and general wave propagation problems.

One of the most interesting features of BEASY is that it allows for
discontinuities in the unknown functions in between elements. This property is a
consequence of having passed all continuities requirements (or derivatives) to
the fundamental solution rather than leaving them on the approximate functions.
The functions only need to be piecewise continuous within an element and are of
special interest for problems in which the variables under consideration are
discontinuous between elements. The convergence of these elements is reasonably
fast in many cases and they have the added advantage that different shape
elements can be more easily combined since there are no continuity conditions to
be satisfied between elements.

FIELD OF APPLICATION

The general architecture of BEASY is shown in Fig. 1. It consists of eight
independent modules for the solution of problems in potential theory - including
temperature - linear isotropic stress analysis and pre- and postprocessing.

Geometrical Field of Application

BEASY can be used to solve two-dimensional, axisymmetric and three-dimensional
solids, the latter including one-dimensional elements for certain range of
problems.

Materials

The code allows for linearly elastic isotropic materials, including thermal
effects. The potential codes also include anisotropic materials.

Analysis Capabilities

The code can analyse linearly isotropic elastic bodies and potential problems.
The structures are two-dimensional, axisymmetric or three-dimensional solids.

Further capabilities under development include elastoplasticity, time-dependent
potential and elasticity problems and elastodynamics. A limited range of options
presently exist for elastoplasticity, (including 2D and axisymmetric options),
time dependent potential (also 2D and axisymmetric case) and a wave propagation
code marketed independently.

Loadings and Boundary Conditions

The boundary conditions allowed for in the potential modules are potential or flux
density prescribed or a linear combination of the two (generally called heat
transfer boundary condition). Nonlinear boundary conditions as specified by
clients for a particular application have also been modelled and this can be
easily done, especially in the field of cathodic protection where the potential

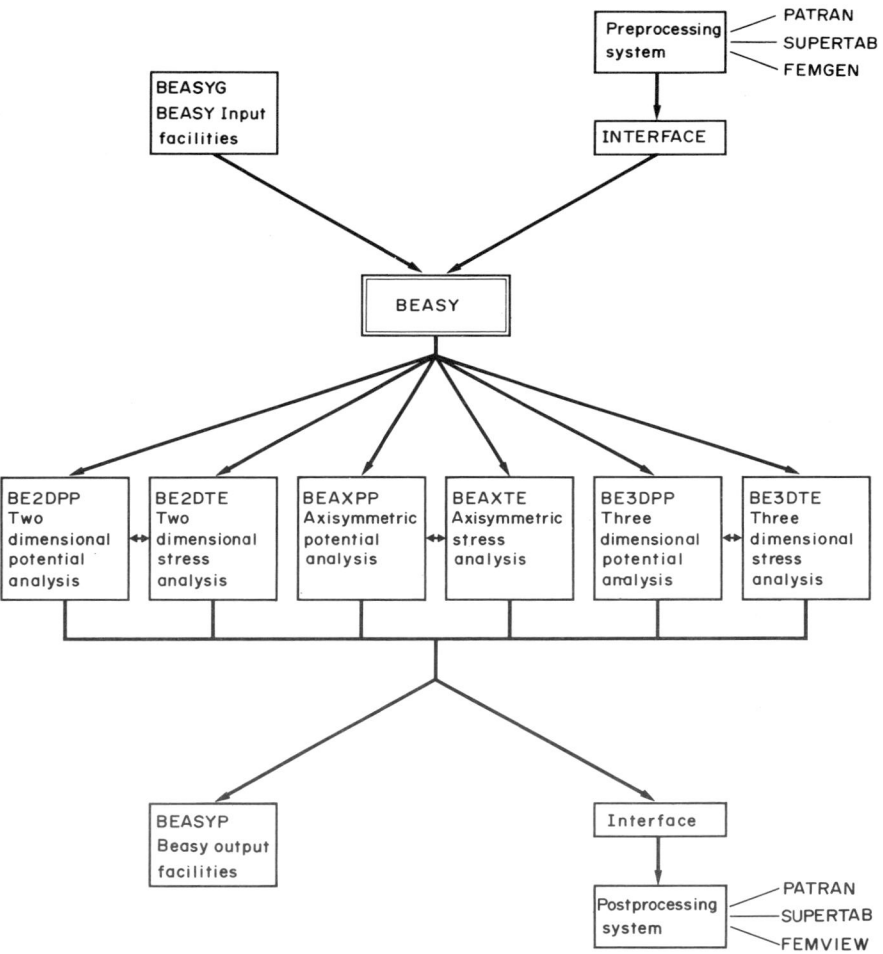

Fig. 1. General architecture of the BEASY system.

and flux (current) density on the cathode is related by a polarization curve. The 2D and 3D potential modules can also model concentrated point and line sources which is a useful feature in many practical applications.

The stress analysis modules allow for prescribed displacement, loads or spring boundary conditions. These can be entered either in the global coordinate system or in a local system, one of whose axes always coincide with the normal to the boundary surface. This load system is not only useful when applying boundary conditions of a single type but is absolutely essential when specifying a mixed type of boundary condition, such as sliding, where the displacement is prescribed normal to the boundary and the loading is given as tangential to the boundary.

Problems with gravitational or rotational loading and problems where the stresses are due to a steady state thermal loading may also be analysed. These problems involving body effects or body forces may be analysed simply by converting domain integrals into boundary terms.[2] Hence even for these cases the user needs to provide the boundary data only.

In the case of thermoelastic problems it is necessary to first solve the potential problem to obtain the temperatures and fluxes at the boundary nodes. This information is then fed to the stress analysis module which then calculates the thermal stress. Exactly the same data file may be used for both analyses.

PROGRAM DESCRIPTION

The code is based on the BEM and uses higher order discontinuous elements and fundamental solutions as influence functions.

Method

The code uses the boundary element method as described in references 1 and 2.

Type of Elements

BEASY uses constant, linear and quadratic discontinuous boundary elements as illustrated in Fig. 2. Their geometry is given in terms of boundary values but

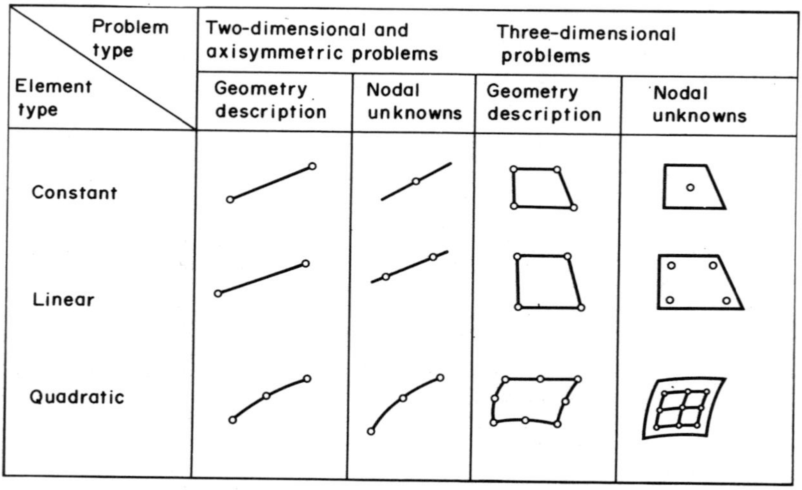

Problem type / Element type	Two-dimensional and axisymmetric problems		Three-dimensional problems	
	Geometry description	Nodal unknowns	Geometry description	Nodal unknowns
Constant				
Linear				
Quadratic				

Fig. 2. Main element types.

the variables - temperatures, fluxes, displacements or tractions - are represented as piecewise continuous. These elements are discontinuous because the nodes are situated within the element boundaries rather than on the edge of the element. The reason for choosing this element is that it is easy to mix discontinuous elements putting quadratic elements in regions of rapidly varying stresses (or temperature) and linear or constant elements elsewhere. It is not necessary for the parts defining the element geometry to be common to several elements. Figure 3 shows part of a mesh for a 3D problem which is perfectly valid mesh for BEASY.

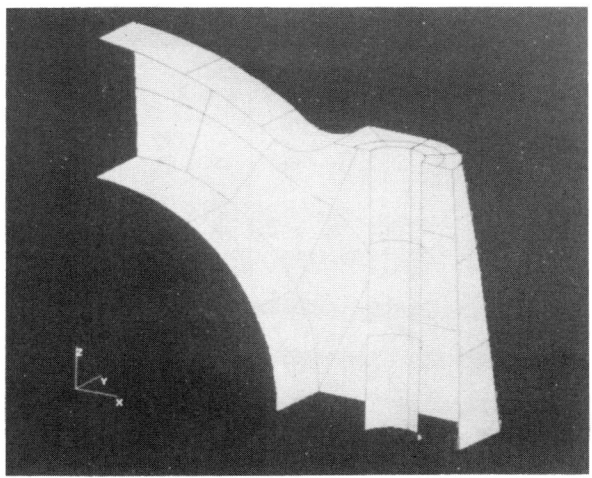

Fig. 3. Bearing cap analysis (Note the use of discontinuous elements).

but would be difficult to handle if continuous elements were used. In this way the problem of "facing out" elements from areas of high density are minimized.

Only quadrilateral elements are available for 3D problems in the current version of the code (version 2). The next version will contain triangular elements as well.

The 3-D quadratic element is a complete bi-quadratic element with 9 nodes as opposed to the commonly used 8 nodes quadrilateral. The reason for choosing a 9-node element is because classical BEM is a collocation technique and 9 nodes give rise to a more even pattern of collocation points.

Program Structure

BEASY is highly user oriented and well suited to CAD applications. The code can be used in conjunction with well-known preprocessors and solid modelling systems such as PATRAN, SUPERTAB or FEMGEN. It also has its own preprocessing facilities, BEASYG which are extremely simple to use as the BEM requires much less data preparation than FEM. BEASYG automatically generates meshes for two-dimensional, axisymmetric and three-dimensional cases. BEASYG carries out some checks on the data.

For postprocessing the user can employ PATRAN, SUPERTAB or FEMVIEW with the relevant interfaces or can rely on BEASY's own facilities, a module called BEASYP. An interface with MOVIE/BYU code is now being developed. Graphics features including colour are possible in all these postprocessors.

BEASY is written in FORTRAN 77 and the codes comprise six distinct stages as shown in Table 1. The analysis may be started or stopped at each step.

TABLE 1 Analysis Steps

Step	Input required	Computation	Output generated
1	Data file	Check on data	
2	Data file	Formation of influence matrices	Influence matrices
3	Data file Influence matrices	Application of the type of boundary conditions to form the system matrix	System matrix $\underset{\sim}{A}$
4	Data file System matrix	Reduction of the lhs of system of equations	Reduced left hand side
5	Data file Reduced left hand side	Application of the *magnitude* of the boundary conditions to form the rhs vector. Reduction and back-substitution to obtain boundary solution	Boundary solution
6	Data file and boundary solution	Computation of results at internal points	Results at internal points

The BEM originates from applying the techniques of finite element discretization to the boundary integral formulation of the problem. This results in the formation of influence matrices (usually cased $\underset{\sim}{H}$ and $\underset{\sim}{G}$) which describes the behaviour at each node due to a unit excitation at another. These matrices are related by the equation,

$$\underset{\sim}{H}\,\underset{\sim}{U} = \underset{\sim}{G}\,\underset{\sim}{P} + \underset{\sim}{B} \tag{1}$$

where $\underset{\sim}{U}$ is the vector of nodal potential or displacements
$\underset{\sim}{P}$ is the vector of nodal fluxes or loadings
$\underset{\sim}{B}$ is the vector which results from sources within the problem or body forces

The $\underset{\sim}{H}$ and $\underset{\sim}{G}$ matrices are analogous to the matrices which one obtains when using the FEM. The fact that there are two matrices and not one is a consequence of the mixed character of the BEM solutions, i.e. one works with potential and fluxes, displacements and stresses, etc. This mixed approach generally makes BEM results more accurate than FEM solutions.

Applications of the boundary conditions enables the elements of the matrices to be rearranged so that one can write,

$$\underset{\sim}{A}\,\underset{\sim}{X} = \underset{\sim}{B}\,\underset{\sim}{F} + \underset{\sim}{B} \tag{2}$$

where $\underset{\sim}{A}$ is the system matrix

$\underline{X}$ is the vector of unknowns
$\underline{B}$ is the complementary matrix
$\underline{F}$ is the vector of prescribed boundary values

The system matrix $\underline{A}$ and the complementary matrix $\underline{B}$ are stored on disc.

Gauss elimination is now applied to the system matrix $\underline{A}$ so that it may be solved.

BEASY uses an out of core solver which enables large problems to be solved efficiently on quite small machines.

The final step calculates the value at internal points using the boundary solution just obtained.

The reason for carrying out the analysis in the manner described above is to minimize the run times required for repeated analyses. The influence matrices are dependent only on the mesh geometry. Once the user has got that right he should never need to form the influence matrices more than once. The analysis for subsequent boundary conditions need only be restarted at step 3 if the *type* of boundary condition has been altered, e.g. a prescribed displacement is specified where previously the load was prescribed. If only the *magnitude* of the boundary condition is altered, or new forces or body forces are added then it is not necessary to repeat steps 3 and 4 but the run may be restarted at step 5. Thus both the time consuming steps of forming the influence matrices and of decomposing the system matrix are avoided.

Once the boundary solution is obtained the solution at internal points is calculated by a fairly simple procedure. No further equation solution is required. If after looking at the results the user decides results are required at some other internal points then only step 6 needs to be repeated.

Symmetry. Symmetry is handled by the simple expedient of reflecting the boundary about the plane of symmetry and continuing the boundary integration around the reflected part of the structure. By making use of the fact that symmetry exists the number of equations is reduced. Furthermore, no elements are required on the plane of symmetry. The time taken to compute the influence matrix is outweighed by the reduced amount of data preparation. Some modules allow also for antisymmetry.

Zones or subregions. Although BEASY does not enable the user to model problems with continuously varying material properties, piecewise constant properties may be handled by considering each part of the problem as a boundary element zone or subregion. This is a very powerful facility as it may also be used in certain types of problems to reduce the total run time and its effect is to produce a banded system of equations.

HARDWARE COMPATIBILITIES

BEASY runs in a large variety of hardware and hardware configurations, starting from simple PC to supercomputers with array processors.

Types of computers. BEASY system is highly transportable and can be run in any computer with FORTRAN 77 compiler. Some machines where BEASY is now operating include: VAX, PRIME, CRAY-1, APOLLO, IBM, UNIVAC, CDC, PDP, and others.

Peripherals. BEASYG and BEAYSYP require a plotter.

Operating system. Any.

Media. As requested by customer.

EXAMPLES OF APPLICATIONS

Test Case

The basic test case selected here is two-dimensional for simplicity. Figure 4 shows the BE mesh used for stress analysis of a fillet when subjected to uniaxial tension load. Notice that no elements are required in this case on the plane of symmetry. The elements used were quadratic.

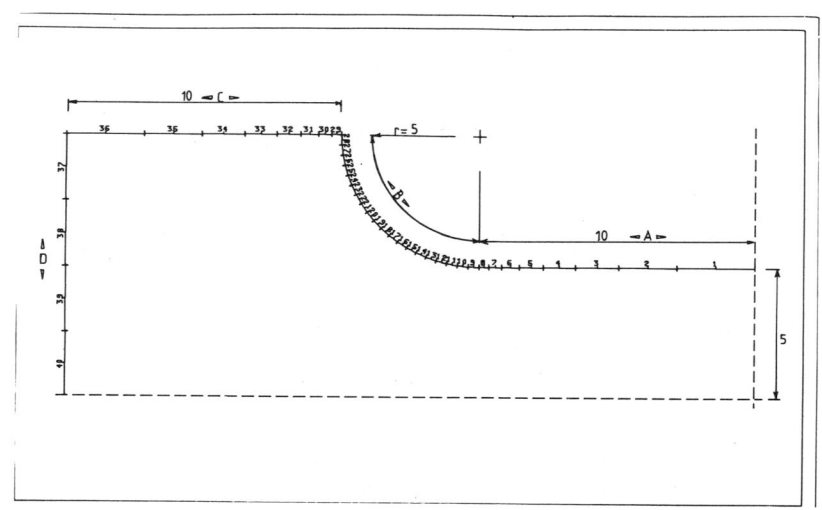

Fig. 4. Plate with fillets subjected to unidirectional pull: boundary element mesh.

The example is a classical test solved by different authors, including Weiber, whose photoelasticity results are quoted by Timoshenko.[3] It is shown that the stress concentration factor for this geometry is $K \cong 1.50$. Figure 5 plots the principal stresses along the external surface of the fillet. It can be seen that the maximum stress represents a concentration factor of 1.53.

This and other examples - some of them three-dimensional - have been used to validate the code. In all cases the program has shown great accuracy.

Industrial Examples

Connecting rod. Figure 6 shows the boundary element mesh used for the stress analysis of a connecting rod when subjected to a load from the crankshaft. Notice that no elements were required on the plane of symmetry. All the necessary commands to create the mesh and loading are given by BEASYG as follows,

```
LE 3
SX
BP 1,0,0
BP 2,35,0
BP 3,35,47.803
BP 4,37,56.75
```

BEASY: A System for Structural Analysis 57

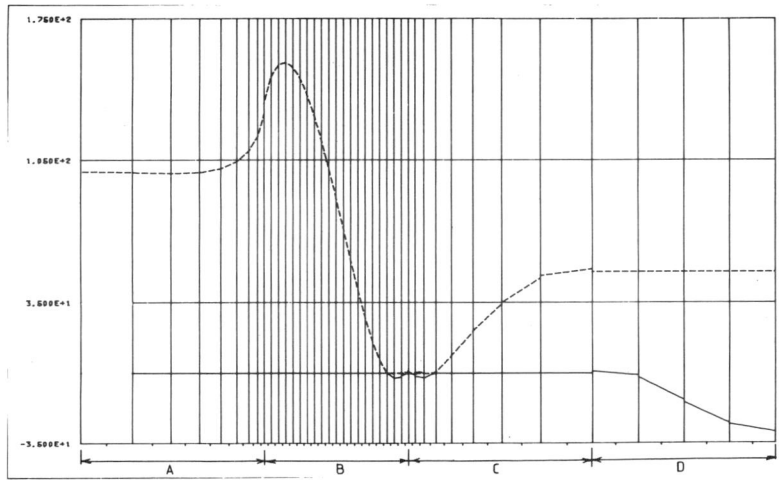

Fig. 5. Plate with fillets subjected to unidirectional pull:
principal stresses on surface.

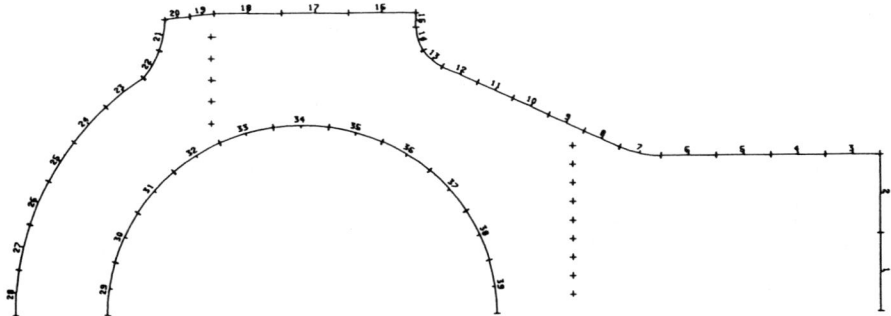

Fig. 6. BE mesh for a connecting rod. Lines of internal points marked
by +'s have been created at sections at which the designer is particularly
interested in the stresses. The example illustrates how easy it is to
prepare a BE mesh.

```
BP  5,55,95
BP  6,63.759,100.56
BP  7,67,100.56
BP  8 67,144.29
BP  9,65,754,155
BP 10,52.846,159.64
BP 11,0,187.5
BP 12,0,167.5
BP 13,0,82.5
BP 14,0,125
```

```
BP  15,56.1,47.803
BP  16,63.759,90.88
BP  17,65.754,175.7
BL  1,1,2,2
BL  2,2,3,4
BC  3,3,4,15,1
BL  4,4,5,5
BC  5,5,6,16,2
BL  6,6,7,1
BL  7,7,8,3
BL  8,8,9,2
BC  9,9,10,17,2
BC  10,10,11,14,6
BC  11,12,13,14,11
ZE  2100
ZP  0.28
ZI  1,4.135,66
ZI  9,37.218,66
ZI  10,42.723,144.29
ZI  14,32.145,144.29
EP  29,-1,-7.802,-7.705",-7.775
EP  30,-1,-7.705,-6.947,-7.424
EP  31,-1,-6.947,-5.533,-6.293
EP  32,-1,-5.533,-3.522,-4.586
EP  33,-1,-3.522,-1.231,-2.442
PD  1,-1,0,0
```

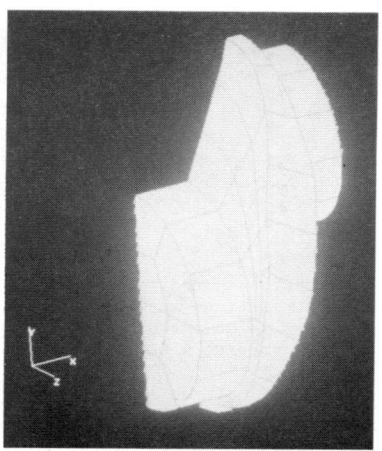

Fig. 7. Mesh for crankshaft analysis.

The discretization was carried out using quadratic elements. The problem has a total of 234 degrees of freedom plus 14 internal points where displacements and stresses were calculated. The problem has only one subregion and the output data is not enclosed to shorten this presentation. The computer time for this case is 15 minutes of CPU time in a VAX11/750.

Crankshaft analysis. Figure 7 describes the mesh used to analyse a typical crankshaft for an automotive manufacturer. The object of the analysis was to calculate the stiffness of the shaft section. Notice that two planes of symmetry have been taken into consideration. The rest of the external surface has been discretized into 51 quadrilateral elements with 1377 unknowns. The module BE3DTE was used in this analysis, the data plots were done using PATRAN for which an interface with BEASY is now available. Results for the displaced shape (Fig. 8) were also obtained using PATRAN.

Bearing cap analysis. The bearing cap shown in Figure 3 was also analysed and the results were displayed using PATRAN. Figure 9 shows a typical stress plot.

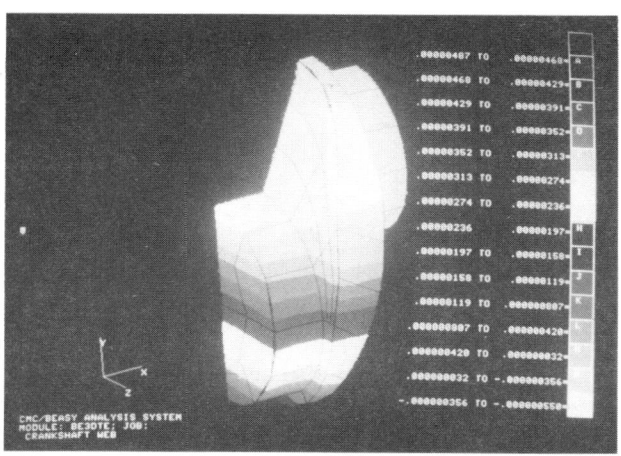

Fig. 8. Contours of displacement.

Because of the way the examples were run, CPU times were not available. More recently another example was run in a CRAY-1 machine consisting of 150 elements, i.e. 4050 unknowns, and its solution took 15 minutes of CPU time.

Potential Users

The BEASY code has a large range of applications in many engineering disciplines. It is of primary interest for mechanical, aeronautical, offshore, civil, naval and other engineering disciplines.

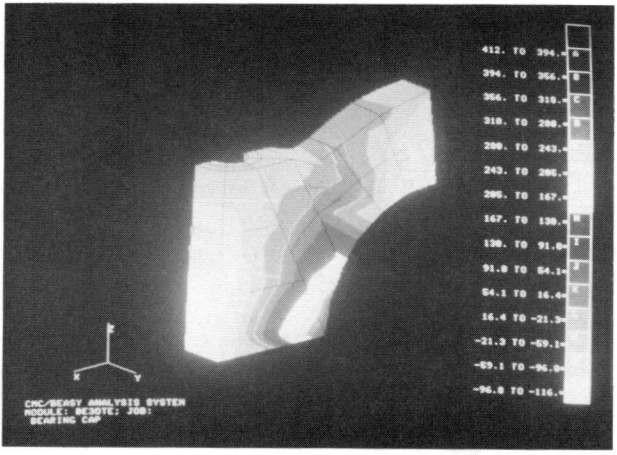

Fig. 9. Bearing cap model - Contours of direct stress.

REFERENCES

1. Brebbia, C. A. (1978). *The Boundary Element Method for Engineers*. Pentech Press, London.

2. Brebbia, C. A., J J. Telles and L. C. Wrobel (1984). *Boundary Element Techniques - Theory and Applications in Engineering*. Springer, Berlin.

3. Timoshenko, S. P. and J. N. Goddier (1970). *Theory of Elasticity*. McGraw-Hill.

CA.ST.OR: FINITE ELEMENTS AND BOUNDARY ELEMENTS ANALYSIS SYSTEM

M. Afzali, A. Turbat, J. F. Billaud and J. Peigney

CETIM, Computation Department, 52 avenue Félix Louat, 60300 Senlis, France

INTRODUCTION

CA.ST.OR is a general purpose finite elements and boundary elements system covering a wide range of applications and computer capabilities.

CA.ST.OR's software has been developed during the last 15 years at CETIM (technical centre of the French mechanical engineering industry).

CA.ST.OR's developments are based on the extensive experience of CETIM in solving the engineering problems and its position in the industrial environment. This allows the software to be well-suited to problems encountered in mechanical engineering.

Two kinds of software are proposed to the users:

- CA.ST.OR software running on mega-mini and main frame computers covering a large number of industrial applications which need the powerful and fast computers;

- CA.ST.OR BE software, running on micro-computers with interactive procedure and graphic display suitable for design offices. The great advantage of CA.ST.OR BE is that the user needs minimum knowledge of computer science and is guided during all steps of computations.

This paper describes CA.ST.OR software running on mega-mini and main frame computers.

This software allows us to perform two-dimensional and three-dimensional stress analysis with linear and nonlinear behaviour. Seismic analysis, fluid structure interaction and also dynamic anlaysis of rotating machinery can be performed. For three-dimensional thermal and elastic analysis the boundary element method is used.

THEORETICAL BACKGROUND

The development of the theoretical background of CA.ST.OR (concerning contact, structure-fluid interaction, seismic qualification, shell formulation etc.) is beyond the scope of this book. For such details, the readier is asked to consult references 5-9. The following section explains some of the algorithms implimented in CA.ST.OR.

Nonlinear Finite Element

The algorithm in nonlinear finite element analysis is incremental. At each step we resolve iteratively the equation

$$[K]\{\Delta u\} = \{\Delta F\}_{mec} + \{\Delta F\}_{th} + \{\Delta F\}_{vp} \qquad (1)$$

where $[K]$ is the *elastic* stiffness matrix, $\{\Delta F\}_{mec}$ the increment of mechanical nodal forces, $\{\Delta F\}_{th}$ the increment of equivalent thermal forces and $\{\Delta F\}_{vp}$ the increment of balance forces due to plastic or viscoplastic strains (unknown):

$$\{\Delta F\}_{vp} = \int_{V_t} [B]^T [D] \{\Delta \varepsilon\}^{vp} \, dV$$

$\{\Delta \varepsilon\}^{vp}$ is calculated from strain rates at t and $t + \Delta t$, with the Euler-Cauchy rule:

$$\{\Delta \varepsilon\}^{vp} = \frac{\Delta t}{2} (\{\dot{\varepsilon}\}^{vp}_t + \{\dot{\varepsilon}\}^{vp}_{t+\Delta t})$$

The nodal displacement vector increment $\{\Delta u\}$ can then be obtained from an iterative solution of equation (1). We stop the algorithm when convergence on $\{\Delta \varepsilon\}^{vp}$ or a maximum number of iterations is reached. The stiffness matrix $[K]$ must be updated only when the elastic coefficients are temperature - dependent.

The behaviour law chosen for each material allows us to express the plastic or viscoplastic strain rates. In the case of an elastoviscoplastic behaviour, we use the Norton-Hoff power law, without yield stress:

$$\{\dot{\varepsilon}\}^{vp} = \frac{3}{2} \lambda^{-n} |\bar{\varepsilon}^{vp}|^{-\frac{n}{m}} (\bar{\sigma})^{n-1} \{s\}$$

where $\{s\}$ is the deviatoric stress tensor (isotropic hardening). For an elastoplastic material, without viscosity, whose yield stress $\bar{\sigma}_y$ depends on temperature and strain hardening we have selected a pseudo - Bingham law:

$$\{\dot{\varepsilon}\}^p = \frac{3}{2} \gamma < 1 - \frac{\bar{\sigma}_y}{\bar{\sigma}} > \{s\}/\bar{\sigma}_y$$

where γ is arbitrary.

Thus, viscoplasticity and plasticity can be treated with the same algorithm. The above behaviour laws follow the Prandtl-Reuss normality rule and can be extended to materials with kinematic or mixed hardening.

Boundary Element

For thermal and elastic analysis of three-dimensional structures the boundary element method is used In elastic analysis, by application of the Betti theorem to Navier equations an integral equation can be obtained:

$$C_{ij}(x) U_j(x) + \int_S T_{ij}(x,y) U_j(y) ds = \int_S U_{ij}(x,y) t_j(y) ds$$

CA.ST.OR: Finite Elements and Boundary Elements Analysis System

where U and T are the kernels and u and t are the displacement and tension unknowns. Because of the singularity of the kernels U and T a special numerical integration procedure has been introduced into the program.[5,6*] By this technique the elements are divided into the subelements depending on its distance from the singular point (r = 0).[5-9]

FIELD OF APPLICATION

The system is composed of four parts:

- CA.ST.OR 2D performs thermal and linear and nonlinear static analysis of plane or axisymmetrical structures with symmetric or non-symmetric loadings.

- CA.ST.OR 3D performs thermal and linear elastic analysis of three-dimensional structures.

- CA.ST.OR SD performs static and dynamic mechanical analysis of 3D bolted and welded structures (plates, shells, beams) and of 3D continuous media with linear elastic behaviour.

- CA.ST.OR MT performs dynamic bending and torsional analysis (natural frequencies and mode shapes) of interconnected shafts.

   All programs are written in Fortran 77 with interactive data input and free formated input for batch, and are run on CRAY XMP, VAX 11/780 and other computers (Micro-VAX-Apollo).

In two-dimensional analyses material nonlinearities such as plasticity or viscoplasticity and cyclic thermoplasticity are included. The contact problems (geometry nonlinear) can be also analysed. Steady state and transient thermal analyses can be performed in two- and three-dimensional modelizations.

The temperature file is used for thermal loading in elastic or plasticity analyses. Other loadings may be concentrated forces, pressure and shear forces, rotational and gravitational forces or a prescribed displacement. Some specific applications, like fracture mechanics in two- and three-dimensional structures, can be also performed. Generally speaking CA.ST.OR software is suitable for the following applications:

engine components, mechanical transmission components, pressure vessels, compressors, gas turbine, mechanical components in nuclear industry, welded and bolted structures in buildings, offshore platforms, nuclear plant and general mechanical components, reservoirs and immersed structures, etc.

PROGRAM DESCRIPTION

CA.ST.OR 2D

Based on the finite element method this allows us to perform 2D thermo-elasticity or thermo-elastoviscoplasticity analysis of plane or axisymmetrical structures.[1-3] The size of CA.ST.OR 2D is about 60,000 statements, but the memory size needed for an analysis depends on the size of the problem, and because the memory is dynamically allocated the size limitations depends on the computer's memory.

<u>Options</u>. *M2D* is an automatic mesh generator for plane and axisymmetric structures. The necessary data are the description of structure contour and the mean size of

---

*Superscript numbers refer to References at the end of the article.

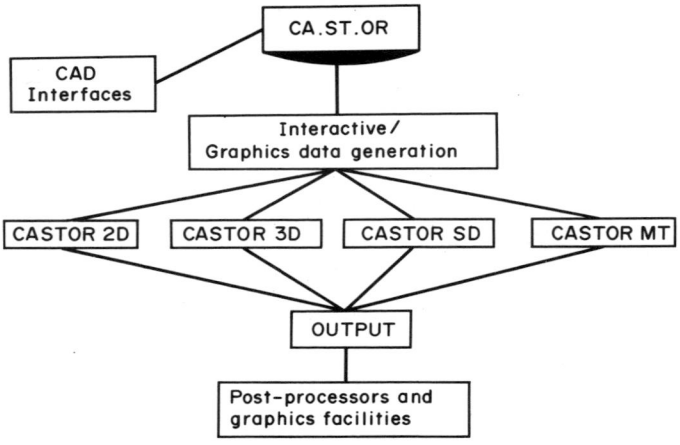

CA.ST.OR software.

the elements. The program generates the 6-node elements with local refinement possibilities. The node renumbering reducing computation time in FEM analyses is included.

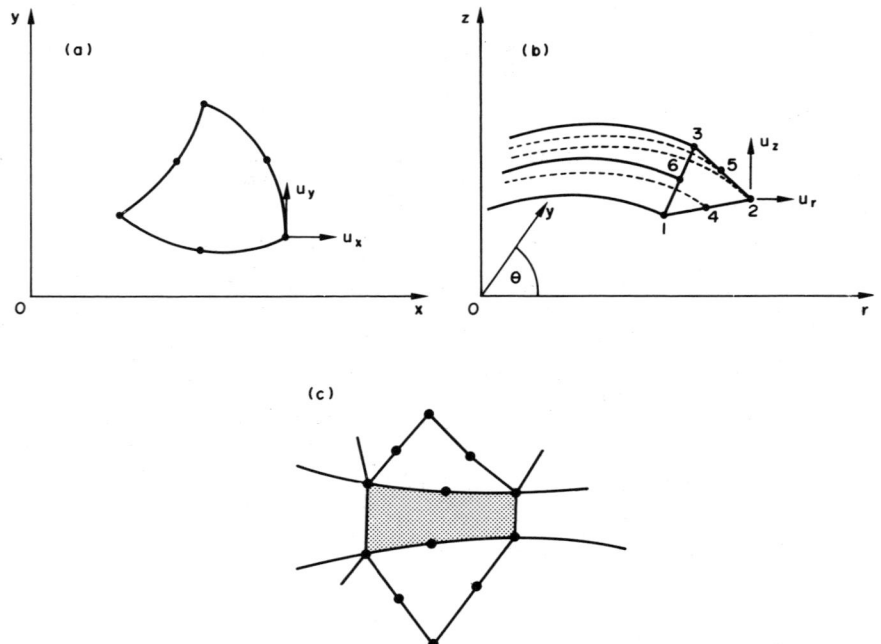

Fig. 1. CA.ST.OR 2D elements: (a) plane, (b) axisymmetric, (c) contact.

CA.ST.OR: Finite Elements and Boundary Elements Analysis System

E2D. This option performs elastic stress analysis of 2D plane or axisymmetric structure. The structure may be composed of one or several isotropic materials. The interpolation functions are quadratic.

The necessary data are:

- description of geometry using the corresponding file generated by M2D or given by the user with semi-automatic generation of nodes and elements;

- material properties;

- boundary conditions (given displacements, elastic supports, perfect contact, perfect sliding or contact with friction between the structure's components). For the contact problems with unknown contact a special contact element (Fig. 1(c)) is used. The algorithm for perfect sliding or perfect contact is incremental in type, but in case of friction an iterative procedure is performed;

- loadings may be concentrated forces, pressure and shear forces, rotational and gravitational forces, thermal loadings (temperature file generated by TH2D).

The results are the displacements, strain and stress tensors, principal stresses and Von Mises and Tresca equivalent stresses at each node. The displacement and stress files are generated for a graphic interpretation of results.

TH2D. The option performs steady state and transient thermal analyses of 2D plane and axisymmetrical structures. The analyzed structure may be composed of one or several isotropic materials.

The data are similar to those in E2D:

- description of geometry (geometry file generated by M2D or by the user);
- material properties;
- boundary conditions (given temperature, given flux, convection, internal power);
- evolution of the boundary conditions with respect to time, and the time-step in transient analysis. The evolution of boundary condition is considered to be linear with respect to time.

  The results are the temperature at each node and for each time-step in transient analysis. The temperature file is also generated for isothermal plotting and/or for thermal loading for stress analysis (E2D).

EF2D. This option performs elastic stress analysis of 2D plane or axisymmetric bodies under non axisymmetric loading. For loading, a fast Fourier transformation is used. The data and results are the same as in option E2D except that the boundary conditions are restricted to given displacements and elastic supports. The angles for which the results must be printed are needed.

EVP2D. This option of CA.ST.OR 2D performs thermal-elastoviscoplastic analysis of 2D plane or axisymmetrical structures in the framework of small strains and large displacements for isotropic and anisotropic materials. The practical problems as creep, plasticity and viscoplasticity may be analysed by this option.

Mechanical and thermal loadings may vary with respect to time. The program takes into account the variation of material properties with respect to temperature.

The following behaviour laws may be considered:

- elastic behaviour (Fig. 2a);
- elastoplastic behaviour with and without work hardening (Fig. 2b);
- elastoviscoplastic behaviour (Fig. 2c).

The input data for description of geometry M2D and boundary conditions are

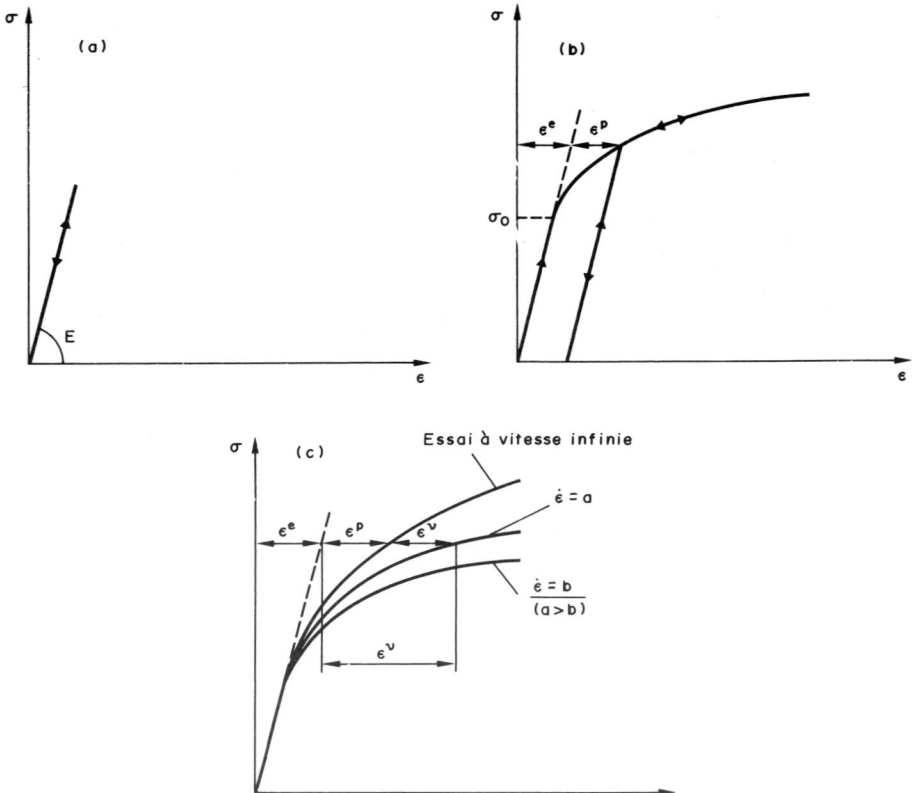

Fig. 2. Behaviour laws in CA,ST,OR 2D: (a) elastic, (b) elastoplastic, (c) elastoviscoplastic.

similar to EF2D. The loadings (similar to E2D) and their evolution with respect to time are needed.

The algorithm in EVP2D is incremental and iterative. At each step the displacements, strains, stresses and plastic or viscoplastic strain increments are calculated; the material nonlinearities (plasticity or viscoplasticity) appear in the second member.

Iteration procedures are adopted for system resolution. The results are the same as those given by E2D for each computation step prescribed by the user. The plastic or viscoplastic strains are also given.

NX2D. This option performs elastic analysis of axisymmetrical shells composed by isotropic or anisotropic materials. The elements are 2-node with 4 DOF per node (radial, axial, circumferential displacements and one rotation), generated by automatic mesh generator.

The necessary data are:

- geometry description (automatic mesh generator);

CA.ST.OR: Finite Elements and Boundary Elements Analysis System

- boundary conditions (given displacement);
- loadings (similar to E2D).

The results are the displacements, rotations, stresses (in elements local base) and forces and at moments each node of mesh. The stresses at the shell surfaces are given.

The results can be plotted by an appropriate graphical postprocessor (TX2D).

PC2D. This option performs the cyclic thermoplasticity based on a simplified approach[4] of the 2D plane or axisymmetrical structures in small strains and small displacements. The materials are considered elastoplastic with linear kinematic work hardening. The initial residual plastic strains may be considered in the structure.

PC2D gives some information about the structure behaviour after a great number of cycles (adaptations or accommodations) without computations at each cycle usual in fatigue simulations.

The necessary data for this option are similar to those in EVP2D. The results are the plastic strains, elastic or inelastic stresses, work hardening parameter field, and inelastic strains.

PP2D. This option is composed of several postprocessors for fracture mechanics application, selective impression of results, etc.

D2D. This option allows us to visualize the geometry generated by the above options before and after loading on a graphic terminal or by a plotter with zoom possibility. The results (displacements, strains, stresses and temperatures) may be displayed or plotted as iso-value curves or printing of results values at each node.

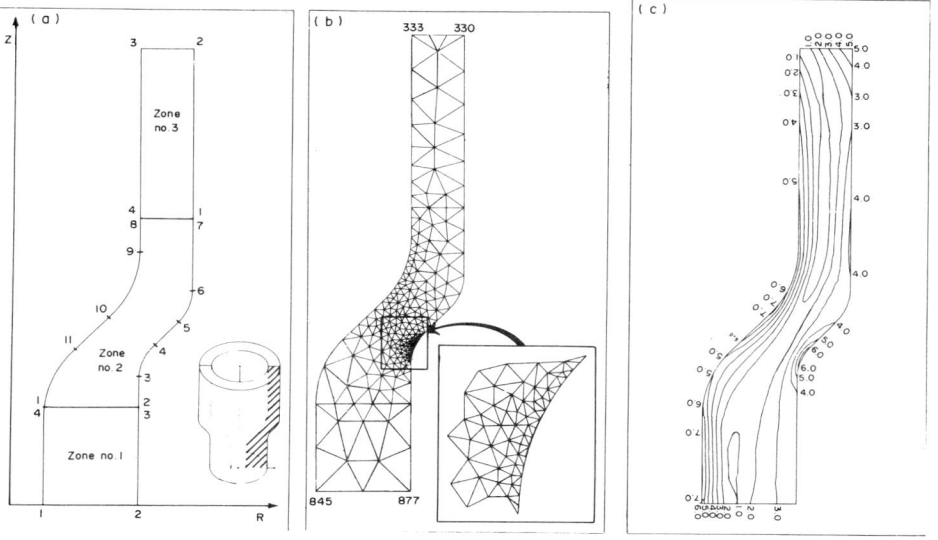

Fig. 3. Transient thermo-elastic analysis of a tube: (a) geometry, (b) mesh, (c) iso-value of Von Mises stresses.

CA.ST.OR 3D

Based on the boundary element method, it performs thermo-elasticity analysis of 3D structures.[1,5,9] The size of the program is about 75,000 statements.

The structures discretized up to 1200 nodes, 400 elements and ten subregions may be analysed by the different options of CA.ST.OR 3D.

Options. *M3D*. Semi-automatic mesh generator for discretization of 3D surfaces into subregions and elements. The surface elements are 6-node triangle and 8-node quadrangular elements (Fig. 4). Node coordinates may be in cartesian, cylindrical or spherical basis.

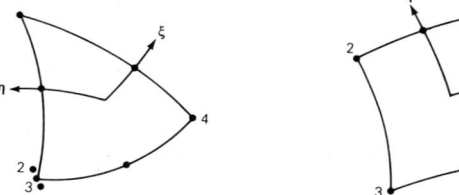

Fig. 4. Elements in CA.ST.OR 3D

Further the algorithm allows the operator to generate a new group of elements by means of rotation and/or translation of existing elements.

E3D. This option performs the elastic analysis of the 3D structures in small strains and displacements. The structure may be constituted by ten isotropic materials.

The necessary data are:

- general information: interpolation functions (linear or quadratic), plane or cyclic symmetries
- geometry description (M3D);
- material properties;
- boundary conditions (given displacements on nodes, lines or elements, sliding on subregions interfaces, elastic support);
- loadings (concentrated forces, distribution tractions on line or element, rotational and gravitational forces, thermal loading).

The results are the displacements, strain and stresses, principal stresses, equivalent Tresca and Von Mises stresses at each node and at any other point on the surface or inside the structure prescribed by the user.

TH3D. This option performs the steady state and transient heat transfer in the 3D structures.

The necessary data are:

- general information (same as E3D);
- geometry description (same as E3D);
- material properties;
- boundary conditions, given temperature on nodes, distributed temperature or flux on elements, heat transfer with exterior domains. In case of transient heat transfer, the evolution of boundary conditions with respect to time and initial temperature should be given;
- output is the temperature, flux, temperature gradient steady state, for the

nodes and requested points. The temperature file for thermal loading in E3D or plotting the isotherms is generated.

PP3D. This option is composed of several postprocessors:

PP3D-POST: calculates the solutions for the points on the surface and inside the structure.

PP3D-COMB: performs the linear combination of elementary load cases.

PP3D-SPECF: allows a selective impression of results.

PP3D-MODIF: performs the same anlaysis as in E3D or TH3D for a structure with local modification compared with the reference structure.

PP3D-JRICE: calculates the J integral of Rice with applications in fracture mechanics.

PP3D-COUPE: generates the solution files in order to plot the iso-value curve of stresses or temperatures on a plane intersecting the structure[7] (Fig. 5).

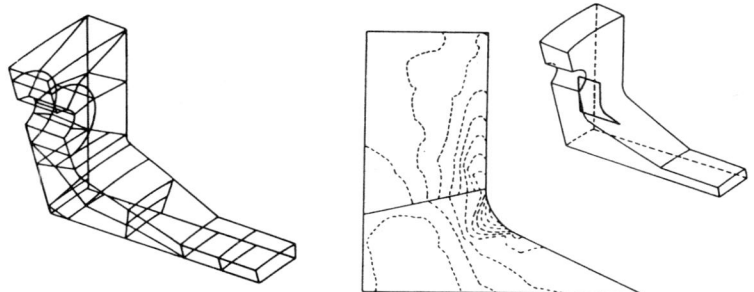

Fig. 5. Iso-value Von Mises stresses on an intersecting plane with zoom possiblity.

PP3D-RECUP: restart of E3D or TH3D.

PP3D-SIGMA: calculates the stress tensor components with best precision for a node on a sharp edge or on a corner.[7]

PP3D-RESTR: generates the solution file of a limited number of subregions.

TPD3D: generates compatible files of results for graphic processor D3D. Can only be used in transient thermal analysis.

D3D. This option allows us to visualize the geometry generated by M3D, E3D or TH3D before and after loading on a graphic terminal or by a plotter with zoom possibility. It allows us also to plot the iso-value curves of results on a 3D surface with hidden lines.

CA.ST.OR SD

CA.ST.OR SD performs static and dynamic analysis of a 3D structure by the finite element method.[1,10-12] The preprocessors and several postprocessors are included. The size of the program is approximately 70,000 statements. The size of the

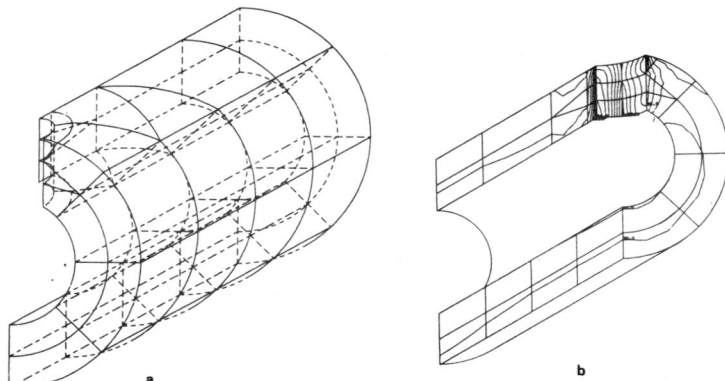

Fig. 6. Thermoelastic analysis of a heat exchanger: (a) mesh, (b) iso-value of stresses.

analysed structures may be very important because of the dynamically allocated memory. Different options are as follows.

Options. *PASD*. This option performs linear elastic, static and dynamic analysis (natural frequencies and mode shapes) of 3D structures with small strains. Several types of elements (plate, shell, transition, brick, beam, etc.) are shown in the element library. Automatic node renumbering in order to reduce the bandwidth and computation costs is included.

The necessary data are:

- general parameters (kind of analysis, node number, element group number);
- geometry description, node and element definition and element groups with corresponding material properties and distributed loadings;
- boundary conditions;
- nodal loads (static) or nodal masses (dynamic).

The results are the generalized forces and displacements in the case of static analysis, critical loadings and mode shape in the case of buckling analysis, and natural frequencies and mode shapes in the case of dynamic analysis.

STRSD. Performs the computation of stresses in beam, plate and pipe elements from the generalized forces. For other elements this option gives the equivalent stress (Tresca or Von Mises) and principal stress tensor.

BASSD. This option allows us to compute the dynamic response (using mode superposition) of a structure submitted to a transient excitation.

CMCSD. Performs the linear combinations of elementary load cases.

SRSSD. Computes the response of a structure to earthquake excitations using the method of the sum of the modal maximum responses (response spectrum analysis).

VENT. This option performs the structure analyses under random wind loadings with computation of fatigue lifetime.

REASD. This option computes the generalized nodal forces (reaction forces).

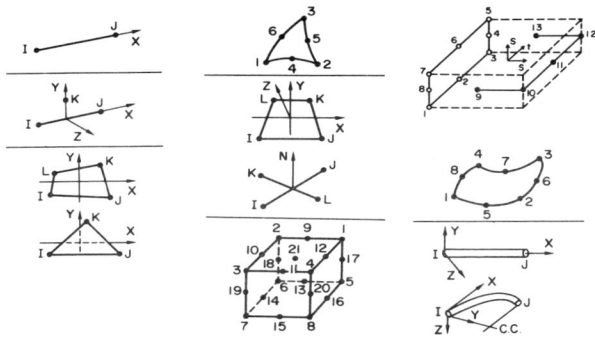

Fig. 7. CA.ST.OR SD: element library.

MASSD. This option performs fluid-structure coupling using the boundary element method for fluid and FEM for the structure.[11] Two kinds of problem may be analysed:

- immersed structures in infinite fluid;
- reservoirs containing the free surface fluid.

EDISD. A selective edition of results.

D3D. Like CA.ST.OR 3D, this option allows us to visualize the geometry before and after loading on a graphic terminal or by a plotter.

CA.ST.OR MT

Based on the finite element method, it performs dynamic analysis of rotating machineries.[1,13] The size of this program is about 10,000 instructions.

Options. *TORMT.* Performs computation of torsional natural frequencies and associated mode shapes for interconnected shafts.

The necessary data are:

- material properties;
- description of geometry (diameter and length or torsional stiffness and inertia);
- rotational speed ratio between the different shafts.

The results are the natural frequencies and associated mode shapes.

DYNMT. This option allows us to calculate the critical speeds, stability threshold and unbalanced response of rotating shafts upon many bearings.

The necessary data are:

- material properties;
- description of geometry;
- mass, diametral and polar inertia of added rigid disks;
- evolution of stiffness and damping coefficients of bearings with rotational speed.

The results are

- evolution of natural frequencies and modal damping with rotational speed of the rotor;
- mode shapes and critical speeds;
- amplitude and plane of the shaft with unbalanced masses.

PLOMT. Performs visualisation and plotting of the mesh and results.

CA.ST.OR - CAD (EUCLID) Interface

An interface between CA.ST.OR 2D and CAD software EUCLID has been developed. By this interface the user introduces only the mesh refinement, boundary conditions and loadings with full interactive input data procedure. A CA.ST.OR 3D - EUCLID interface is under development.

CA.ST.OR - GIFT Interface

In order to use GIFT as a preprocessor for 3D mesh generation, an interface between CASTOR and GIFT has been developed. This interface allows the user to introduce input data interactively.

HARDWARE COMPATIBILITIES AND CA.ST.OR AVAILABILITY

CA.ST.OR software is commercialized by CETIM all over the world. The English versions of CA.ST.OR 2D, and 3D are also available. Those who wish to use CA.ST.OR on their own computers may obtain information regarding detailed capabilities and prices from CETIM. Some international commercial software centres are preparing the CA.ST.OR version on the following computers:

- Mini 6 (CII-HB)
- Apollo
- Fujitsu
- Prime
- IBM

The interfaces between CA.ST.OR and GIFT and CAD software (EUCLID) are under development. Full interactive data preparation is already included in CA.ST.OR BE and will be included in the other CA.ST.OR software. The versions of CA.ST.OR on VAX 11/780 and CRAY, Apollo, Micro-Vax and those of CA.ST.OR BE on HP materials are supplied by CETIM.[15]

For using the CA.ST.OR software different possibilities are adapted:

- CETIM's engineers may perform your structure analysis using CA.ST.OR software. A detailed report will be supplied.

- CA.ST.OR software may be used by your own staff on CETIM's computer (self-service).

- CA.ST.OR software is available on the CISI, CGG and INVECTOR networks.

- CA.ST.OR software may be supplied and implemented on your own equipment.

Documentations

A complete documentation of CA.ST.OR which is regularly updated is available:

- user's manual,
- theoretical manual,

- test manual,
- documentation of the programs.

INDUSTRIAL APPLICATIONS

CA.ST.OR 2D Applications

Gear. Stress analysis and the effect of gear geometry on mechanical behaviour of a transmission component.

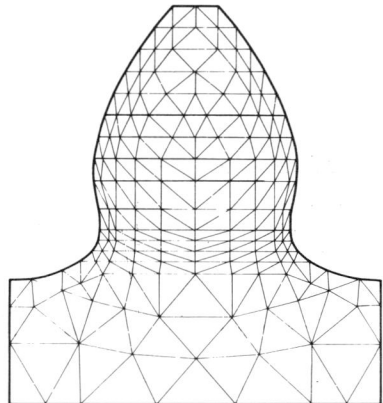

Fig. 8. Gear and its discretization.

Different geometry has been studied. The results have been presented as charts giving maximum stresses versus gear geometry and applied loads. These charts are a valuable help in gear design.

<u>Compressor casing</u>. Stress analysis of a compressor casing under thermal and mechanical loading. The compressor casing was made of two bolted half-shells. A finite element for axisymmetric geometry has been carried out for the evaluation of temperature and stress distribution.

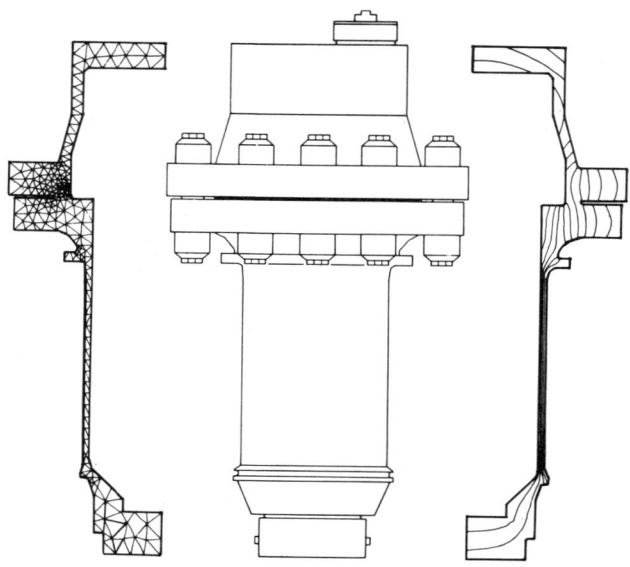

Fig. 9. Compressor casing, mesh and isotherm curves.

<u>Notch</u>. The study of a notch under mechanical loading and shape optimisation in order to decrease the stresses.

This problem has been carried out using CA.ST.OR 2D with an automatic shape optimization algorithm which is under investigation. For discretization, 6-node elements are used. The following figures show mesh, stress and isostatic stress distribution before and after optimization.

<u>Crack propagation</u>. Study of crack propagation and evaluation of stress intensity factor of a plate. The plate has been discretized by 6-node elements with a refined mesh around a crack front by using an automatic mesh generator (M2D).

<u>CA.ST.OR 3D Application</u>

<u>Connecting flange</u>. The operating safety of equipment such as pressure vessels depends upon the behaviour of their connecting flanges. A 2D finite element analysis in elastoplasticity has been carried out in order to estimate the working limits of the flange. A 3D elastic analysis by the boundary element method with a small number of elements compared with the finite element gave a good agreement between computation and experimental results.

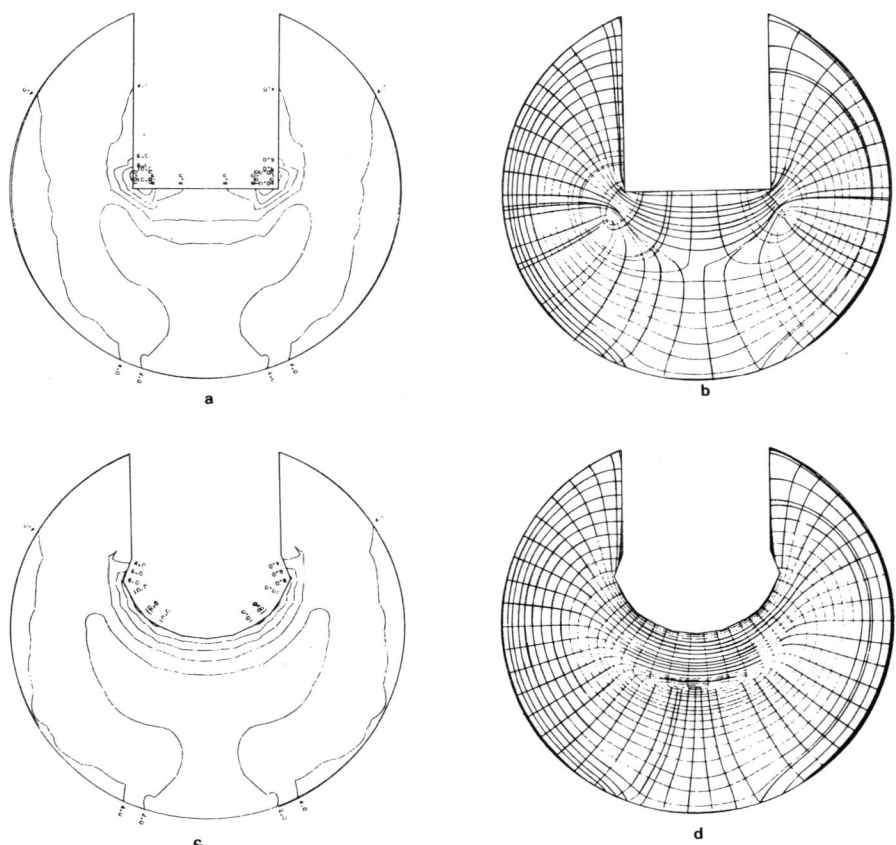

Fig. 10. Notch: (a,b) stresses and isostatic stresses before optimization, (c,d) stresses and isostatic stresses after optimization.

Acceptability of defects in turbine discs. This problem was to study the 3D crack propagation inside a disc of a high-power turbine. The boundary element technique is very well adapted to fracture mechanics and CA.ST.OR 3D was used for this problem. This work has been carried out under the heading of safety in nuclear plants and the critical size of the defects has been determined.

Crank-arm. CA.ST.OR 3D has been used for stress analysis of the crank-arm of a diesel engine. The problem was to study the behaviour of the crank-arm under mechanical loadings, especially the bolting forces. The influence of a fillet radius near the bolt has also been studied in order to decrease the stress level. The 6-node and 8-node boundary elements are used for the modelization. Stress distribution and deformed structure has been displayed on a graphic terminal with zoom possibility.

CA.ST.OR SD Applications

Statue of Liberty. For the 100th anniversary of the Statue of Liberty CETIM has

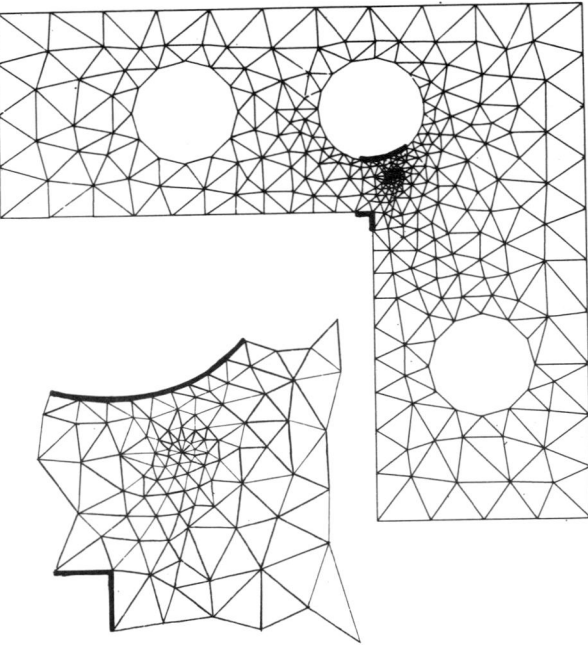

Fig. 11. Crack propagation in a plate.

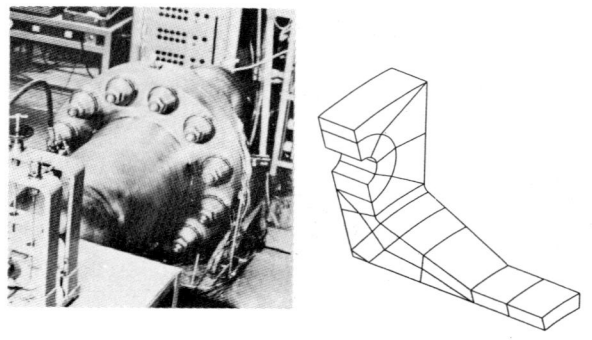

Fig. 12. Connecting flange and its modelization.

modelized the structure using CA.ST.OR SD. The aim of the study was to
investigate the lifetime of the structure under random wind loading of more than

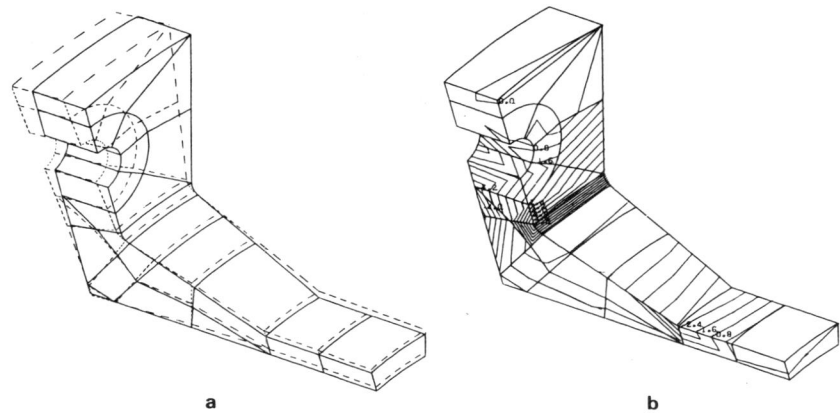

Fig. 13. Connecting flange: (a) mesh before and after loading, (b) stress distribution on the 3D surface.

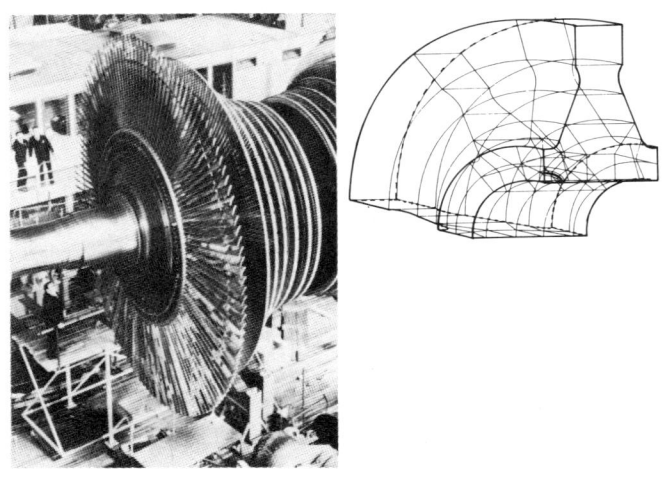

Fig. 14. Study of 3D crack propagation in turbine discs by the boundary element method.

120 km/hr.

The statue was modelized by the beam and plate elements with 12,000 degrees of

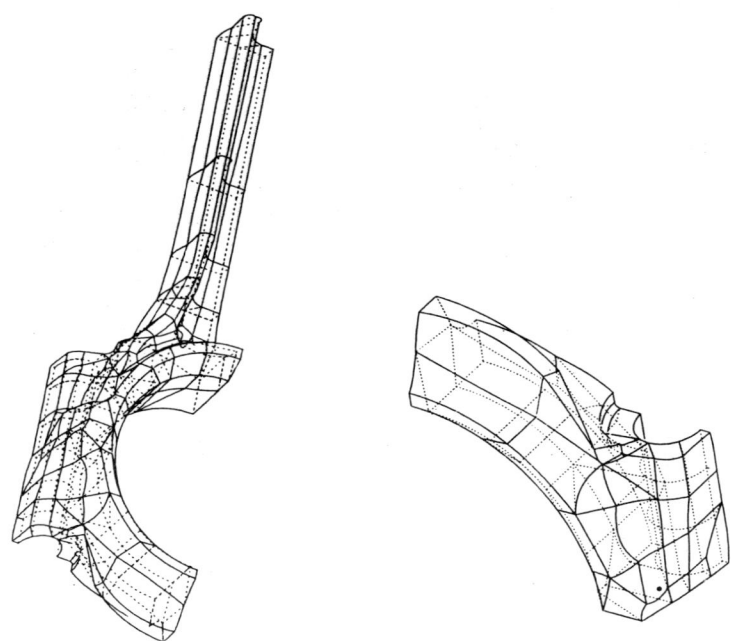

Fig. 15. Mesh of crank-arm with zoom possibility (SEDEMS Company).

freedom. The stresses in static and dynamic analyses have been evaluated and are used for estimation of lifetime by fatigue analysis.

<u>Water tank</u>. Static, dynamic and seismic analysis of a water tank has been carried out by CA.ST.OR SD.

The discretization has been done by shell elements. The following figures show mesh, first to fourth mode shapes.

<u>Valve casing</u>. In this problem stresses and displacement of a valve casing under mechanical loading have been calculated by using CA.ST.OR SD. For discretization, shell 3D solid and 3D solid-shell transition elements have been used. The loading cases are pressure, applied bending and torsional forces.

CA.ST.OR MT Application

<u>Air compressor</u>. The gear-branched system, as shown below, consists of:

- an air compressor (driven machine),
- a fly wheel,
- two couplings (flexible),
- gear wheels (reduction),
- a turbine.

The problem is to calculate natural frequencies and associated mode shapes of the system in torsional vibration.

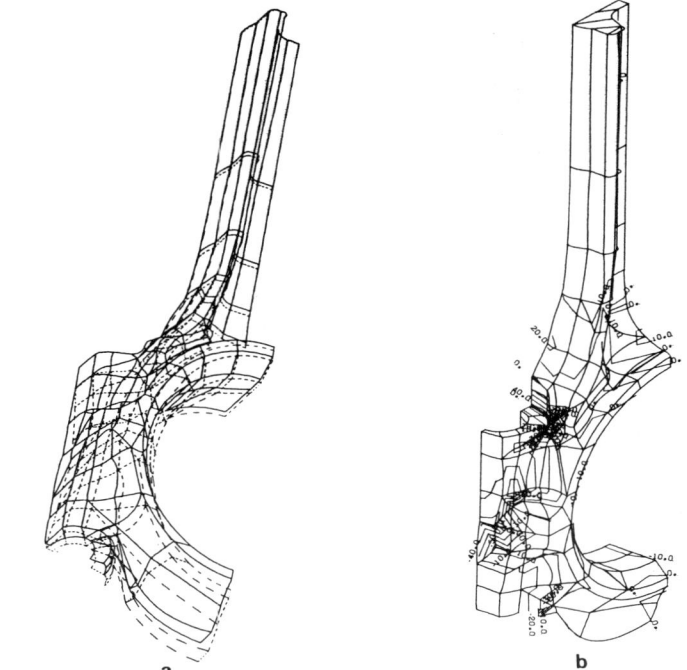

Fig. 16. Crank-arm: (a) mesh before and after loading, (b) iso-value curve of stresses.

A finite element discretization is used with 2-node elements (8 elements).

This calculation is performed using TORMT, and an example of the output is shown below.

Air ventilator. This rotor, as shown in Fig. 23, consists of a shaft with a rigid disk, hinged on two flexible bearings.

We want to calculate:

- the critical speeds,
- the unbalance response,

of the complete system in bending vibrations.

A finite element discretization, using a special "rotor element" (2-node beam element with shear and gyroscopic effects), is applied.

The calculations are performed with the option DYNMT of CA.ST.OR-MT.

Different levels of output are available:

- influence of rotational speed on the natural frequencies of the rotor bearing system and calculation of critical speeds;

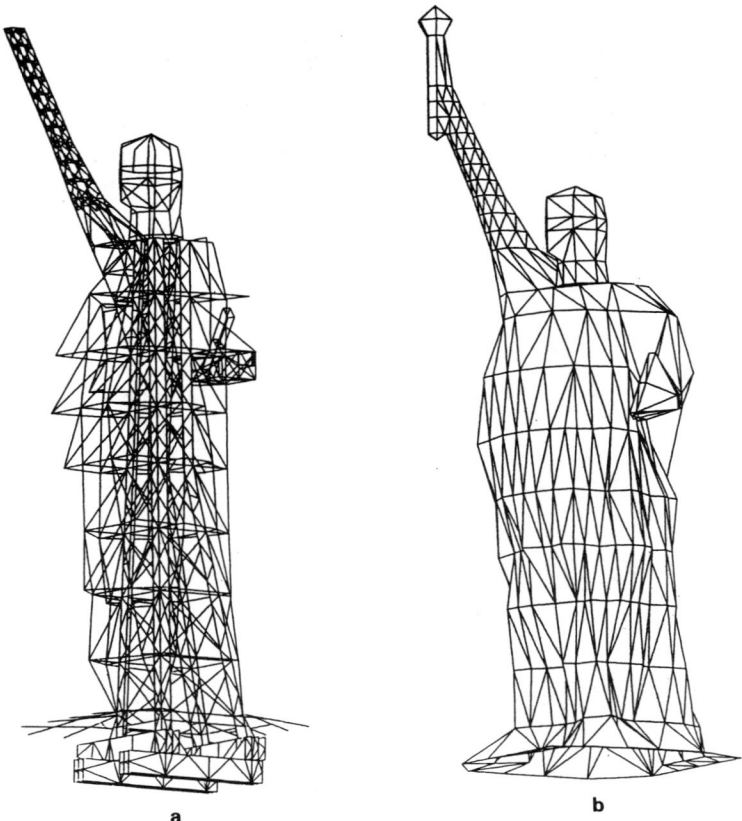

Fig. 17. Statue of Liberty: (a) inside structure, (b) skin.

- the mode shape evaluation for each critical speed (backward and forward whirl);
- evolution of amplitude and phase with rotational speed at different nodes.

CA.ST.OR USERS

CA.ST.OR software is used frequently by the engineering industries (general mechanical, energy, marine, material processing, aerospace, civil, nuclear, etc.). Some of the users are:

- General Electric
- Thomson-Brandt
- Potain
- Renault Car Industry
- Stein
- Cnexo
- Alsthom Atlantique
- Ateliers et Chantiers de Bretagne (ACB)
- EDF

CA.ST.OR: Finite Elements and Boundary Elements Analysis System

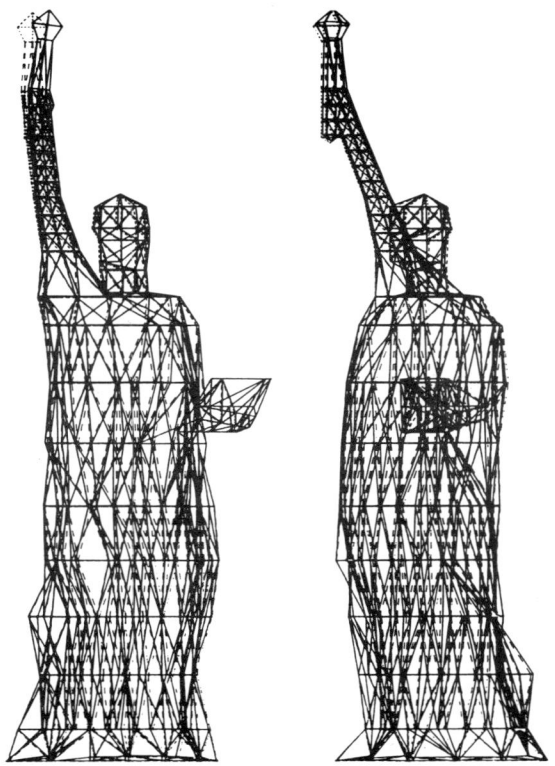

Fig. 18. Statue of Liberty: mode shapes.

- Creusot-Loire
- Messier Auto Industrie
- Technofan
- Sereg-Schlumberger
- C.E.A.
- C.D.F. Chimie
- SEP
- SACM
- Turbomeca
- CCSA (Army Computation Center)
- Fives-Cail Babcock
- Merlin Gerin
- Clecim
- Sedems (RVI - Renault)
- Coflexip
- SNPE
- AMRI
- SAGEM

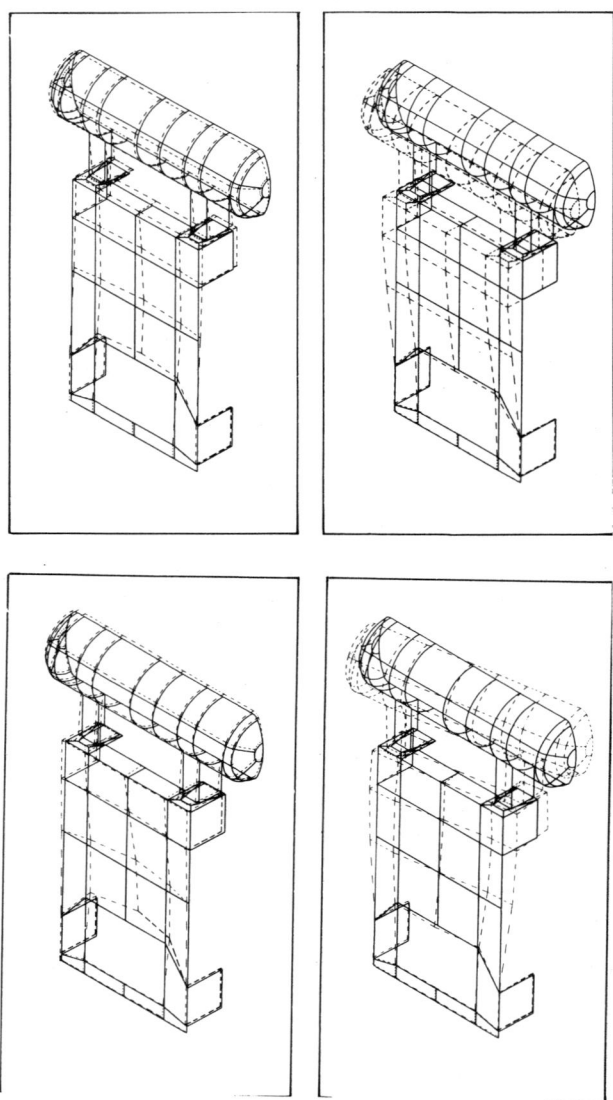

Fig. 19. Seismic qualification of a chassis supported water tank (SteDamois et Cie) - Doël III nuclear plant (Belgium).

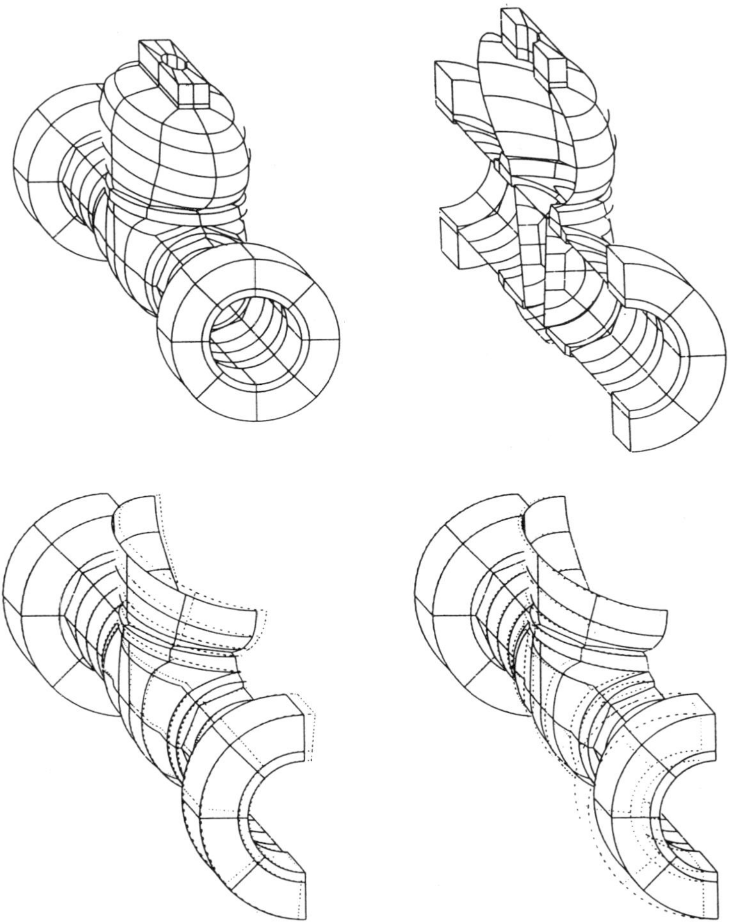

Fig. 20. Modelization of valve casing by finite elements before and after loading.

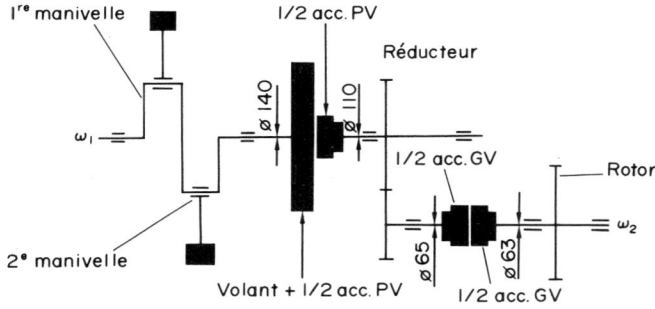

Fig. 21. Air compressor.

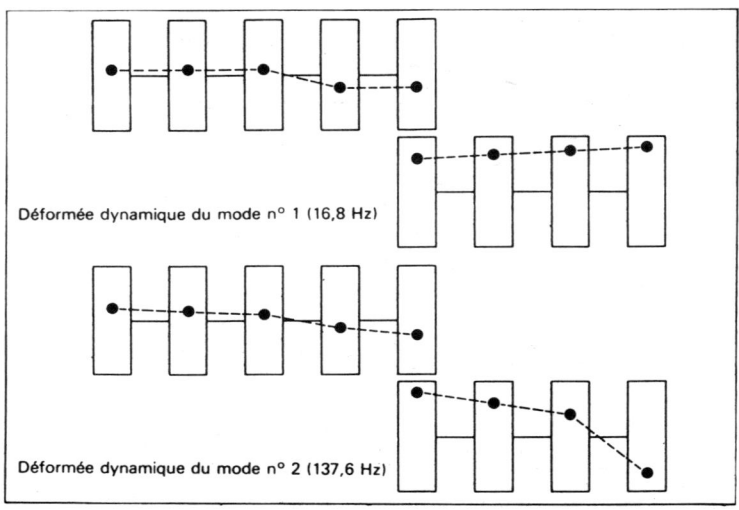

Fig. 22. Mode shapes.

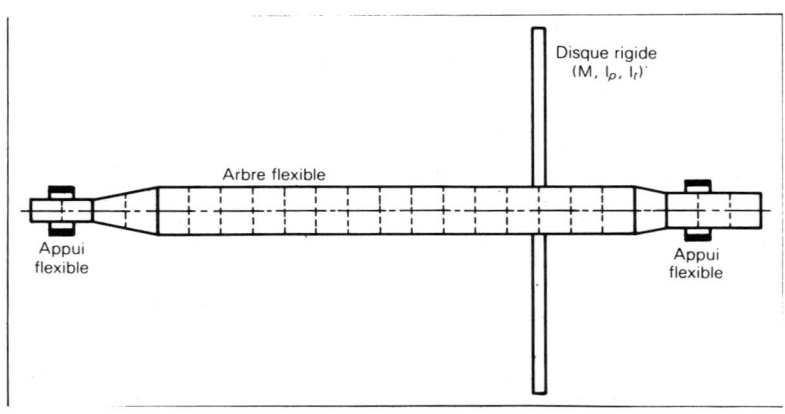

Fig. 23. Modelization of air ventilator.

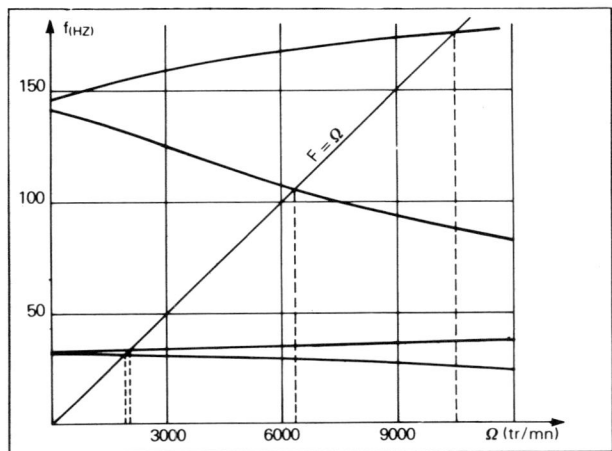

Fig. 24. Variation of natural frequency with speed.

REFERENCES

1. CASTOR software, CETIM documentation 1982.

2. M. Dubois, M. Cristescu and A. Turbat 1983. Non linearités physiques et géométriques par la méthode elements finis. *CETIM Report*.

3. A. Turbat, F. Convert and N. Skalli (1983). Prediction of thermal residual stresses by the finite element method. Effect of a phase change. *Proceedings of the 3rd Int. Conf. on Numerical Methods in Thermal Problems*.

4. J. Zarka and J. Casier (1979). Elastic-plastic response of a structure to cyclic loading: practical rules. *Mechanics Today*, Ed. Nemat-Nasser, vol. 6, pp. 93-198.

5. J. C. Lachat and J. O. Watson (1976). Effective numerical treatment of boundary integral equation. *Int. J. Num. Meth. Engng*.

6. J. C. Lachat and J. O. Watson (1977). Progress in the use of boundary integral equations illustrated by examples. *Comp. Meth. Appl. Mech. Engng*.

7. M. Afzali and A. Chaudouet (1982). A plane intersection of a three dimensional boundary elements mesh and stress; displacement contour plotting. *Bound. Elem. Meth. Engng.*, Springer Verlag, pp. 594-606.

8. J. M. Boissenot *et al.* (1978). Application de la méthode des équations integrales à la mécanique. *CETIM Report*.

9. A. Chaudouet and M. Afzali (1983). CA.ST.OR 3D: Three-dimensional boundary element analysis computer code. *Proceedings of the 5th Int. Conf. on BEM*, Japan.

10. A. Bonnefoy, P. d'Anthouard and J. F. Billaud (1982). Seismic analysis of equipment subjected to multiple support response spectra input. *7th Symposium on Earthquake Engineering*, Roorkee, India.

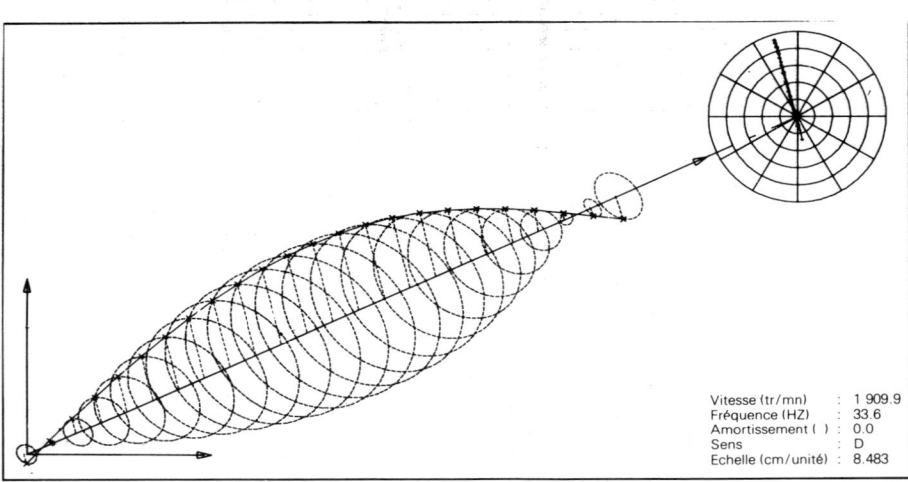

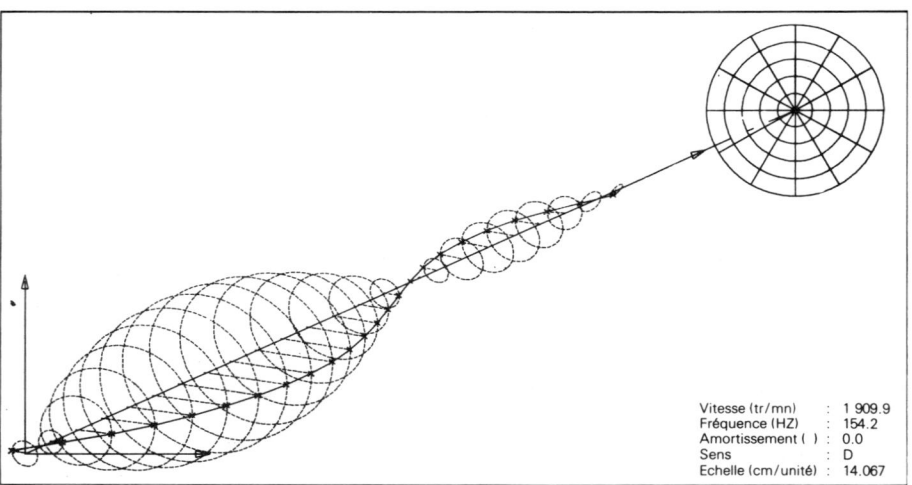

Fig. 25. Computation of critical speed.

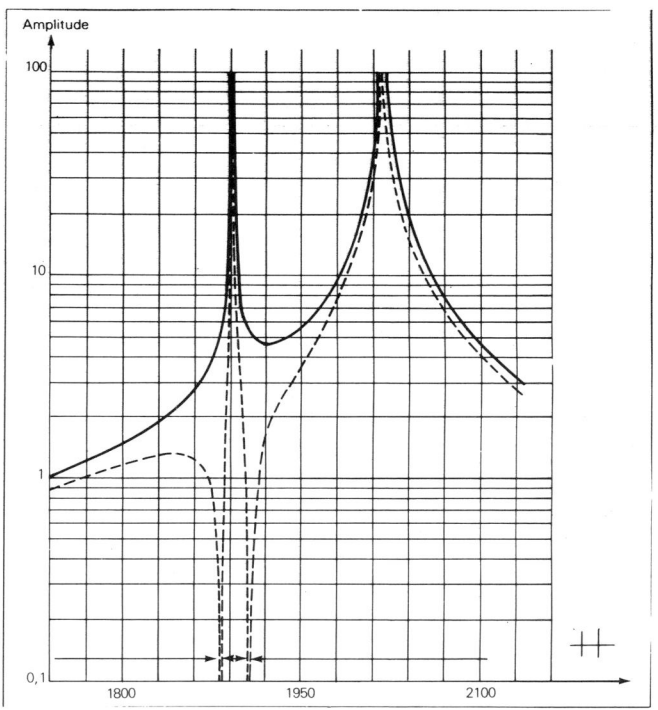

Fig. 26. Evolution of amplitude with speed.

11. Y. Ousset and M. N. Sayhi (1983). Added Mass computations by Integral Equations Methods *Int. J. Num. Meth. Engng.*, vol. 19, pp. 1355-1373.

12. P. d'Anthouard (1983). About seismic qualification of equipment by the multi-spectra method. *Transactions of the 7th Int. Conf. on SMIRT*, Chicago.

13. J. Peigney (1980). Prediction of dynamic properties of rotor supported by hydrodynamic bearings using FEM. *Int. J. of Computers and Structure*, vol. 12, no. 4.

14. A. Bonnefoy, T. Belbin and M. Accoley (1984). Statue of Liberty wind loading analysis. *CETIM Report* 8 442.00/8214.

15. M. Afzali and P. Devalan (1984). Structure analysis software on mega-mini and micro-computer. *5th Finite Element Systems Seminar*, Southampton, England.

Note: All CETIM reports can be obtained directly from the authors.

DAPST: A FINITE ELEMENT PACKAGE FOR THE DYNAMICAL ANALYSIS OF STRUCTURES

H. Sol, M. Van Overmeire and W. P. De Wilde

Free University of Brussels (V.U.B.), Faculty of Applied Sciences, Pleinlaan, 2, B-1050 Brussels, Belgium

ABSTRACT

Designed as an educational tool, DAPST (Dynamical Analysis Program for Structures) incorporates interesting features, such as complete modularity, normalization of varibles and programming technique, documentation of the program structure and detailed user manuals with the theoretical background of each module.

SHORT PROGRAM DESCRIPTION

At the Free University of Brussels, the finite element method is considered to be important enough to become an obligatory course in the education of civil and mechanical engineers. The course is divided into a theoretical and a practical part.

The theoretical part involves the mathematical principles underlying the method as well as the different fields of applications. In former days the students were asked to develop a small finite element program as the practical part of the education. This task was found to be not completely satisfactory because of two reasons; firstly, a large amount of time was spent in debugging program errors and secondly, the student gained no practical experience with the use of large finite element programs for the solution of engineering problems.

In order to achieve the double goal, education of students as *users* and as *programmers*, it was decided to develop a large finite element package with specific educational purposes. Therefore the package had to satisfy several conditions:

(a) A very strong modularity, with a detailed description of every module (theoretical background, used numerical methods, variable names, common blocks, input-output-stream). This demand was the reason why a commercially available program could not be used. Indeed, commercial programs provide poor - or sometimes no - information, necessary to allow the program to be extended or modified by students.

(b) A reasonably large element library.

(c) Different analysis types: statical analysis, eigenvalue problems, dynamical analysis.

(d) A clear manual for use as a black box system with sufficient examples.
With these specifications in mind, "DAPST" was developed.

As a consequence of its purposes, the *modularity* of the program was always set prior to *performance*.

Indeed, the computer memory and time consumption for one big routine is often smaller than the consumption for a group of routines which perform the same function.

But in DAPST the main concern is program transparency, rather than computer savings. Although the program DAPST was designed as an educational tool, it was soon discovered that it could be used to solve special research and industrial applications. Indeed, some applications ask for special options which are rarely available in multiple purpose PE programs. Because of its transparency such special demands can easily by implemented in DAPST (special element types, special analysis types, etc.).

THEORETICAL BACKGROUND

DAPST is a finite element program written in FORTRAN IV. Linear isotropic or orthotropic material behaviour and small deformations and strains are assumed.

All used elements are isoparametrical displacement or hybrid elements.

(1) *Statistical analysis*: $[K]\{u\} = \{F\}$ (1)

  $[K]$: structure stiffness matrix

  $\{u\}$: structure displacement vector

  $\{F\}$: structure force vector

  To solve the statistical problem (1), a frontal solution method is used. As a consequence no assemblage of the structure stiffness matrix is necessary. The element matrices are first calculated and stored on disc memory and afterwards returned in core memory during the solution procedure.

(2) *Eigenvalue extraction*: $([K] - \lambda [M])\{u\} = \{\phi\}$ (2)

  $[M]$: structure mass matrix

  A conservative eigenvalue problem (2) is solved based on the principle of Sturm sequences and an inverse power iteration. For the isolation of the eigenvalues, a bisection method is used.

  Each bisection demands the factorization of the matrix $[K] - \lambda [M]$. The factorization is carried out using a frontal method. Therefore no assemblage of the stiffness or mass matrix is necessary.

(3) *Dynamical analysis*: $[M]\{\ddot{u}\} + [C]\{\dot{u}\} + [K]\{u\} = \{F\}$ (3)

  $[C]$: structure damping matrix (orthogonal)

  . : time derivative

  To solve the dynamical problem, a modal superposition method is used.

  - As a first step a limited number (m) of eigenmodes is calculated.

  - These eigenmodes are used as a base for the displacements as a function of time.

The transformations:

$$^T[\phi]\ [K]\ [\phi] = \lfloor k \rfloor \tag{4}$$

$$^T[\phi]\ [C]\ [\phi] = \lfloor c \rfloor \tag{5}$$

$$^T[\phi]\ [M]\ [\phi] = \lfloor m \rfloor \tag{6}$$

$$^T[\phi]\ \{F\} = \{f\} \tag{7}$$

$[\phi]$: Matrix containing the eigenmodes

$\lfloor \ \rfloor$: diagonal matrix

deliver a set of m uncoupled equations.

- The uncoupled equations are solved in discrete timesteps.

FIELD OF APPLICATION

DAPST is a multipurpose program for structural analysis. The structure can be a beam - truss composition, a two-dimensional plane stress-, plane-strain-, or axysymmetrical structure, a plate or a shell structure or a three-dimensional solid structure.

The material of the structures can be isotropic or orthotropic or anisotropic (e.g. laminate composites). Linear material behaviour is assumed.

As already mentioned, DAPST can perform a statical- or eigenvalue- or dynamical analysis. It is also possible to calculate the sensitivity of the eigenvalues for parameter changes in the elements.

Finally, it is possible to calculate the bending stiffness and the torsional stiffness of arbitrary cross-section of isotropic beams.

The loading is limited to concentrated loads, concentrated masses, gravity loads and pressure loads.

PROGRAM DESCRIPTION

DAPST is a finite element program of approx. 20000 lines with a reasonable large element library:

- beam-, truss-elements,
- plain strain, plane stress, axisymmetrical-elements,
- plate-general shell elements,
- PR20 - PR21-elements,
- Poisson-element.

STRUCTURE

The most transparent and modular way to structure a program is the Overlay-segmentation. Each overlap has its own specific function. At this moment 8 overlays are implemented in DAPST:

OVERLAY 0 : master overlay
OVERLAY 1 : INPUT + datachecking
OVERLAY 2 : element formation
OVERLAY 3 : statical analysis
OVERLAY 4 : eigenvalue analysis
OVERLAY 5 : dynamical analysis
OVERLAY 6 : sensitivity of the eigenvalues for parameter changes
OVERLAY 7 : stiffness calculations of beam cross-sections
OVERLAY 8 : OUTPUT + plotting

Each of these overlays are divided into suboverlays which on their turn are divided into a number of subroutines and functions. As an example, the structure of overlay 1 is given (Fig. 1).

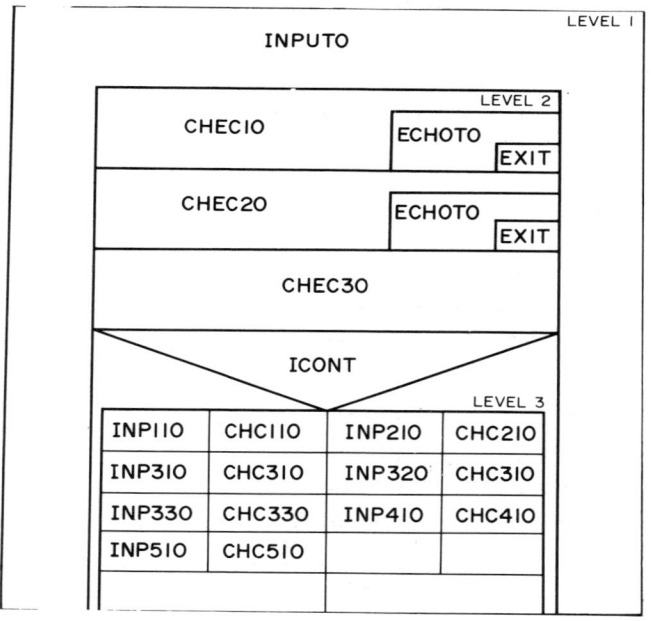

Fig. 1. Structure of overlay 1.

LEVEL 1

INPUT0 : input of problem independent data (1)

LEVEL 2

CHEC10 : control of general information (2)
CHEC20 : control of specific information
CHEC30 : determination of requested dimensions for memory management (3)
ECHOT0 : in case of error, printing of the not yet controlled information

LEVEL 3

INPnm0 : input of problems dependent information (4)
CHCnm0 : control of problem dependent information

Comment: (1) General information, material properties, geometrical data boundary conditions.

(2) The checking routines are only used in datacheck mode.

(3) The program has dummy variables for the dimensions of vectors and matrices. In datacheck mode these dummies are set to 1. The value of the dummies for execution mode is calculated.

(4) The control variable ICONT determines the kind of problem.

PROGRAMMING RULES FOR DAPST

The subroutines and functions are written according to strict programming rules. Some examples:

VARIABLES - each variable has 5 characters
- a flag variable starts with an I
- a dummy variable ends with an X
- a variable containing a maximum value starts with an M
- a counting variable starts with a K
- a variable containing a tape number starts with NT
- implicit FORTRAN-variable declarations remain:
  I, J, K, L, M, N : integers
  rest: real

COMMON BLOCKS   labelled with a 6 character name

LABELS   1 -   99 : reserved for GOTO statements
          100 - 8999 : reserved for DO-LOOPS
         9000 - 9999 : reserved for FORMATS

A detailed instruction set and a description of all the variable names, common blocks and tapes are added to the user manual.

USER FACILITIES

- free format input,
- postprocessing of results,
- graphical output of deformations of/and stresses,
- failure criterion on stresses,

- easy to adapt or to extend the program,
- the software can easily be transferred to different computers. Both source and compiled code are available.

HARDWARE

DAPST runs on a CYBER 170 from CDC. A calcomp-plotter or a graphical terminal (Tektronics) is necessary if graphical output is desired.

EXAMPLE OF APPLICATION

DAPST has been used for an investigation on the dynamical behaviour of tennis rackets.

The investigation was sponsored by the Belgian tennis racket manufacturers (Donnay-Snauwaert-Browning).

The goal of the study was to establish a design and optimization method for tennis rackets using the finite element method.

The performance and comfort of a tennis racket is determinated by items like rigid body properties (total mass, centre of gravity, centre of percussion, mass inertia) and the spectrum of eigenvalues and corresponding eigenmodes (only the four lowest values are important).

During large-scale experiments on a large series of different tennis rackets, with the co-operation of top class tennis players, the ideal values for the above-mentioned items were established.

The design problem was to develop tennis rackets with those experimentally determinated properties.

To solve this problem, DAPST was adapted in such a way that the program produced all those properties in the output section. For the finite element modellization of the frame of a tennis racket, simple beam and plate elements were used.

The strings were modelled as a tensioned membrane. This special purpose element was programmed and added to the element library of DAPST. The eigenfrequencies and eigenvalues produced by the program DAPST (Fig. 2) were checked by an experimental modal analysis set-up. Excellent agreement of the results has been reached (Fig. 3).

The calculations of three eigenfrequencies and eigenvalues of a finite element model of a tennis racket with 800 d.o.f. and a frontwidth of 128 take 450 CPU-seconds on a CYBER 170 computer.

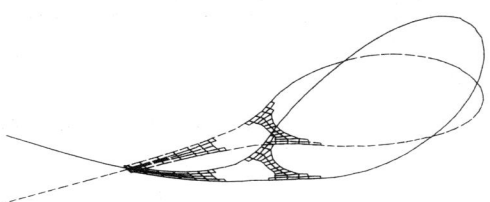

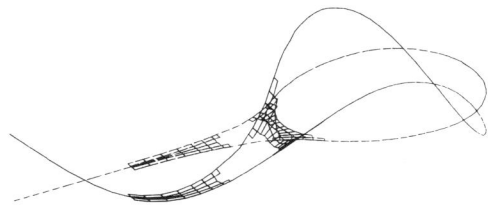

Fig. 2. First and third eigenmode of a tennis racket.

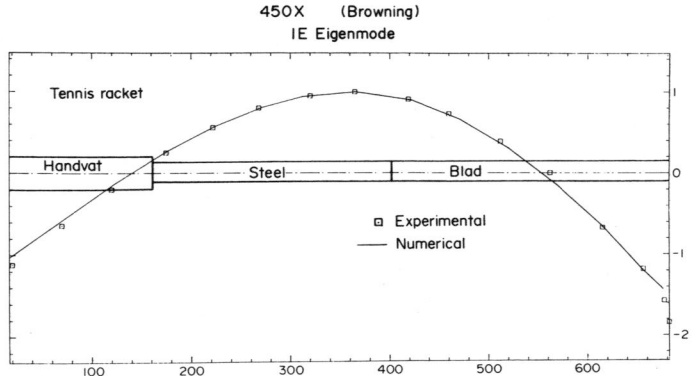

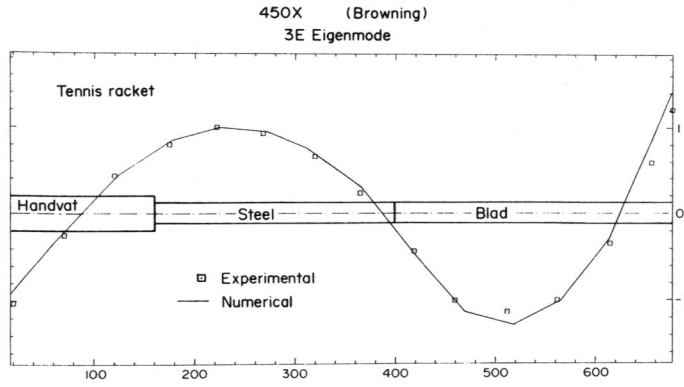

Fig. 3. Experimental verification of the numerical results.

DEFOR: PROGRAM FOR STATICAL ANALYSIS OF STRUCTURES COMPOSED OF ONE-DIMENSIONAL ELEMENTS

V. Kolář and I. Němec

Technical Institute Dopravoprojekt, 658 30 Brno, Leninova 17, Czechoslovakia

ABSTRACT

The program DEFOR can perform linear and geometrically nonlinear analysis of elastic, statically loaded structures composed of one-dimensional beam elements following the finite element analysis method. The elements can be of any technical type which can be represented by a centroidal axis and analysed as line elements. The structure can be composed of both prismatic and nonprismatic, straight or curved slender members, which can be divided into elements or regarded as one element, generally also as a substructure with two ends (stiffness or flexibility given) The program can solve the structures which extend in one (continuous beams), two (plane trusses, frames and grids) or three (space trusses and frames) dimensions. The members can be pinned or rigidly connected at any joint. Hybrid connections are admissible. The solution provides the information about joint and support (nodal) displacements and rotations, internal forces in all members (at the ends or in more cross sections) and external nodal forces (checking of the given loads and moments, unknown reaction components). The algorithms are based on the modern finite element ideas and employs recently developed techniques in structural analysis, matrix and network formulations. The program uses the stiffness method and in the simple cases of the common technical praxis the solution is exact. In more complicated cases, dividing some members into elements which model them approximately, the solution converges to the exact solution when refining the division. Nonlinear problems of large displacements and still small strains, as it is rule in engineering, can be solved too.

THEORETICAL BACKGROUND

The simplest case of program element is a straight line beam element respecting all six internal forces (i.e. normal force $N = X$, shear forces $Q_y = Y$, $Q_z = Z$, torsional moment $M_x = (X)$, bending moments $M_y = (Y)$, $M_z = (Z)$) in the potential energy. The bending follows the Bernouilli-Navier hypothesis of plane cross-sections without warping, the shear stresses are calculated by the Grashof-Zhuravski law, the torsion by St. Venant hypothesis. The physical properties are described by the Young modulus of elasticity E and the shear modulus G. The geometry of cross section is represented by six constants F_x (area), F_y, F_z (effective shear area), I_x (torsional rigidity), I_y, I_z (moments of inertia). The position of the element in the space is given by the x, y, z coordinates of its ends and the angle β, defining the direction of the main centroidal axis y. In plane structures (planned x,y or x,z) loaded only in their plane (frames, trusses or perpendicular to that plane (grids) some of data are omitted.

The nonprismatic or curved elements can be given through their stiffness matrix or flexibility matrix which can be calculated separately by the program DEFOR itself or by the formulas from references.

The leading variational principle of the solution is the Lagrange principle of minimum of total potential energy. The corresponding equations are the equilibrium conditions in each structure node. The unknowns are nodal displacement components u, v, w and rotation components ψ_x, ψ_y, ψ_z in global structure coordinates x, y, z. All six components are relevant only when solving space frames. Space trusses possess only three nodal unknowns u, v, w, plane frames u, v, ψ_z, plane grids w, ψ_x, ψ_y, plane trusses u, v.

The program capacity is from the theoretical point of view unlimited. Up-to-date (1984) information for computer of the type PDP 11: the number of nodes should not exceed 1000, which means about 6000 unknowns in the most general case. The capacity depends on the extent of internal and external memory (e.g. PDP11: minimum 32 KB and minimum 2.5 MB, see further).

FIELD OF APPLICATION

Geometrical. 1D-, 2D- and 3D- structures composed only of beam elements, i.e. continuous beams plane trusses, frames and grids, space trusses and frames. Large displacements admissible. Simple elements are given by their cross section properties the other elements by their stiffness or flexibility matrix.

Materials. Linear elastic.

Analysis capabilities. Static loading problems, thermal expansion, special distorsions, influence lines and functions.

Loadings. Any loading concentrated in nodes pertaining to the nodal deformation parameters. In the general case of space frames: Nodal force components P_x, P_y, P_z and moment components M_x, M_y, M_z. A very great class of element loadings: concentrated loads or moments in any element section including the ends (primary loads and moments in more complicated cases), uniformly or linear distributed loads on the whole element or on a part of it. Any concentrated load impulse is connected with the possibility of inputting the appropriate deformation impulse: the given nodal displacements and rotations, the displacement and rotation singularities (dislocations) in any cross-section (e.g. generating the influence lines etc.).

Input KOMB can arrange the calculation of the effect of any combination of the previous load cases including the previous KOMB cases.

PROGRAM DESCRIPTION

Method

Finite element method with 1D elements, Lagrange variational principle of minimum total potential energy leading to the equilibrium conditions in the nodal points of structure division. Unknowns: deformation parameters, i.e. displacement and rotation components in the nodal points.

Type of Elements

Beam elements. Classical beam elements governed by the theory of Bernouilli-Navier, Grashof, Saint Venant with the influence of all six internal forces on

DEFOR: Program for Statical Analysis of Structures

the potential energy, i.e. normal force and shear effect included. Non-prismatic and curved elements entered through their stiffness or flexibility matrix which can be in any case calculated by the program DEFOR itself when the references are not available. In that case only the triangular part of a nodal submatrix is sufficient for the input, i.e. six terms in a plane case and maximum 21 terms in a space case.

PROGRAM STRUCTURE AND USER COMFORT

Data can be entered through traditional modes (card deck, disk) or interactively through a terminal console in DIALOG without the use of a manual. Both input and output data can be visualized graphically on screen, hard copy or plotter drawing. The postprocessor can arrange the form, content and extent of data output according the user demands, including the colour graphs, etc. The language of the program is FORTRAN.

HARDWARE COMPATIBILITIES

Minimum configurations. 32 KB internal and 2.5 MB disk memory, a terminal console with an editor.

Type of computers and other hardware. IBM, DEC, HP.

Peripherals. Plotter, digitizer graphic display, printing console. Colour graphic display, terminals.

Operating system. OS, RSX-11 MS DOS.

Media. Availability: source code complete listing and cards, complete program on disc, user's manual, theoretical manual. Services: regular updates, new capabilities, free consultation, technical help (implementation, user's training etc.).

EXAMPLES OF APPLICATION

Test Cases

The solution of problem is exact. Test cases can be performed only from the modelling point of view in the complicated members replaced through a system of simple elements. In engineering practice dividing into ten elements should be sufficient. For more information see References.

Practical Examples in Industry

Plane frame (Fig. 1). Six store and three field plane frame with wind load in its plane.

Application area: civil engineering

Type of problem: static

Drawing: see Fig. 1. Dimensions: horizontal: 6 + 8 + 6 m, vertical: 6 x 3 = 18 m

Discretization: one beam = one element

Type of elements: beam elements (1D-elements)

Number of elements: 42

Number of nodes: 28, number of supports: 4

Number of degrees of freedom: 76

Band width: 15

Subregions: not used

Program DEFOR - plane frame part used. Computer PDP 11, plotter Calcomp 836

Input: interactively in DIALOG

Output: printing: displacements u,v, rotations ψ_z, internal forces N, T, M, nodal forces and reactions. Graphic output: deformation, see Fig. 1

Computation time: 3 min including output

Costs: 3 x 7.50 = 22.50 (Czechoslovak Crones)

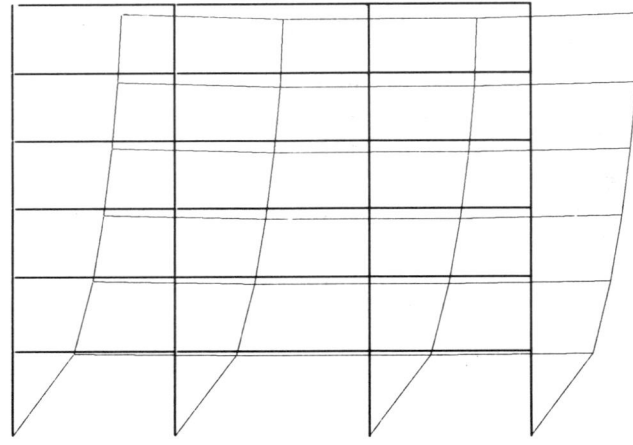

Fig. 1.

Space frame (Fig. 2). Cooling tower 100 m high, concrete structure in the 1st stage of construction (60 m high), symmetry conditions (a half on Fig. 2), horizontal loads pertaining to the montage cable forces.

Application area: civil engineering, power stations

Type of problem: static

Drawing: see Fig. 2

Discretization: see Fig. 2, grillage

Type of elements: beam elements modelling the real structure

Number of degrees of freedom: 2016

Number of elements: 660

Number of nodes: 349

DEFOR: Program for Statical Analysis of Structures

Band width: 90

Subregions: not used

Program DEFOR-space frame part, computer PDP 11, plotter Calcomp 836

Input: interactively in DIALOG

Output: all six displacement and rotation components in 349 nodes, internal forces $N, T_y, T_z, M_x, M_y, M_z$ in all 660 elements, nodal forces and reactions, two load cases. Drawing on plotter

Computation time: 40 min including output

Costs: 40 x 7.50 = 300 (Czechoslovak Crones)

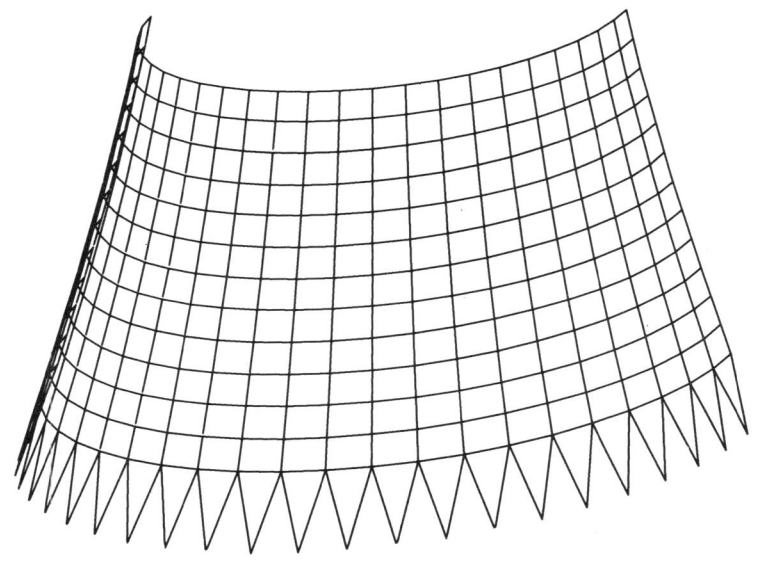

Fig. 2.

Tension tower (Fig. 3). Steel structure of the tension tower 300 kV, rigid connection of beams in joints (space frame), loaded by cable forces and wind.

Application area: civil engineering, power stations

Type of problem: static

Drawing: see Fig. 3

Discretization: one beam = one element

Type of elements: beam elements

Number of elements: 460

Number of nodes: 183

Number of degrees of freedom: 1098

Band width: 100

Subregions: not used

Program DEFOR - space frame part, computer PDP 11, plotter Calcomp 836

Input: interactively in DIALOG

Output: displacement and rotation components in 183 nodes, internal forces in 460 elements, reactions, drawing on plotter

Computation time: 25 min including output

Costs: 25 x 7.50 = 187.50 (Czechoslovak Crones)

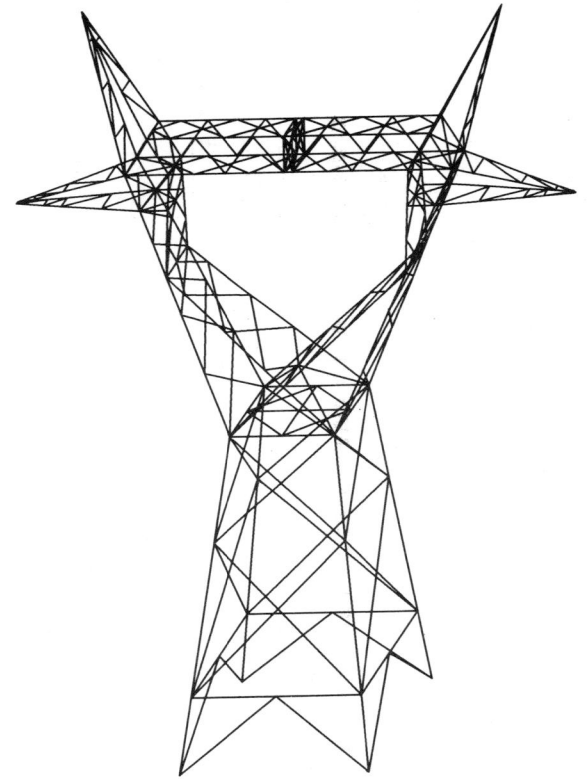

Fig. 3.

REFERENCES

1. V. Kolář et al. *Design of Two- and Three-dimensional Structures by the FEM*. Springer, Vienna, pp. 1-68. 1975 (in German).

2. Kolář, V. and I. Němec. Finite element analysis of structures. *United Nations T.E.M. Workshop CAD*, Prague-Laussane, 1985.

FLASH: AN ANALYSIS AND DESIGN TOOL FOR ENGINEERS

D. Pfaffinger and U. Walder

Walder and Partners, Computer Centre and Softwarehouse, Bern and Zürich, Switzerland

ABSTRACT

The finite element program FLASH analyses beam structures, folded plates, plates in bending and stretching and shells under static loads. In addition FLASH permits dynamic eigenvalue and buckling analysis. The program uses hybrid finite elements including a newly developed elastically supported element for foundation plates or embedded shells as well as for flat slab structures to model the columns. The input is described by simple and concise syntax diagrams. In addition to its high numerical efficiency FLASH offers extensive means of postprocessing the results. It has also been interfaced with the CAD System CAD-B (Computer Aided Design of Buildings). Due to its modern elements, high numerical efficiency, the simple and user-oriented input language as well as the postprocessing facilities and the interface with CAD, FLASH is a powerful and cost effective tool of the engineer.

INTRODUCTION AND THEORETICAL BACKGROUND

The original version of the program FLASH has been developed at the Swiss Federal Institute of Technology (ETH) in Zurich. It was subsequently completely rewritten and expanded to meet most of the needs of the practising engineer. The program first of all analyzes by linear elastic theory shells, folded-plates, plates in bending, plates in stretching (plane stress or plane strain), ribbed slabs, plane and space frames or trusses under static loads. Special elements allow the investigation of plates and shells on elastic foundations. These elements can also be used to model columns in flat slab structures to avoid moment singularities at the support points. Taking into account the shear deformation in plates and shells also relatively thick structures (for example arch dams) as well as sandwich plate problems can be analysed. Stiffeners of plates and shells can be modelled by using eccentrically connected beams. This offers the possibility of treating stiffened plates as plane problems. For the modelling of the structure as well as for the representation of some results FLASH permits the introduction of additional coordinate systems (cartesian, cylindrical or spherical). Thus structures or parts of a structure with special geometries can be modelled in the appropriate coordinate system. FLASH comprises many options to graphically display the model or special properties of the model (e.g. thicknesses, support conditions etc.). The results of the analysis can be obtained in the joints or in the centre of the elements for single loadcases, loading combinations as well as for the envelopes. Practically all results can be obtained either numerically and/or graphically in the form of contour lines or other adequate graphic representations. In addition the user will obtain deformations, reactions and the support pressures of the elastically supported elements as well as bending moments and membrane stresses, the principal moments and stresses and the reinforcement moments. It is possible to obtain the required amount of steel in the different parts of the structure. The program also permits definition of sections across the structure for which integral section forces can be determined.

With the exception of the beam element FLASH uses an improved hybrid stress model for its elements. Its mathematical formulations are described in detail in references 2, 3 and 4. The model can be shortly described as follows:

- For each element two functions are formulated namely one for the interior stress field and the other for the edge displacements.

- The assumption for the interior stresses satisfies the inhomogeneous differential equation of equilibrium, but leads to discontinuities along the element edges.

- The functions for the edge displacements are such that kinematic compatibility along the element boundaries is guaranteed.

- The stiffness-, stress- and load-matrices are formulated by applying an extended principal of minimum complementary energy such that the hybrid stress model too leads to the matrix deformation method.

- All element integrations are done numerically, which allows the use of arbitrarily shaped triangular and quadrilateral elements.

- The convergence of the elements is proved. The results lie always in between the solution of a compatible deformation model with equivalent edge displacements, which is always too stiff and the solution for a pure equilibrium model with the same stress assumptions, which is always too flexible. Hence, even coarse meshes can yield very accurate results.

FIELDS OF APPLICATION

With regard to the geometry, FLASH permits the analysis of two-dimensional and three-dimensional structures. At the moment, however the program does not contain volume elements. Using the shell elements with shear deformation on the other hand permits the analysis of rather thick structures. The user specifies at the beginning of the input data the type of his structure (membrane, plate or shell). The number of degrees of freedom per node is then chosen accordingly.

The material properties are linear elastic. The program supports isotropic and orthotropic materials.

FLASH permits first of all the static analysis of three-dimensional structures under arbitrary loadings. The loads may consist of nodal loads, gravity loads, uniform pressures, centrifugal loads, residual stresses, singularities for influence lines or influence surfaces or for thermal loads. For beam structures the loadings may consist of nodal loads (forces and moments) or of concentrated or linearly distributed loads in between the nodes. In addition the second order effects can be considered either in the static or in the vibration analysis. It is also possible to determine the buckling load of the structure. FLASH permits the calculation of eigenvalues as well as related dynamic quantities of the structure such as mass participation factors, effective masses and so on.

The support conditions in FLASH are very general. The user may fix special degrees of freedom either rigidly or elastically. He may also rotate the support nodes. One special modelling technique in FLASH is the use of constraint equations. The program offers rigid connections as well as general linear constraint equations. By means of such equations many special properties of a structure such as polar symmetry can be easily modelled.

The program FLASH till now has been extensively used in structural engineering work (buildings, towers, dams etc.) and in the machine building industry.

PROGRAM STRUCTURE

The program has been designed to work with high numerical efficiency and also to offer great user-friendliness. One of the means to obtain efficiency is to calculate the element stiffness, stress- and load-matrices only once for elements with the same shape and material properties. The global matrices can be automatically optimized before decomposition.

The handling of the data is fully dynamic. The maximum problem size has been fixed to approximately 30,000 degrees of freedom. In special cases this upper limit can be extended.

The input of FLASH is free format and consists of a problem-oriented input language. All input statements are represented in railroad diagrams as shown in Fig. 1. The experienced user can prepare his input data exclusively by means of the few pages describing these diagrams. The output comes in DIN A4 paper size. The output and also the plots can thus directly be used in reports.

The program offers restart capabilities to enable the user to reenter the program at specified restart points. By means of restarts, for instance, additional loadcases can be calculated or additional results can be requested.

FLASH contains extensive error checking features. The input data are scrutinized for logical and syntax errors without entering into any time consuming algorithms. The program contains hundreds of error messages which give clear indications of any input problems. In addition, extensive graphic representations permit the user to display practically all properties of the structure he is interested in. The graphic representations comprise the plot of the mesh, the numbering of nodes and elements, the marking of thicknesses and element types, the graphic representation of boundary conditions, local and global coordinate systems and so on. In the plots multiple colours are used for additional clarity. With these capabilities in addition to the option of data check runs the user can completely verify his model before the analysis is done.

FLASH has recently been interfaced with the CAD System CAD-B. By means of this approach it is possible to use the data from the geometric data base of the CAD-B System for an automated mesh generation. Once a proposal of a mesh is obtained, the user can then interactively change this mesh until it satisfies all requirements. In the same way the material properties, the boundary conditions and the load cases are defined in an interactive way. As a result the input data for FLASH are generated. After the analysis specific results in the form of plots can be fed back into the CAD System and can be used further on in the design process. This permits the user for instance to design his reinforcements interactively on the bases of the FLASH results. He then not only gets the drawings for the reinforcements but also the list of materials.

FLASH is completely written in FORTRAN IV. It is thus highly portable and can be installed on various computers.

The program is fully supported and is steadily further developed by professional engineers.

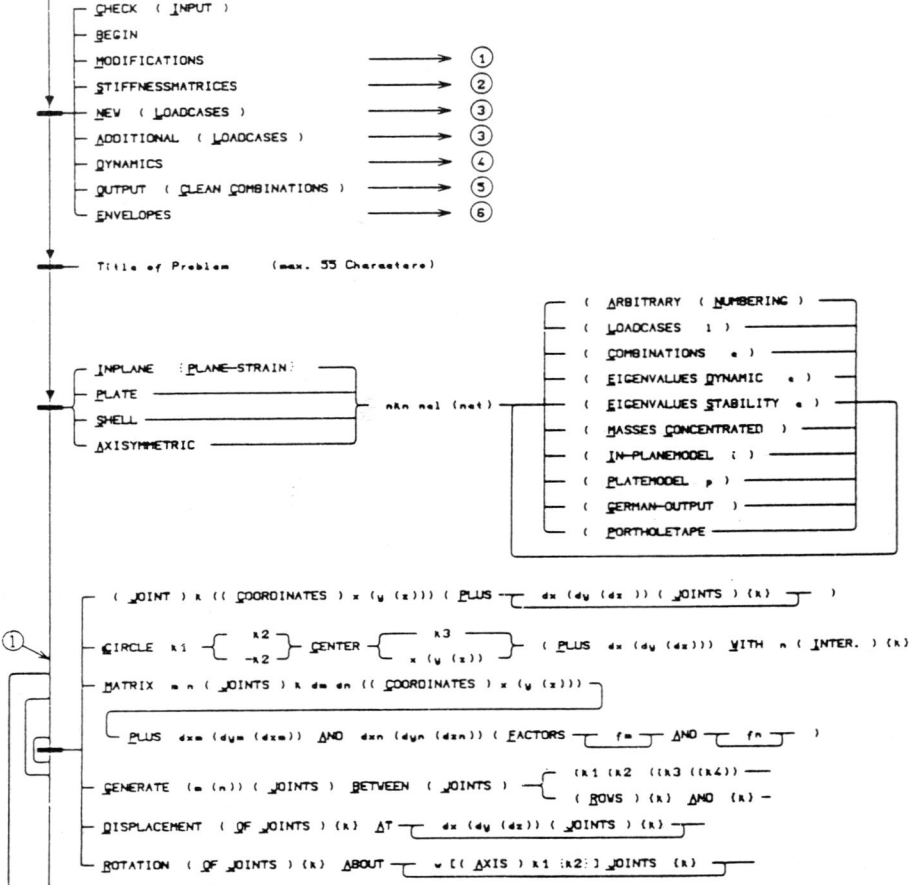

Fig. 1. Example of railroad diagram.

HARDWARE COMPATIBILITIES

FLASH is available in versions for numerous computers including PRIME, DEC, CDC, IBM, UNIVAC, DATA GENERAL, SIEMENS, HP9000, APOLLO and others. To run the program satisfactorily it is necessary to have a graphic terminal, a printer and a plotter. The operating system should support a segmental loader.

EXAMPLES OF APPLICATION

Demonstration Example

Figures 2 and 3 show the complete input data and the mesh of a slab of a tall building. The input could be further reduced by using for each word its first letter only. The structure is modelled by plate elements and by elastically

FLASH: An Analysis and Design Tool for Engineers   107

```
BEGIN
FLAT SLAB OF A TALL BUILDING
PLATE 424 381 2 LOADCASES 2 COMBINATION 1
MATRIX 18 15 JOINT   1   1 18 COORD  0.  PLUS 0.  1.  AND 1.
MATRIX 14 11 JOINT 271 1 14 COORD 15.  4. PLUS 0.  1.  AND 1.
*
ISOTROPIC 3.E6  .167   .24   TYPE  1 2
FLEXIBILITY 1.2E-6 TYPE 2
*
MATRIX 17 14 ELEMENT   1   JOINTS 19  20    2    1  TYPE 1
MATRIX 13 11 ELEMENT 239 JOINTS 271 272 258 257 TYPE 1
*
TYPE 2 ELEMENTS  4 10 17 123 129 136 244 251 335 342
*
NOT-FREE FREE      NOT-FREE JOINTS 271 TO 411 STEP 14
NOT-FREE NOT-FREE NOT-FREE JOINT   257
N  N  F 253 TO 256 $ F F N 412 TO 424 $ F N F 1 TO 235 S 18
*
PLOT SHRUNKEN ELEMENTS WITH BOUNDARYCONDITIONS
PLOT SHRUNKEN ELEMENTS WITH JOINTNUMBERS /
      WITH ELEMENTNUMBERS
*
*
LOADCASE DEADLOADS
GRAVITYLOAD 2.5
LOADCASE LIVELOADS   3 TO/M2 POSITION 1
UNIFORM -.3 ELEMENTS   1   MATRIX 3 7 1 17 11   MATRIX 7 7 1 17 /
123 MATRIX 7 7 1 17   245 MATRIX 7 7 1 13 330 MATRIX 6 4 1 13
*
LOADCASE 1
DEFORMATIONS $ MOMENTS JOINTS 1 TO 270
LOADCASE 2 $ COMBINATION 1   5.   2   -.1
*
LOADCASE 1 AND IF RELEVANT 2 AND 1001
ENVELOPES POS.REINFORCEMENT X NEG.REINFORCEMENT X /
P  Y  N  Y  DRAW   SCALE 100. ELEMENTS
*
```

           Fig. 2. Complete input data.

supported elements for the columns. Figure 4 shows the envelope of the moment M_x.

Miscellaneous Applications

FLASH has been used in the analysis of many civil engineering structures. The following list contains only some of the more unusual structures which have been analysed with the program:

 - John Hancock tower in Boston.
   Reanalysis on the basis of a frame structure.

 - Plaza tower in Houston, Texas.

 - Turntable of a crane.
   Static analysis of a shell model.

 - Rotor of a turbine.
   Static and collapse analysis.

 - Supporting structure of an offshore platform.
   Static reanalysis on the basis of a model of beam and shell elements.

Selected List of Clients

 - Allusuisse
 - Elektrowatt

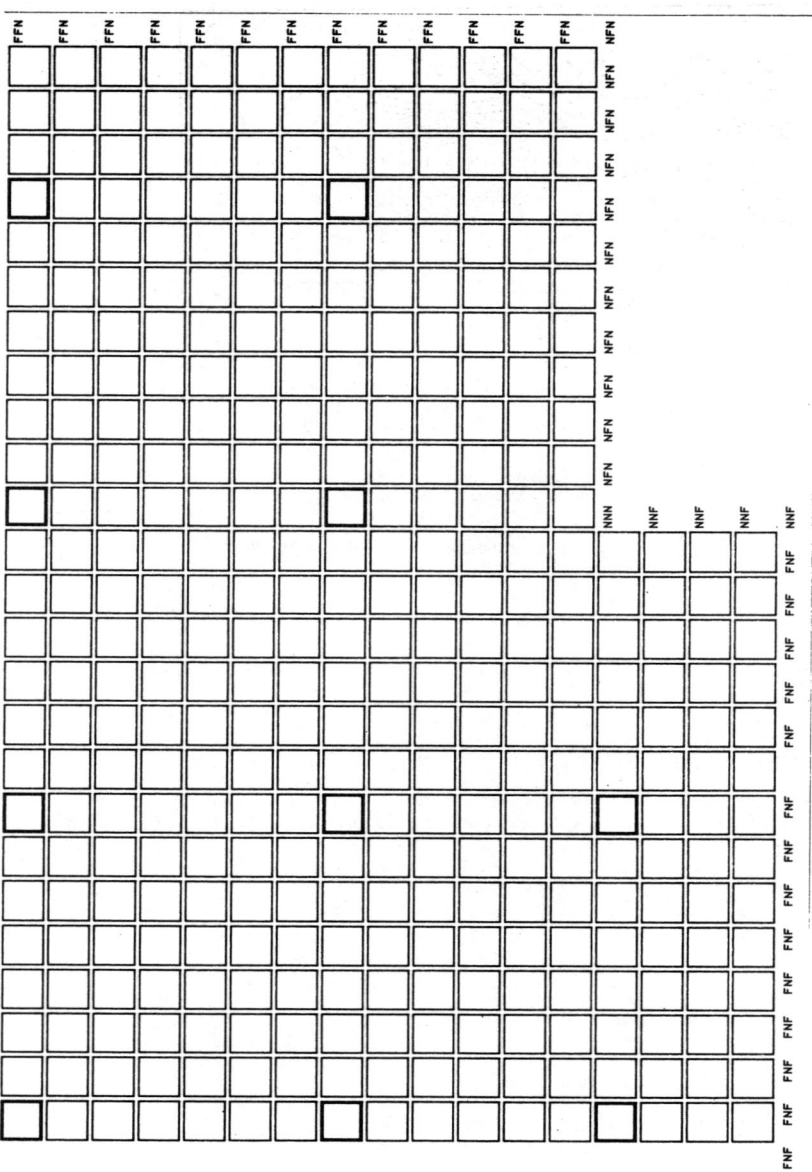

Fig. 3. Mesh with support conditions of slab.

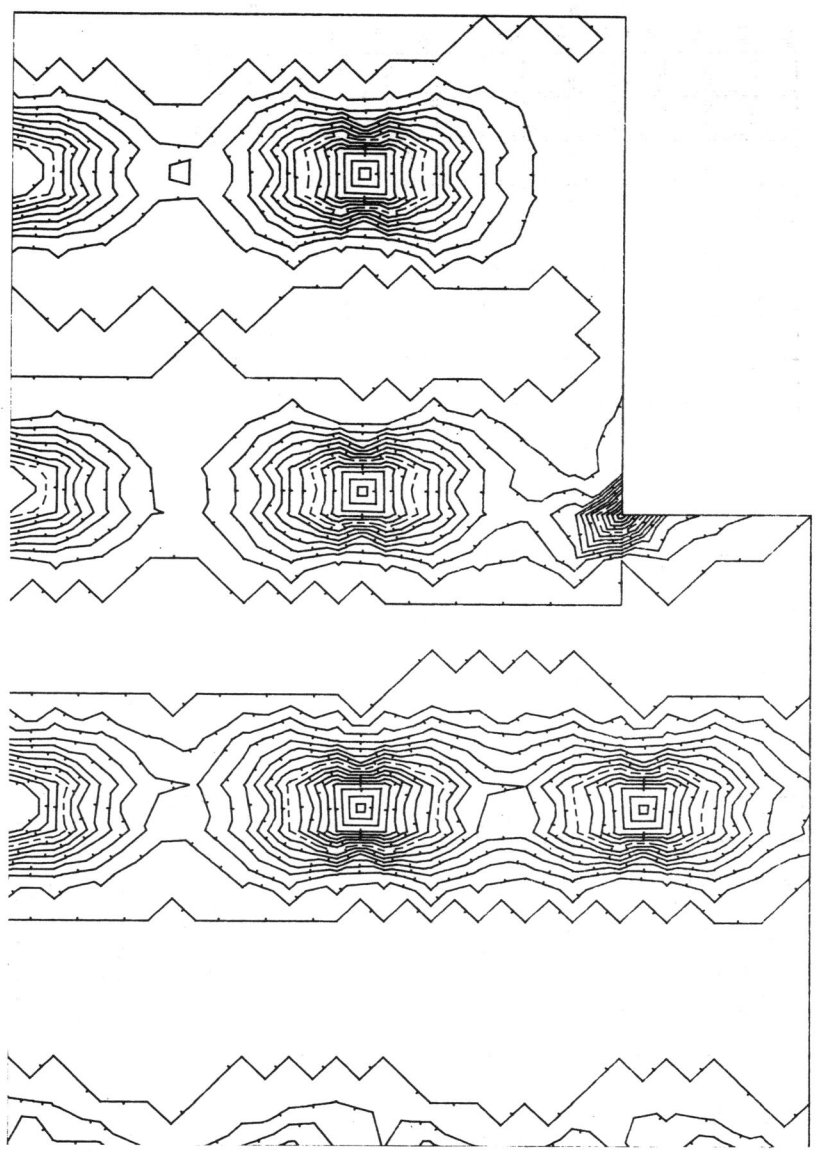

Fig. 4. Design moment Mx.

- Emch & Berger
- Basler & Hofmann
- Fides
- Control Data
- Liebherr
- Universale
- ETH Zurich and Lausanne
- Mass. Institute of Technology
- Universities: Aachen, Bangkok, Berlin, Brisbane, Brussels, Calgary, Graz, Hong-Kong, Peking, Trondheim

REFERENCES

1. Pfaffinger, D., and U. Walder. The use of CAD in finite element analysis. *12th IABSE Congress*, Vancouver, Sept. 1984.

2. Pian, T. H. H. Elements stiffness-matrices for boundary compatibility and for prescribed boundary stresses. *Proc. 1st Conf. Matrix Methods in Structural Mechanics*, Dayton 1965, AFFDL-TR-66-80, Nov. 1966.

3. Wolf, J. P. *Generalized Stress Models for Finite Elements Analysis*, Institut fuer Baustatik und Konstruktion, ETH, Zurich, Bericht No. 52, 1974.

4. Walder U. *Beitrag zur Berechnung von Flaechentragwerken nach der Methode der finiten Elemente*, Institut fuer Baustatik und Konstruktion, ETH, Zurich, Bericht No. 77, 1977.

5. Walder, U., D. Pfaffinger, D. Green. *FLASH User's Manual*, 8th edn., May 1985.

IBA: INTERACTIVE BUILDING ANALYSIS

F. Braga*, M. Dolce**, C. Fabrizi* and D. Liberatore*

Istituto di Scienza delle Costruzioni, Via Eudossiana 18, 00184, Rome, Italy
**Istituto di Scienza delle Costruzioni, Roio (L'Aquila), Italy*

ABSTRACT

The program is oriented to the analysis of framed buildings subjected to vertical and horizontal actions (wind and earthquake); the static and/or dynamic analysis is performed directly on the spatial frame, treated as an assemblage of plane frames connected by diaphragms (floors, roofs) which are infinitely stiff in and infinitely deformable out of their own plane.

Particular attention is devoted to the input-output operations; they are performed by suitable pre- and postprocessors, which provide graphic representations of the input data and of the results. Corrections are allowed at any stage of the analysis, without redefining any of the previously given data.

At present, the program is implemented on Olivetti M20, M24 (personal computers 64 kb user's RAM and two $5\frac{1}{4}$ inch floppy disks) and on Olivetti M40 (microcomputer multitask multiuser with 1.5 Mb user's RAM and one 18 Mb Hard Disk) and is codified in Microsoft BASIC and FORTRAN 77; in the lesser configuration buildings with up to 20 stories and 40 vertical column lines can be analysed.

THEORETICAL BACKGROUND

Processor IBA makes use of the substructuring technique and of the direct stiffness method to assembly the stiffness matrix of the pseudo-three-dimensional model. The various substructures are two-dimensional (frame, frame-wall, single wall, coupled walls) and are connected through diaphragms infinitely rigid in and infinitely deformable out of their plane (usually one diaphragm for each floor). Each diaphragm has three degrees of freedom: the two translations and the rotation in its plane. Any kinematism of the model is described by only these degrees of freedom. In comparison with plane models, the pseudo-three-dimensional model has a wider field of applicability, which includes buildings with torsional effects, without any increase in user and computer time. In comparison with fully three-dimensional models (six degrees of freedom per node) a drastic reduction in computer time and storage is gained, thus allowing microcomputers to treat even bigger size structures.

The stiffness matrix **K** of the model is calculated by means of a set of operations which transform and condense the nodal degrees of freedom to diaphragm degrees of freedom. The main steps are as follows.

(1) Making of the element matrices R_e in the element coordinate systems u_e.

(2) Transformation of the stiffness matrices of the elements from the element coordinate systems u_e to the substructure coordinate systems v_s ($u_e = T_{es} v_s$):

$$R_{es} = T_{es}^T R_e T_{es}$$

(3) Assemblage of the transformed element matrices R_{es} into the stiffness matrix K_s of the relevant substructure:

$$K_s = \Sigma_e R_{es}$$

(4) Partitioning of K_s (separation of nodal v_{ns} from floor v_{fs} degrees of freedom):

$$K_s = \begin{bmatrix} K_{s1} & K_{s2} \\ K_{s2}^T & K_{s3} \end{bmatrix}$$

and condensation with respect to the substructure coordinate v_{fs} directly connected to the diaphragm coordinates x:

$$C_s = K_{s3} - K_{s2}^T K_{s1}^{-1} K_{s2}$$

(5) Transformation of the condensed substructure matrices from the substructure floor coordinate system v_{fs} to the pseudo-three-dimensional or diaphragm coordinate system x ($v_{fs} = T_{sp} x$):

$$C_{sp} = T_{sp}^T C_s T_{sp}$$

(6) Assemblage of the transformed substructure matrices C_{sp} into the pseudo-three-dimensional stiffness matrix.

$$K = \Sigma_s C_{sp}$$

The load vectors p of the pseudo-three-dimensional model are obtained through analogous operations, for each load condition.

The mass matrix M of the pseudo-three-dimensional model is calculated directly from the translational mass and the rotational mass about the vertical axis through the barycentre of the diaphragms. Only the horizontal translational and rotational masses are considered, even for inclined diaphragms; in this case a suitable transformation is performed. The governing equation of the static problem is therefore:

$$K x = p$$

and the solution is obtained by means of the Choleski-Banachievitz method, which is based on the factorization:

$$K = L D L^T$$

The governing equation of the dynamic problem for the seismic load condition is:

$$M \ddot{x} + K x = -M r \ddot{x}_G$$

and the eigensolution is obtained via the transformation of the $M^{-1/2} K M^{-1/2}$

matrix to the tridiagonalized form (Householder method) and the Q-D method; the convergence is greatly quickened through the shift of the origin and deflation operations. The dynamic analysis for the seismic actions is performed through the response spectrum method.

FIELD OF APPLICATION

IBA is oriented to the analysis of framed buildings and idealizes a three-dimensional building structure as an assemblage of plane frames connected by diaphragms. Each frame lies on a vertical plan not necessarily parallel to the coordinate plans and is composed of beams, columns, structural walls, bracing elements, infilled panels. Beams and diaphragms are not necessarily horizontal, nodes out of the diaphragms are allowed (to idealize stairs and so on), columns and structural walls are necessarily vertical. Beams, columns and structural walls are treated as beam elements with flexural, axial and shear deformability; diaphragms are treated as if they were infinitely stiff in and infinitely deformable out of their plane.

The constitutive law of the materials is linear elastic; large displacements and deformations are not taken into account. Any kind of structural material can be treated as long as creep and relaxation phenomena can be neglected.

Static and dynamic analyses can be performed. For static analyses several types of load can be considered:

(a) gravity loads,
(b) wind loads,
(c) equivalent seismic loads,
(d) concentrated nodal forces with any direction.

The procedure for dynamic analyses is oriented to seismic excitations and is divided in three steps:

(1) solution of the eigenproblem,
(2) evaluation of the modal displacements from the given design spectrum,
(3) evaluation of the effective values of displacements and stresses.

In step 3 several formulas can be adopted to take into account eventual modal couplings, and any direction of the seismic action can be defined by the user.

Any linear combination of stresses resulting from static and dynamic analysis can be made by using suitable combination coefficients.

PROGRAM DESCRIPTION

IBA works by using the matrix analysis techniques for finite elements. Beam, truss as well as boundary elements are available: all the three elements can have infinitely rigid arms which allow for idealization of beams, columns, reinforced concrete shear walls, infilled brick masonry walls, steel bracing, deformable foundations. In particular the rigid arms permit the high stiffness of the beam-column joint panels to be taken into account as well as the offset of the axis of a member with respect to the axis of the adjacent aligned member, due to a change of the cross-section geometry (e.g. caused by tapering of columns along the height of the building); finally they permit the location of nodes at the intersection points between the input alignments rather than between the member axes, thus greatly simplifying the structural idealization. In Fig. 1 the flow-chart of processor IBA, with a clear identification of the three parts, Assemblage, Solution and Translation, in which the processor is divided is reported. As can be seen the procedures of Assemblage and Translation are so organized that equal frames (substructures of the pseudo-three-dimensional model) are treated in the

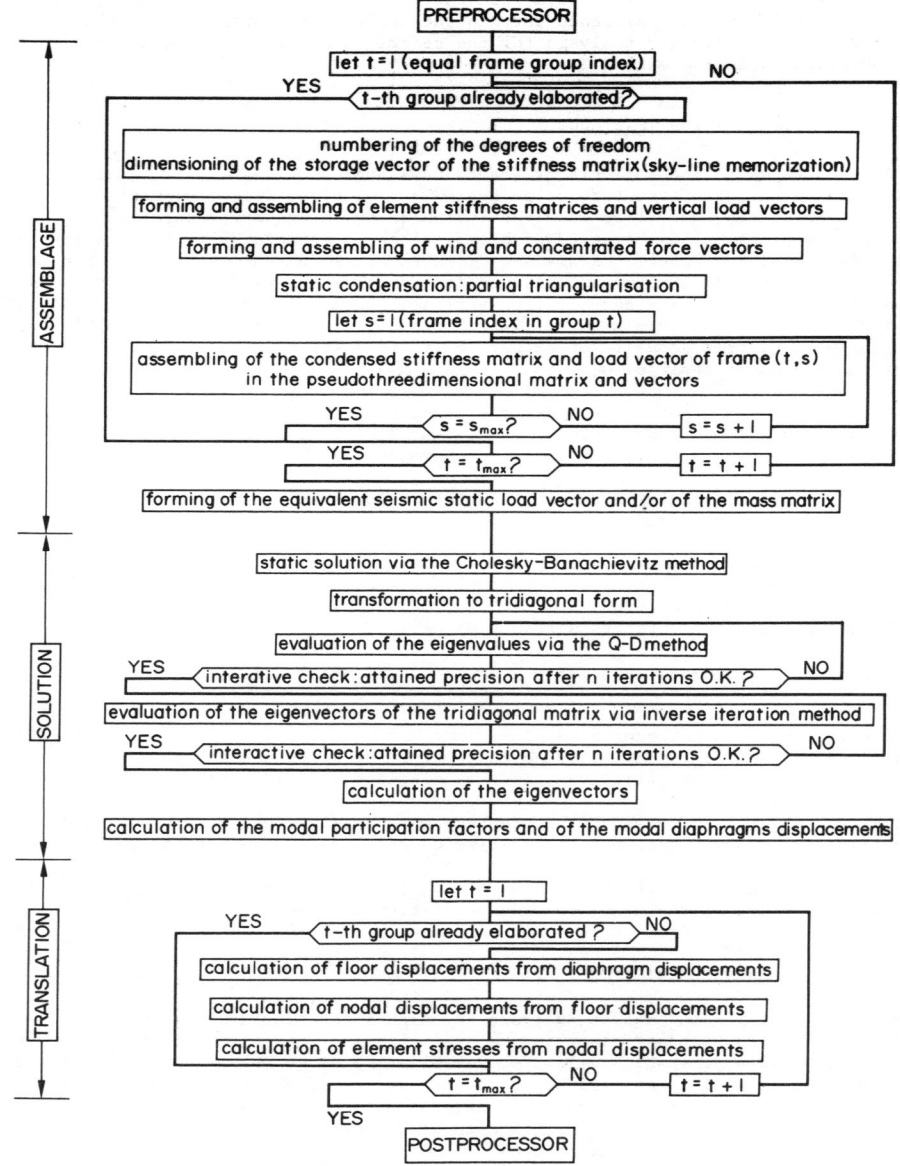

Fig. 1. Flow-chart of IBA processor.

same step, with consequent advantages in terms of time and storage. In particular the structure is assembled by treating a group of equal frames at time, starting from the forming of the stiffness matrix and of the load vectors of each element and arriving at the calculation of the contribution of each of the frames of the

IBA: Interactive Building Analysis

current group t on the pseudo-three-dimensional stiffness matrix and load vectors; this permits the system to keep in memory only the data relevant to each single group, thus saving memory occupation. To the same end, an algorithm of static condensation of Gauss-type, which does not require any additional matrix to be stored in central memory and saves mass memory occupation too, is implemented.

As shown in Fig. 1, the processor is preceded by a preprocessor and followed by a postprocessor.

The main purpose of the preprocessor is to reduce the "user time" required to prepare data; besides this function, which will be called "Quickening", there is the exigence of reducing mistakes, by making the operations of "Diagnostics" easier. In particular, as regard "Quickening", preprocessor IBA defines the structural scheme, with main and secondary elements, directly from the architectural alignments and from the cross-section geometry of beams and columns. It also calculates the offsets with respect to the nodes of the axis line of the elements, to account for change of cross-section geometry and for the high stiffness of beam-column intersection panels and numbers elements and nodes, in a way that saves computer time and storage. Vertical loads on beams are directly defined from the unitary loads and from the texture of floors and roofs, and floor and roof masses and/or seismic equivalent loads are directly calculated from vertical loads. The automation of all such operations, besides improving the input rate, practically eliminates any risk of schematization mistakes.

Preprocessor IBA allows for corrections and modifications, without losing any of the previously input and/or elaborated informations which are not affected by the modifications, but in the same time guaranteeing the automatic correction of all the data actually affected.

As regards "Diagnostics", the identification of mistakes is usually performed "by sight" for other programs, i.e. by examining the input data printed in tabular form; this operation is very time-consuming and, consequently, expensive. Mistakes are generally due to a bad interpretation of the user's manual of the program or to oversights wrong typing, etc. The IBA preprocessor prevents mistakes of the first kind by directly explaining and interactively requiring the type of data to be input in the current step and the mistakes of the second kind by immediately plotting on the screen, the consequences of each operation. The great capability of graphical representation to show up errors is thus exploited. Furthermore, many checks of consistency are performed, so that input of data which are not consistent with the previously input ones, are prevented.

In Fig. 2 the most important input operations and elaborations of preprocessor IBA are illustrated by means of the pictures of the screen which appear during the various steps. Figures 2 a (request) and 2b (visualization) are relevant to the input of the geometrical characteristics of the various beam and column sections. Figures 2c (request) and 2d (final visualzation) are relevant to the input of the alignments in the plan; these alignments define, along with alignments in elevation, the base network in which the various main and secondary structural elements will be positioned (Figs. 2f, 2g, 2h). Figure 2e shows the assonometric view of the previously input foundation plan; the same representation is used for roofs. Figures 2f, 2g 2h show the way in which beams, columns and outside walls are positioned in the plan (f) and the elevation (g, h) networks, by means of the cursor X, and the way in which their section characteristics are described, visualized and positioned (f, g) with respect to the alignments. Figures 2i, 2j are relevant to the input operations of the vertical distributed loads on slabs (i) and on beams (j). Figure 2k shows the numeration of nodes and elements, automatically performed by the program, and Fig. 2l visualizes the geometry of one of the rigid beam-column panels, automatically evaluated by the program; they are proposed on the screen to allow the engineer to make any modification he wants.

The main tasks of the postprocessor are:

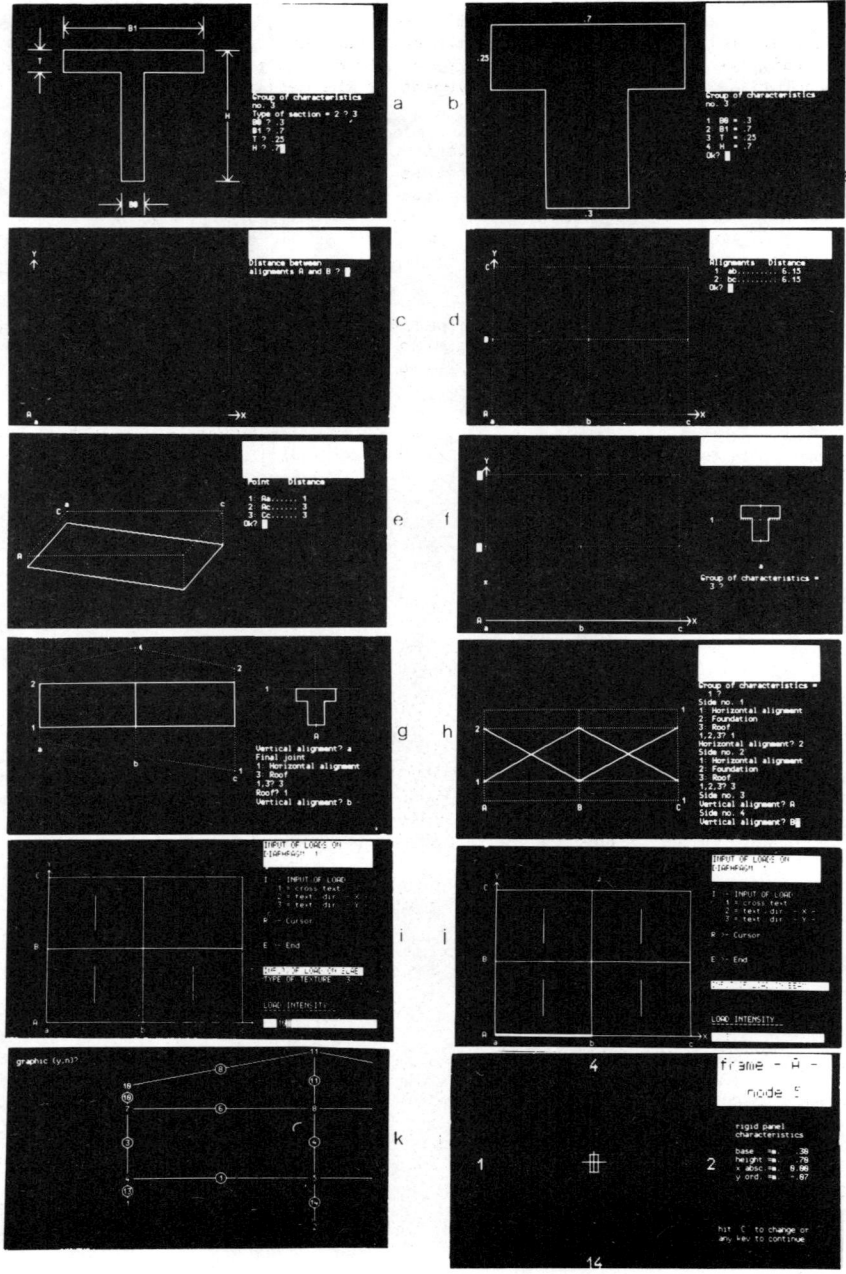

Fig. 2. Interactive graphic input operation.

IBA: Interactive Building Analysis

- "Synthetization" of the results obtained by the processor and representation of them in a graphical and/or numerical form.
- "Utilization" of the above results, through further elaboration; the purpose is to obtain the final product (drawings) of the design process or, at least, all the quantities which make the achievement of the design process as easy and fast as possible.

Besides these two fundamental tasks it is opportune that the postprocessor performs diagnostic operations and provides the relevant results in a suitable form, so that the approximation level connected to round-off errors and to the type of structural model can be easily evaluated. Such operations must be devised bearing in mind the structural model and the type of analysis. According to that, postprocessor IBA checks the overall equilibrium of forces.

After the output of errors, program IBA provides graphic representations of the diaphragm displacements, through assonometric representation of the entire model, and of the nodal displacements, through plan representation of each single frame. These graphics allow the engineer to make a fast and effective control against gross errors.

As regards the synthetic representation of stresses in members, it would be very onerous if it were performed graphically, moreover it is of little use when the results are further elaborated by the program. This is the case of postprocessor IBA, which is oriented to the design of R/C structures. Therefore it provides all the resultant stress states of the members in tabular form and subsequently calculates the reinforcements of beams and columns. The algorithm for this task is composed by three sub-procedures, i.e.:

(1) interactive graphic input, of all possible reinforcement devices of each type of section of beams and columns;
(2) automated search for the optimal, among the previously input ones, reinforcement of every section of every member of every preselected frame;
(3) interactive modifications, with immediate stress verification, of the reinforcements selected in point 2.

It is clear that this procedure is a good compromise between a completely interactive procedure and a completely automated one; it has been studied to eliminate the drawbacks of interactivity and automation, i.e. the large user's time required and the short control on the results respectively, while maintaining their respective advantages.

Preprocessor, processor and postprocessor are codified in BASIC Microsoft and FORTRAN 77; thus it is extremely easy to add new finite elements or new subroutines.

At present a postprocessor oriented to the steel structures is at a developing stage.

HARDWARE COMPATIBILITIES

The minimum required hardware configuration is:

(a) microcomputer with 60 kb user's memory (available for data and program);
(b) two drivers for floppy disks of $5\frac{1}{4}$ inch;
(c) graphic screen and printer.

IBA can therefore be implemented on all the most world spread personal computers (IBM, Apple, etc.).

The compatible operating systems are: PCOS (M20 version), MS-DOS (M24 version), MOS (M40 version).

118   F. Braga et al.

IBA is available on $5\frac{1}{4}$ inch and 8 inch diskettes.

EXAMPLE OF APPLICATION

An application relevant to the structure shown in Fig. 3 is presented. It is a three-dimensional frame made up of beams and columns and with deformable foundations. It has been used to test the effectiveness of the eigensolution procedure (seismic dynamic analysis), since some of its vibrational modes are coupled. The discretization is clearly shown in Fig. 3; beam, column and boundary elements have been used. If treated with a full three-dimensional model

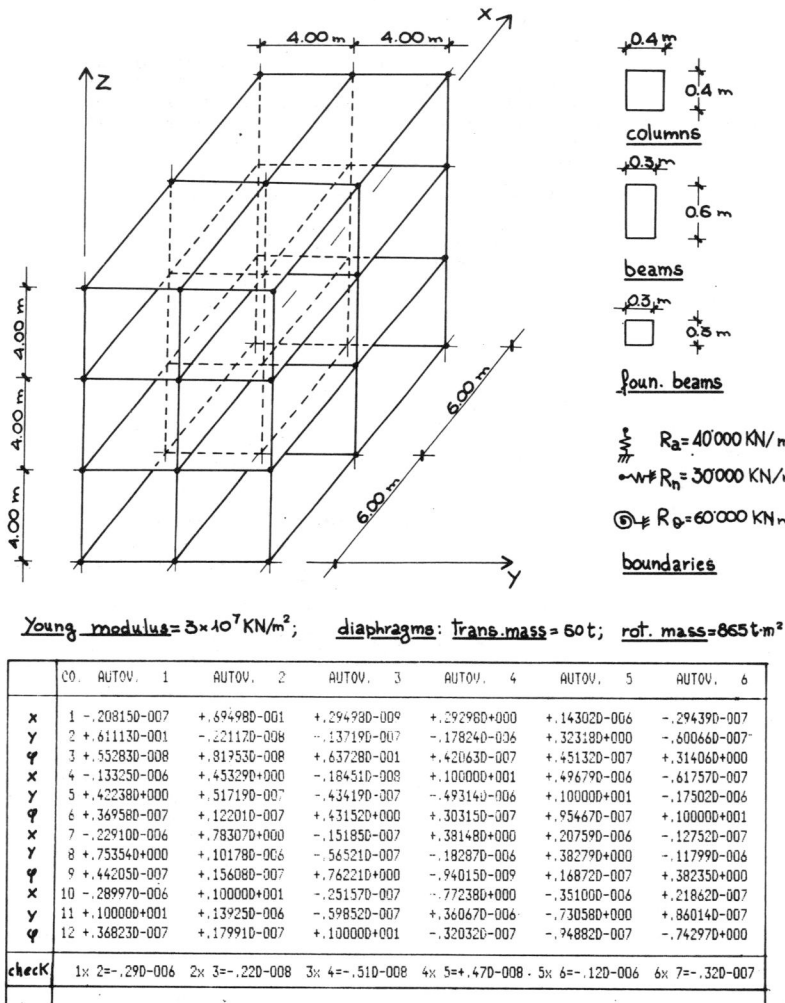

Fig. 3. Example of application.

there would be 36 nodes, 27 column elements, 48 beam elements, 216 degrees of freedom as a pseudo-three-dimensional model there are 6 substructures (plane frame), each being composed of 12 nodes, 9 column elements, 8 beam elements, 9 boundary elements, 28 degrees of freedom (average bandwidth = 8.25 with sky-line memorization), and 4 diaphragms each having 3 degrees of freedom (x, y, ψ). In Fig. 3 there are shown the first six eigenvectors (autov.) and eigenvalues (squared circular frequencies). A comparison with the results obtained by using program SAPV on the same model gives errors on the fourth digit (Olivetti M20 consider seven digits only for single precision variables). Computation times are about 15 minutes for assemblage and 3 minutes for solution (language BASIC Microsoft interpreted).

POTENTIAL USERS

Potential users are all civil engineers who are involved in seismic design of buildings. Thanks to its ease of use, its self-explaining and graphic input-output operations, its numerous precision checks, IBA can be easily used even by not computer-trained professionals.

KYOKAI: A USER-FRIENDLY BEM.FEM SOLVER

Y. Ohura, K. Obata and K. Onishi

Department of Applied Mathematics, Fukuoka University, Fukuoka 814-01, Japan

ABSTRACT

The KYOKAI is a menu-driven small turn-key command system of boundary element and finite element methods performing the numerical solution of two-dimensional potential and elasticity problems. The system is coded in FORTRAN IV, and especially dedicated for the interactive use on a super-mini. The program and the manuals are fully documented. KYOKAI has hierarchal interrelation of subprograms. Modules are classified into preprocessing, analysis, and postprocessing. The preprocessor automatically generates boundary element-finite element meshes for an arbitrary geometry. The input file once generated can be easily modified and quickly updated when the numerical modeling is to be restarted. The postprocessor displays pictorial results on a graphic terminal. New features involved in the pre- and postprocessors to increase the productivity are described. Automatic data conversions between boundary element and finite element analyses, and between constant and linear boundary elements are incorporated. Multiple user simulation is possible under the shared use of c.p.u. The output numerical data file can be used subsequently as a potential input for design and manufacturing.

THEORETICAL BACKGROUND

The first version of the KYOKAI was announced in 1982 to academic circles and industries.[1-4]* With the growing interest of boundary element methods in engineering, the need for an augmented edition has been increased.[5,6] KYOKAI has established a wide use among universities and industries for introductory course of numerical modeling in computational mechanics. Recently, KYOKAI was publicized in a review article,[7] and it is represented in a seminar[8] and symposium[9] on boundary element methods. KYOKAI for three dimensions was presented for boundary solution of potential and elastostatic problems.[10]

The problems covered in KYOKAI are two-dimensional potential and elasticity problems for zoned inhomogeneous, orthotropic materials.

We consider the unknown potential u in the rectangular coordinate system x_i ($i=1,2$). The potential is related to the flux components q_i ($i=1,2$) by:

---

*Superscript numbers refer to References at the end of the article.

$$q_i = -k_i \frac{\partial u}{\partial x_i} \qquad \text{in } \Omega \tag{1}$$

in which k_i is the orthotropic coefficient and Ω is the domain of the material. The conservation law is generally written in the form:

$$\sum_{i=1}^{2} \frac{\partial q_i}{\partial x_i} = S \tag{2}$$

in which S denotes some kind of sources or sinks. The potential problem is to find the unknown function u of Equations (1)-(2) subject to the boundary conditions:

$$u = \bar{u} \qquad \text{on } \Gamma_u \tag{3}$$

$$q = -\frac{\partial u}{\partial \nu} = \bar{q} \qquad \text{on } \Gamma_q \tag{4}$$

in which $\bar{u}$ is the given value of the potential on a part of the boundary Γ_u, and $\bar{q}$ is the given value of the boundary flux in the direction of unit conormal ν on the rest of the boundary Γ_q.

For the two-dimensional problem of stress and strain, materials are assumed linearly elastic and the displacement is assumed infinitesimally small.

Let us denote by σ_{ij} ($i,j=1,2$) the stress components. Force equilibrium can be expressed in terms of the stress components as:

$$\sum_{j=1}^{2} \frac{\partial \sigma_{ij}}{\partial x_j} + b_i = 0 \tag{5}$$

in which b_i is the i-th component of some body force.

Let us denote by ε_{ij} ($i,j=1,2$) the strain components. The strains are related to the displacement components u_i ($i=1,2$) as:

$$\varepsilon_{ij} = \frac{1}{2} \left(\frac{\partial u_i}{\partial x_j} + \frac{\partial u_j}{\partial x_i} \right) \tag{6}$$

Hooke's law representing linear stress-strain relationships including the initial stresses σ_{ij}^0 can be written by:

$$\sigma_{ij} = \lambda \delta_{ij} \varepsilon_{ll} + 2\mu \varepsilon_{ij} + \sigma_{ij}^0 \tag{7}$$

where δ_{ij} is the Kronecker's delta and λ, μ are elastic constants. In the case of thermal stress, the initial stresses and strains are related by:

$$\sigma_{ij}^0 = -\beta \delta_{ij} \varepsilon_{ij}^0 \tag{8}$$

in which β is constant, $\varepsilon_{ij}^0 = \alpha \Delta T$, where α is the delatation coefficient, and ΔT is the variation of the temperature.

The boundary conditions are given by:

$$u_i = \bar{u}_i \qquad \text{on } \Gamma_u \tag{9}$$

$$p_i = \bar{p}_i \qquad \text{on } \Gamma_p \qquad (10)$$

where $\bar{u}_i$ is the prescribed displacement and $\bar{p}_i$ the prescribed surface traction.

FIELD OF APPLICATION

Two dimensional field problems governed by equations (1)-(4) and plane thermoelasticity problems governed by equations (5) (10) can be solved by KYOKAI. Materials for field problems are allowed to be orthotropic and zoned inhomogeneous. Materials for elasticity problems are linearly elastic and zoned inhomogeneous. Composite materials can be modelled by the use of the technique for subregions. Only static problems can be dealt with. The thermal loading effect is incorporated in the analysis. Temperature variations as a calculated result from the potential problem prior to the elastic analysis can be used as the thermal load. Pressure load, initial strains and stresses can be considered.

Physical problems described by the quasi-harmonic equation can be solved by KYOKAI. Heat conduction in solids, concentration diffusion in liquids, gas diffusion, seepage of ground water, irrotational ideal fluid flow, electromagnetostatics, torsion of prismatic bars, Reynolds film lubrication are typical examples. Plane stress and plane strain problems can be solved by KYOKAI.

PROGRAM DESCRIPTION

The boundary element method and the finite element method are integrated in KYOKAI, but the combined use of the method within one problem in such a way that a part of the material is modelled by boundary elements and some other part is modelled by finite elements is not supported.

For boundary element analysis, constant elements and linear elements are used, but they cannot be intermingled. A problem modelled by one of these boundary element types can be automatically remodelled by the other type of elements. For finite element analysis, 2D solids are used in terms of linear triangular elements. Triangular axis-solids are partially delivered.

The language of the program is FORTRAN IV. Higher versions of the FORTRAN language have been forbidden in KYOKAI programming. KYOKAI has a modular program structure depending on the type of problems and the type of analysis. Modules are classified according to the functions involved into three categories: preprocessor, analyzer, and postprocessor. About seventy subprograms are contained in the program. They are interrelated in a tree structure. The tree has seven levels, from the root segment to the seventh branch. The tree up to the second level is shown in Fig. 1. Each subprogram is designed to be monofunctional. New subroutines can be incorporated easily into KYOKAI in an interactive environment.

The KYOKAI system is driven by a series of commands. Two sets of commands for boundary element analysis and finite element analysis closely resemble each other. The list of commands is shown in Table 1.

Input data are created in an interactive mode. The outline of the geometry of the domain is defined by the combination of blocks. The block is an eight-noded quadrilateral. Once the block mesh is constructed, the blocks are subdivided into either boundary element or finite element mesh automatically. Elements and nodes are numbered successively. The nodes can be renumbered in order to minimize the band-width in the coefficient matrix of the resulting linear system of finite element equations.

Multiply connected domains are dealt with. A string of boundary node numbers is

```
KYOKAI---MESH---BLOCM   : generates block mesh.
    |       |-GRAPH     : digitizes graphic screen.
    |       |-BCON      : defines boundary conditions.
    |       |_FINEM     : generates fine meshes.
    |
    |-CAL----LOAD       : specifies load conditions.
    |       |-BEM       : invokes BE solver.
    |       |_FEM       : invokes FE solver.
    |
    |-LIST              : displays numerical list data.
    |
    |_BFPLOT---PREPB    : plots the block mesh.
            |-PREPF     : plots the fine mesh.
            |-PCONTR    : plots contours.
            |-PPERSP    : plots perspective projection.
            |-PFLUX     : plots vector arrows of fluxes.
            |-PDISPM    : plots displacement.
            |_PSTRES    : plots stresses.
```

Fig. 1.   KYOKAI subprogram tree.

TABLE 1   KYOKAI Commands

Preprocess	Analysis	Postprocess	
MESH	CAL	LIST	PLOT
-BLOCK	-LOAD	-POTN	-PBLC
-FINE	-BEM	-ELAS	-PFIN
-END	-FEM	-END	-POTN
STOP	-SOL	STOP	-ELAS
	-CHAN		-END
	-END		STOP
	STOP		

made for each boundary component automatically. A list of internal node numbers is obtained. For the boundary element analysis, double nodes are always taken along the interface of adjacent subregions.

Preprocessor

In the preprocessor, a conventional section paper is displayed on a screen of the graphic display terminal. Size of the section paper is virtually infinite. Users can draft the block mesh on looking the screen. The coordinates of the nodes are digitized by a cursor on the section paper. The coordinates can also be input in terms of local coordinates. Linear sides of the block can be defined by two edge points. The middle node point is calculated automatically. Arcs can be defined by specifying the central point, the radius, and the angle in degrees.

A set of subcommands are available when the block mesh is being generated. The mnemonic subcommands are listed in Table 2. They are self-explanatory. HELP command is used for the self-instruction for users. The block mesh is therefore created directly on the graphic screen. This saves the time required to get the outline of the geometry and reduces the input error to a minimum. Nodes and lines once generated can be erased interactively, and the corrected lines are instantly displayed. Block data can be modified by DELETE, CHANGE, and INSERT subcommands.

TABLE 2   Preprocessor Subcommands

BLOCK	FINE	Editing
GRAP	MAKE	DELETE
CORD	REST	CHANGE
MATE	PFIN	INSERT
BCON	CHAN	PRINT
PBLC	END	GRAPH
CHAN	STOP	
END		
STOP		

Boundary conditions are defined in the block mesh. However, the local boundary conditions which cannot be appropriately modelled on the block mesh can be given after the fine mesh has been generated. Boundary values may be constant. Linear and parabolic variations of the values can be specified by a few prompts.

The fine meshes for boundary elements and finite elements are automatically generated after the number of subdivisions and relative spacings has been specified in two local directions within each block. By increasing the number of subdivisions, the user can refine the element mesh. For the finite element analysis, the node numbers are given always in such a way that the band width of the coefficient matrix is to be minimum. The fine mesh can be modified interactively by using DELETE, CHANGE, and INSERT subcommands. The list of fine mesh data can be displayed on the screen by PRINT subcommand.

Load conditions must be specified prior to the solution procedure. Point load and pressure load are given as the boundary condition, but distributed loads due to gravity and any other forces derived from the potential must be given at node points. The intensity of sources or sinks in the potential problem is given node by node. The same value can be specified for a number of successive nodes by a return key.

BEM.FEM Analyzer

Boundary element and finite element methods are invoked by the CAL command. Two-dimensional potential problems and thermoelasticity problems are solved. The problem once solved by one of the two methods can be resolved by the other method under automatic internal data conversion. Therefore the input data for analysis seems similar from the user's point of view. The mesh data are shared by the two methods, and boundary conditions and loadings are reformatted between these methods with minimum effort. After the solution, the input data can be modified by DELETE, CHANGE, and INSERT subcommands for the improvement of the numerical model.

Postprocessor

The calculated results are displayed on a graphic screen by the postprocessor. Table 3 shows the list of subcommands for the postprocessor.

For potential problems, contours, three-dimensional perspective of the potential surface, and fluxes can be plotted automatically. The maximum and minimum values of the contours can be specified. The number of lines and the increment can be also specified The value assigned to each contour is indicated on the screen. The perspective projection can be translated and rotated in three dimensions.

TABLE 3   Postprocessor Subcommands

Potential	Elasticity
CONT	DISP
PERS	STRE
FLUX	TRAC
	STCL
	STPS

The frame size of these plots is arbitrary. Part of the plot can be selected by a zooming feature.

For elasticity problems, displacement, principal stresses, and traction can be plotted automatically. The displacement for the outline of the total geometry is depicted. Also, the distorted mesh plotted by solid curves is overlaid on the original mesh having been plotted by the dotted lines. Principal stresses are shown by arrows representing compression and tension distinguished by arrowheads. Contours for two components of the principal stress are also plotted.

Calculated results are listed in numerals with suitable headers and item names on the screen. The intermediate files are updated only when the same type of analysis is undertaken. Therefore, the computation can be restarted through any stage of calculation. The execution of KYOKAI can be terminated at any command level.

HARDWARE COMPATIBILITIES

KYOKAI is so designed as to be machine independent and be modular to facilitate the maintenance quickly and to include new developments easily.

The major design constraint on the KYOKAI arose from a requirement for interactive computer graphics environment. If this could not be maintained, there would be some difficulties in model definition and result examination. Without graphic capability, the data check would be error prone.

KYOKAI was developed on a PRIME 250 under the PRIMOS operating system with 512 kilobytes central memory, 96 megabytes peripheral disc storage, and a TEKTRONIX 4010 graphic display terminal, and a TEKTRONIX 4113 hard copy unit. The system configuration is shown in Fig. 2. The minimum architecture required can be much

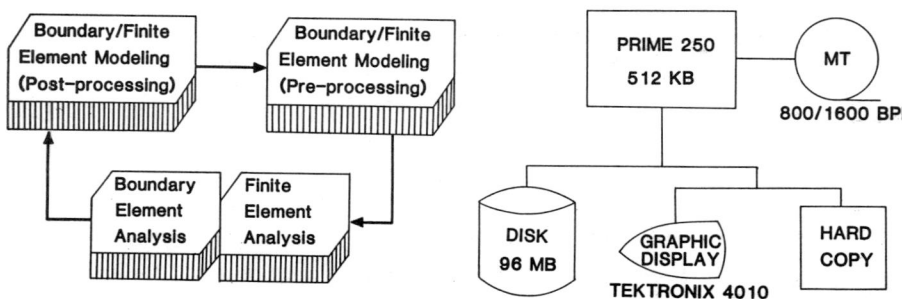

Fig. 2.   Software and dedicated hardware.

smaller. KYOKAI runs successfully on smaller mini-mainframe computers using three megabytes virtual memory. The installation of the KYOKAI system on other IBM-compatible type computers has been attempted. Among them, installation on computers: PRIME 50 series, VAX 11 series, SIEMENS, IBM, HITACHI, FUJITSU machines has been successfully implemented. Multi-user simulation is possible for a number of terminals by the time-shared use of the c.p.u.

Perhaps the only difficulty encountered in the installation of KYOKAI on other computer systems is the different standards of graphic routines. Minimal use of graphic and plotter routines is intended in the KYOKAI source program. Those routines are: OPEN, DEFINE ORIGIN, ERASE SCREEN, PLOT, DRAW NUMBERS, and CLOSE.

The correlation of subsystems and data files is shown in Fig. 3. There are nine intermediate data files: IGI-IG9. During the execution of KYOKAI, the program flow is watched by the supervisor, and the intermediate data are accumulated in the print-image file IW1 in order to trace back the user's activity.

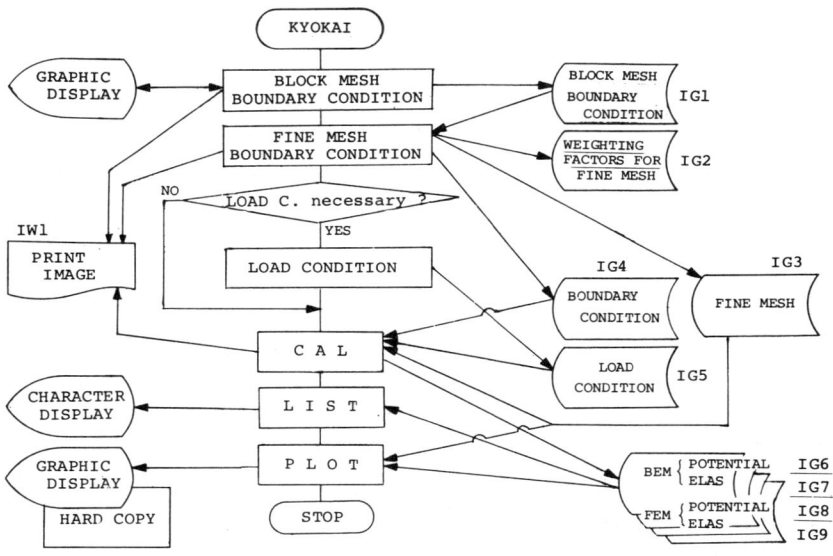

Fig. 3. Correlation of data files and subprograms.

The KYOKAI program is shipped on a magnetic tape. A complete source program together with input data of simple examples is duplicated on a tape in a volume in multiple files. The input data of the examples can be used for diagnostic purpose in the shift operations of the program. The magnetic tape follows the format: 9 tracks, at 1600 bpi, in EBCDIC code, with blocking factor = 20, 1 record = 80 bytes, and unlabeled. Source manuals can also be delivered on the tape.

EXAMPLES OF APPLICATIONS

Basic Test Analysis

For a potential problem, steady heat conduction in an infinitely long seven-rod hexagonal bundle is considered. One seventh of the cross section is shown in

128   Y. Ohura, K. Obata and K. Onishi

Fig. 4. The total geometry is modelled by four blocks. Although the material is assumed to be homogeneous and isotropic, the blocks are taken as different subregions in the boundary element analysis.

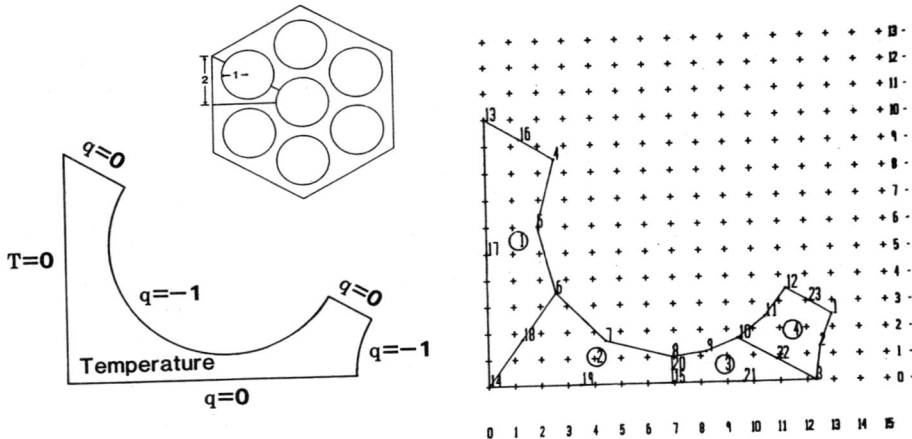

Fig. 4.   Steady heat conduction.

Calculated isotherms are shown in Fig. 5. Constant elements are used in the boundary element analysis.

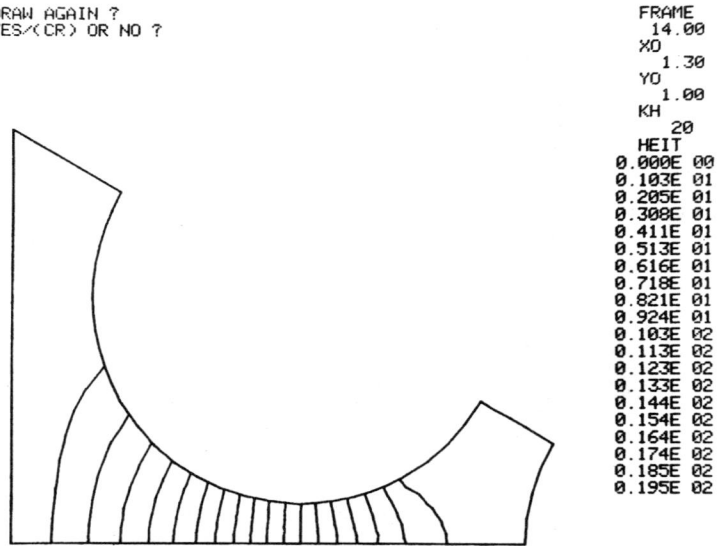

Fig. 5.   Isotherms.

Next, we consider an illustrative problem of plane stress in the elasticity. A square thin plate with a circular hole is under tension. Constant elements are used in the boundary element analysis.

Calculated traction and displacement are shown in Fig. 6. Distorted shape is shown by solid curves, and undistorted shape shown by dotted lines is overlaid.

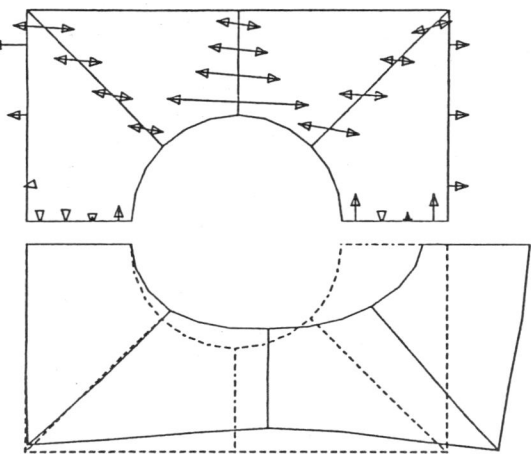

Fig. 6. Traction and displacement.

Industrial Application

Plane stress analysis of a reinforced concrete plate as shown in Fig. 7 was conducted. The plate was used for the wall of the residential room in construction engineering. The purpose of the boundary element analysis was to examine the stress concentration and the displacement subject to a seismic loading. The seismic load was accounted for by the equivalent static load. Material constants are indicated in the figure.

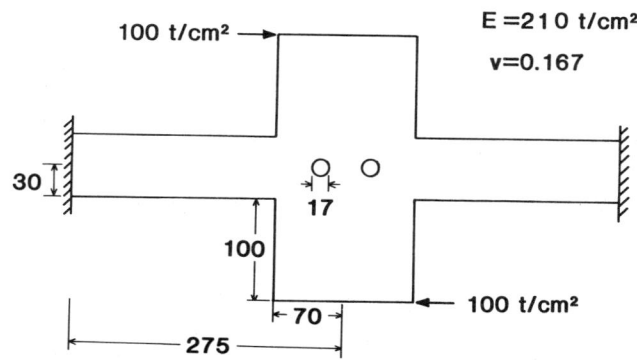

Fig. 7. Reinforced concrete plate.

The geometry of the plate is modelled by twelve blocks that are served for subregions in the boundary element analysis. Constant elements are used. Block and fine meshes are shown in Fig. 8.

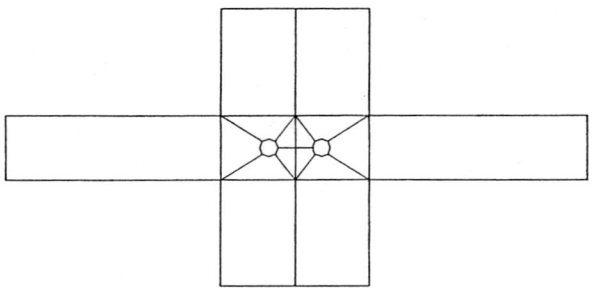

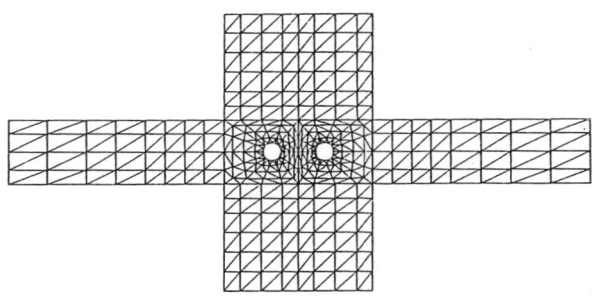

Fig. 8. Block and fine meshes.

The boundary element subsystem of KYOKAI is used for the solution. Mesh and input data are generated automatically in an interactive mode, and the output data are plotted by the postprocessor on the graphic terminal. Calculated traction and principal stresses are depicted in Fig. 9. Stress is concentrated at four re-entrant corners. Calculated displacement is shown in Fig. 10.

REFERENCES

1. Obata, K., Y. Ohura and K. Onishi. Computer programming in interactive FEM-BEM. *Proc. Japan Society of Civil Engineers, Western Branch*, 319-(13), Ryukyu University, February 1982, pp. 65-66.

2. Onishi, K. Introduction to boundary element methods. *SOFT-GIKEN*, Text of the Two-days Seminar, Tokyo, July 1982.

3. Onishi, K., K. Obata and Y. Ohura. KYOKAI - user's guide. *Materials of the 5th Seminar on Software, Association of Computer Education for Private Universities in Japan*, Tokyo, October 1982, pp. 21-40.

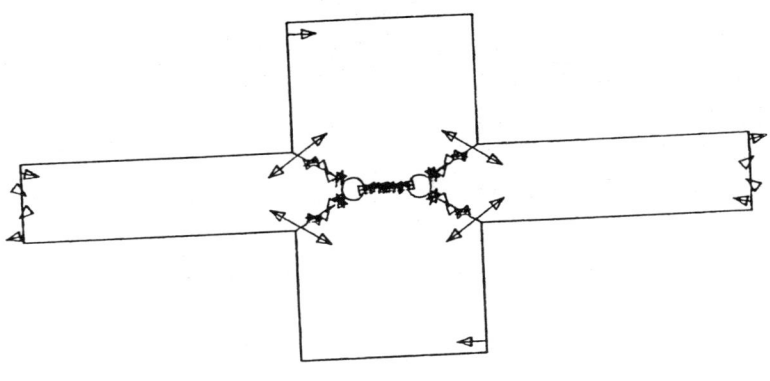

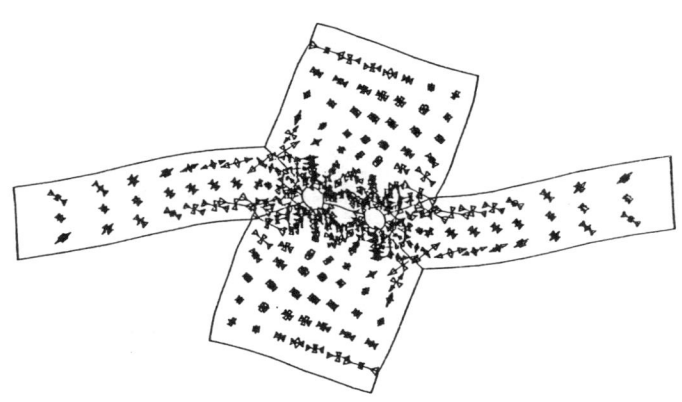

Fig. 9. Traction and principal stresses.

4. Obata, K., Y. Ohura and K. Onishi. KYOKAI - An interactive FEM.BEM program. *Proc. Symp. Computer Usage*. Japan Society of Civil Engineers, Tokyo, October 1982, pp. 5-8.

5. Onishi, K. Developments and applications of interactive FEM.BEM program "KYOKAI" (in Japanese). *Bridge Engineering*, 19 (5), 33-39, 1983.

6. Onishi, K., T. Kuroki, Y. Ohura, K. Obata and T. Ito. KYOKAI - An interactive BEM.FEM solver. *Bulletin of The Central Research Institute*, No. 68, Fukuoka University, 1983, pp. 1-179.

7. Mackerle, J. and T. Andersson. Boundary element software in engineering. *Advances in Engineering Software*, 6 (2), 66-102, 1984.

8. Tanaka, M., K. Onishi, T. Kuroki and G. Aramaki. BEM Seminar, Nagano, August 1984.

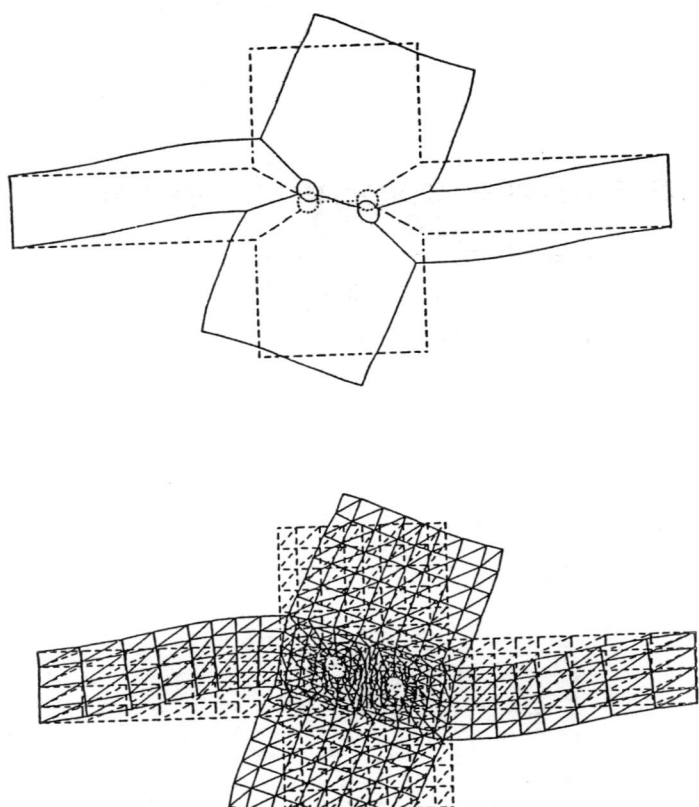

Fig. 10. Displacement.

9. Ohura, Y., K. Obata and K. Onishi. A new version of the BEM.FEM program - KYOKAI. *Proc. 1st Symp. Boundary Element Methods, JASCOME*, Tokyo, November 1984, pp. 127-132.

10. Obata, K., Y. Ohura, K. Onishi and T. Kuroki. KYOKAI.3D - An interactive three-dimensional BEM solver. *Proc. 8th Symp. Computer Usage*. Japan Society of Civil Engineers, Tokyo, October 1983, pp. 49-52.

MICRO STRESS: STRUCTURAL ANALYSIS PROGRAM

D. Nardini

C. M. Consultant Ltd., Ashurst Lodge, Ashurst, Southampton, U.K.

ABSTRACT

Micro STRESS is a computer program for the analysis of framed structures using a microcomputer. The input format of the program was adapted to be similar to the STRESS program developed for the mainframe computers by the M.I.T. in the sixties, with which practical engineers became very familiar. Around this basic idea, a powerful analysis tool has been developed, which includes extensive input generation capabilities, as well as postprocessing and graphics.

INTRODUCTION

The rapid advance of microcomputers into various fields of business and industry is making a tremendous impact on the whole society. It is however interesting to note that the engineering community is still somewhat suspicious at the power of these machines, and many engineers still like to stick to the mainframe computers or the minis.

There is, of course, still a big difference in the performance of a mainframe or super mini in comparison to a microcomputer, but considering the price and availability of the latter, there are significant advantages which must be taken into account.

The approach of various software developers in this respect has been different. One smaller group started writing and modifying their structural analysis software to run it on microcomputers which some 5 years ago were still in their infancy. This was an extremely hard job, considering the very limited capabilities of those machines. Nevertheless, they proved that a microcomputer with a modest processing power could do formidable jobs for a structural engineer, in certain classes of problems. Micro STRESS was one of the first general structural analysis programs with the first full version (in 1980) running on a Z80 computer with a mere 48k of RAM.

The second (larger) group of software developers had the approach of waiting for a microcomputer to develop to a power of a mini, and then to just transfer the existing software to it. Although this has started to happen, there will still be, for some time to come, microcomputers which will not meet these standards. Therefore today one should distinguish between the very powerful microcomputers, often referred to as workstations, with the processing power per user no less than that of a mini computer, and the less pretentious (and much less costly) ones, in the category of personal computers, to say nothing of the new breed of truly portable micros with astonishing performances.

To conclude this introduction, it is quite easy to foresee that the constant emergence of the new, smaller, and less expensive microcomputers will for some time to come keep a demand on the specialized structural analysis programs which could be run on them, and that, for software developers, it will be important to maintain a spectrum of products ranging from simple linear static analysis tools to the nonlinear and time dependent codes.

PROGRAM DESCRIPTION

Micro STRESS performs linear statical analysis of plane or space structures composed of beams and columns. The method employed is the matrix stiffness procedure, with joint displacements being the primary unknowns.

The geometry of a structure analysed, being defined by positioned of joints and members, may be arbitrary. Joint degrees of freedom are determined by the type of the structure, ranging from two for a plane truss to six for a space frame. Supported joints are considered to be restrained from moving in all the global directions, but deviations from this may be introduced by use of joint releases capabilities.

Members (beams or columns) can be either of a constant cross section, or alternatively defined by supplying a local stiffness or flexibility matrix. Deformations due to shearing stresses may be taken into account using the effective shear areas values of which are computed from the shape of a cross section.

There is a possibility of introducing rollers and hingers at individual member ends using member releases.

Material constants may vary through a structure, which allow structures composed of two or more different materials to be analysed with equal ease.

Loadings on the structure may be of various types, namely:

Joint loads - forces or moments acting at specified joints

Concentrated member loads - point forces along members

Uniformly distributed loads - partial or total along members

Linearly distributed loads - partial or total along members

All of the member loads may act either in the local or the global coordinate directions. In addition, loads can be given in form of prescribed support displacements or initial member deformations.

The primary analysis module assembles the relevant matrices, solves the equilibrium equations and backsubstitutes for member forces. The primary output consists of member end forces, support reactions and joint displacements. Any of these may be suppressed for certain or all of the loading cases.

POSTPROCESSING

Results of the primary analysis can be saved on a mass storage, to be later accessed for postprocessing. Micro STRESS Postprocessor is an interactive program, with aid of which a user can obtain detailed analysis of internal forces and stresses in numerical or graphical form.

The program supports the following features.

Section forces. Internal forces at any points due to any loading cases are computed and displayed in the form of diagrams.

Section force envelope. Extremes of internal forces are computed and output for any load combinations. Loading cases may be included into different categories, as described later.

Normal stresses. Stresses due to the combined action of axial force and bending moments are computed and displayed for any members due to any load case.

Normal stress envelope. Extreme stresses are computed due to a defined set of loading cases, which may be combined in various ways, as described further on. The results are also displayed graphically.

Loadings. Extensive load combination possibilities are available in postprocessing, these being:

- adding unconditionally
- adding when unfavourable
- defining groups of mutually excluding load cases.

Any load or loading combination, defined in the primary analysis may be further combined in the postprocessing. The specified options are invaluable when performing analyses with a large number of load cases which may be combined in numerous ways.

Further graphical capabilities include display of structural geometry which may either be partial or total, and joint and member numbering display. Space structures are shown in axonometry with user defined viewing angle.

Deformations of the structure due to any loading case are superimposed on the undeformed shape in a different colour or dashed. Deformation scale is computed automatically, but may also be explicitly defined by the user.

EXAMPLES OF APPLICATION

In order to demonstrate the scope of Micro STRESS, two distinct application examples will be shown.

Pylon. A reticulated space structure composed of 88 joints and 314 members is plotted in Fig. 1. This is a version of a standard pylon used for carrying a fair number of high voltage power lines. The main problem in analysing such a complex structure is making sure that the described topology is right. Micro STRESS can plot axonometric views of the whole structure (as shown in Fig. 1a), or selected parts of it (Fig. 1b). Member and joint numbers may be included or suppressed in order to verify the geometry.

Loads applied to the structure are:

- Self weight
- Vertical pull of cables
- Horizontal pull of cables
- Horizontal wind load in two directions

Motorway Bridge. This example demonstrates use of Micro STRESS for the analysis of a topologically very simple structure, (Fig. 2) but which is subjected to a complex loading pattern (as shown in Fig. 3). Bridge design codes of most

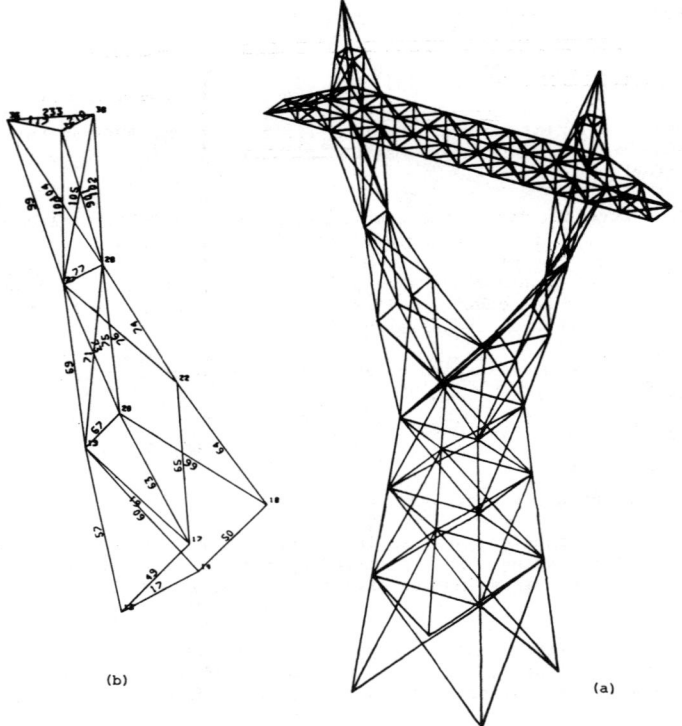

Fig. 1. A reticulated pylon - geometry.

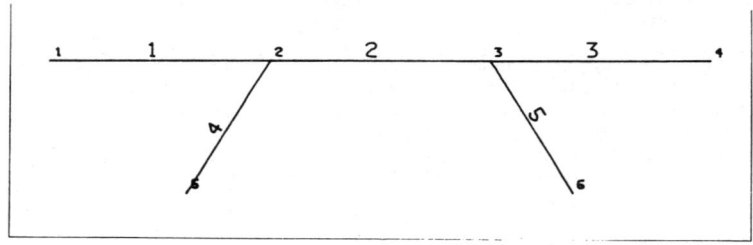

Fig. 2. Motorway bridge.

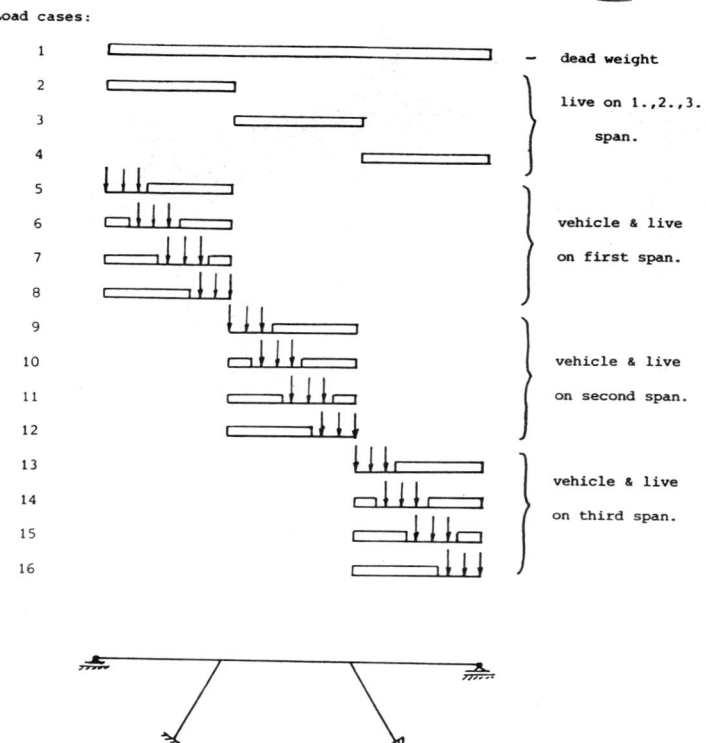

Fig. 3. Load cases for motorway bridge analysis.

countries include usage of a special heavy vehicle, which may be placed in any position along the spans. The rest of the deck may be filled with live load (pedestrians) when unfavourable). For multispan bridges, the number of relevant loading conditions becomes very high indeed. However, Micro STRESS has facilities not only to combine independent loading conditions, but also to assign load attributes to them and act accordingly (as was described for load specification in the postprocessor)

Figure 3 shows all the primary load cases that have to be specified. Dead weight is obviously always present, so that attribute Add (or A) is assigned to it. Loads due to various positions of the special vehcile may not act simultaneously, i.e. they exclude each other. Therefore they are assigned attribute exclude (or X). Live load due to pedestrians on other spans than that occupied by the vehicle may be present if it is unfavourable. Therefore these loads are assigned the attribute Unfavourable (or U).

With that, our load definition will be

```
LOADINGS   (A1 X(2 5 6 7 8) U(3 4)
           (A1 X(3 9 10 11 12) U(2 4))
           (A1 X(4 13 14 15 16) U(2 3))
```

Some of the resulting extreme internal forces and stresses are displayed in Fig. 4.

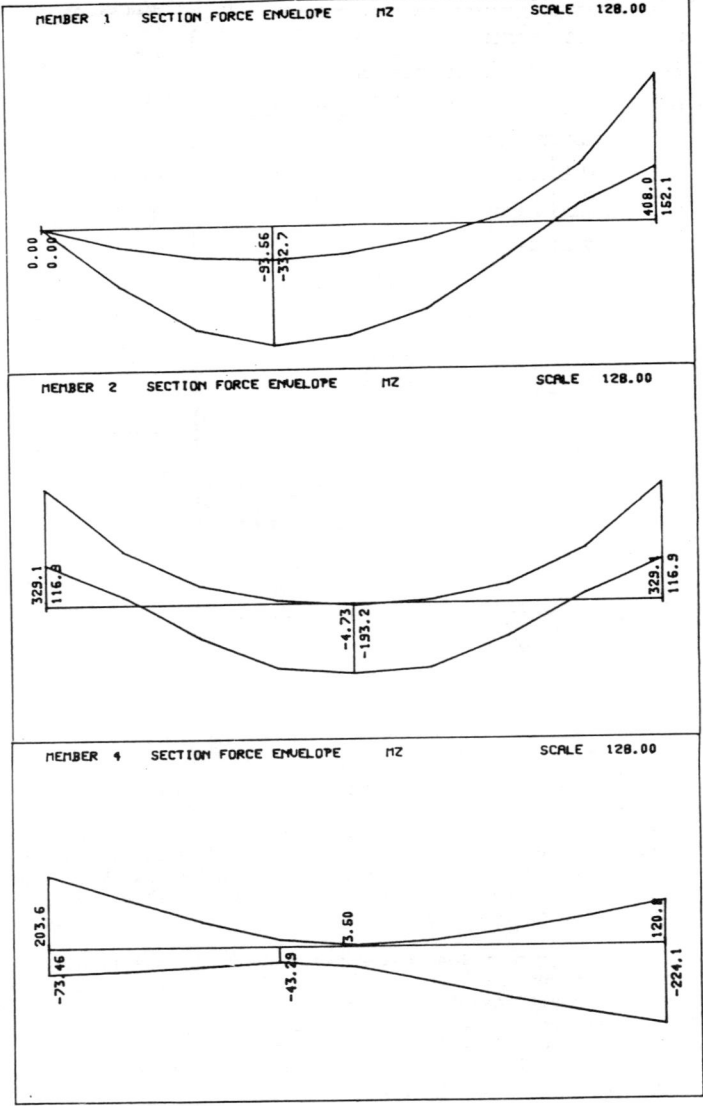

Fig. 4. Diagrams of extreme bending moments for motorway bridge analysis.

CONCLUSION

Micro STRESS program has been implemented on a wide range of microcomputers, including the CP/M operating system machines and the PC-DOS, MS-DOS machines. The program can also be used without any problem on mini computers.

Micro STRESS constitutes a powerful analysis tool for structural design in the areas of civil engineering, mechanical engineering, offshore design, ship building and other selected areas. The main advantages of the package are that it can be used on a very modest computer installation but that it also includes advanced graphical and postprocessing techniques. Since the recent appearance of more powerful microcomputers, it has become feasible to upgrade the capabilities of the program, and the following version will also include geometrical and material nonlinear analysis.

SUPPORT

Micro STRESS is supported and distributed by:

Microcomp Ltd., Ashurst Lodge, Ashurst, Southampton, SO4 2AA, Hampshire, England. Telephone (042 129) 3233. Telex CHACOM G 47388 Attn. COMPMECH.

NE-XX: A FINITE ELEMENT PROGRAM SYSTEM

V. Kolář and I. Němec

Technical Institute, Dopravoprojekt, 658 30 Brno, Leninova 17, Czechoslovakia

ABSTRACT

NE-XX is a general purpose elastic finite element program package for static structural analysis. It consists of ten programs for partial problems such as thin and thick plates, plane stress and plane strain, axisymmetrical stress, shells, shells with ribs of the other stiffeners, if needed in a generalized elastic medium, Saint Venant torsion, heat conduction, plates on a multilayered subsoil etc.

Each program can be implemented and run separately. The shape of solved structure, loadings and boundary conditions can be arbitrary. The structure can work in interaction with an elastic medium or subsoil, which is often used in solution of the foundations or underground structures.

The new very efficient model of the structure-soil interaction is used. The eigenfrequencies and free vibration shapes can be solved too.

The system has been developed for bridges, highway structures, road slabs and modern exacting structures in traffic and civil engineering, but now it is used for the statical analysis of any structure, e.g. buildings, machines, cars, airplanes etc. All the programs have both numerical and graphic outputs and can run in batch or dialogue mode. Some outstanding features are:

(a) ease of use,
(b) full compatibility between all elements of a structure,
(c) soil-structure interaction,
(d) a new efficient subsoil model,
(e) both Kirchhof (thin plates or shells) and Mindlin (thick plates or shells) theories of bending, including the shear effect,
(f) implementation on mini and microcomputers possible,
(g) interactive pre- and postprocessors and many forms of the graphic outputs.

The system has been frequently used since 1974 and represents the first widespread Czech finite element program package, connected with the lectures of Professor V. Kolář on the Technical University of Brno and the pioneer work done by the Director of the Computer Centre of DOPRAVOPROJEKT, I. Němec. The full theory can

be followed by the appropriate user's and theoretical manuals of the NE-XX
system. Regarding the similar known program packages the quoted outstanding
features, firstly the very efficient expression of the structure-soil interaction,
may be of user's interest.

THEORETICAL BACKGROUND

All programs are based on the Lagrange's variational principle of the minimum of
total potential energy and use the pure deformation variance of the finite element
method. The unknown quantities are displacements, rotations or their derivatives.
The sense of all solving equations is similar as the sense of the equilibrium
conditions generalized for any virtual displacements pertaining to all deformation
parameters introduced in the solution.

The full theoretical background including convergence conditions, interpolation
polynomials etc. has been published (see references 1 and 2).

The new very efficient model of soil-structure interaction has been published
firstly in English[3] and recently[4] in Czech with English inscriptions and summary.

The main significance of the new theoretical ideas incorporated in the NE-XX
program package is: The solution of any structure on a subbase or subsoil
(generally in an elastic medium) can be performed with the same number of unknown
deformation parameters and the same bandwidth as in the case without soil-structure
interaction despite the general nature of the subsoil behaviour. The model can
express all known models (Winkler, Boussinesq, Pasternak, Vlasov-Leontjev,
Gibson, Awojobi etc.) and works with the general idea of infinite elements.
Infinity in z-direction, perpendicular to the element plane, and in x- or y-
direction, perpendicular to the foundation boundaries (generally in any m-
direction in the foundation plane) is expressed separately. The user can choose
the exactness of modelling according to the geotechnical input data. The
settlements can be solved without restricting hypothesis.

FIELD OF APPLICATION

Geometrical. 1D, 2D, 3D, Minlin's beams, plates, walls, axisymmetric solids,
shells, thin or thick shells with any arbitrary ribs or the other stiffeners
including piles etc., any subsoil region.

Materials. Linear elastic, composite materials can be modelled, orthotropy and
general anisotropy admissible.

Analysis capabilities. Static, thermal, free vibrations (modes and
eigenfrequencies).

Loadings. Nodes: Any load component pertaining to the nodal deformation
parameter (force or moment or bimoment component). Elements: Uniformly
distributed load with all components, admissible in the problem, generally with
all three components in the planar coordinates x,y,z. Modelling of thermal loads
(temperature changes in middle element plane, temperature gradients along the
element thickness). Steady state temperature field solution. Torsion of any
cross section. Prescription of the deformation parameter values.

Other. Displacements can be prescribed in the arbitrary internal point of
subsoil mass. Influence of caverns in the subsoil. Elastic bonds between two
chosen deformation parameters. Introduction of substructure by their connecting
stiffness matrices.

NE-XX: A Finite Element Program System

PROGRAM DESCRIPTION

Method. Finite elements, deformation parameters, Lagrange variational principle.

Type of Elements.

NE-01: 2D-thin Kirchhof plate element, rectangular or skew, bicubical polynomial base function, 16 deformation parameters of the type w, w_x, w_y, w_{xy}, effective subsoil with the three physical constants C1, C2X, C2Y.

NE-02: 2D-thick Mindlin plate element (Lynn-Dhillon type), triangular, 9 deformation parameters of the type w, ψ_x, ψ_y, material orthotropy, shear effect, subsoil with the three physical constants C1, C2X, C2Y.

NE-03: 2D-thin plate element as in the NE-01 program, line hinges in the x-direction, doubling of the w_y and w_{xy} parameters in a node, orthotropy of the plate and the subsoil as well, plates composed of precast parts.

NE-04: Plane stress or plane strain, triangular element with linear displacement components, six deformation parameters of the type u,v, full anisotropy admissible.

NE-05: Axisymmetric solid annuloidal element with the triangular cross section similar to the NE-04 element, transversal orthotropy of material.

NE-06: Triangular 3D-shell thick or thin element (Mindlin or Kirchhof) with quadratic displacements u,v,w and linear rotation components ψ_x, ψ_y, ψ_z in the planar element coordinates x, y, z. Twenty-seven deformation parameters of the type u,v,w, ψ_x, ψ_y, ψ_z. All six degrees of freedom of any element point are respected (Cosserat's geometry). Full compatibility in space in any case of connection. Physical anisotropy of orthotropy or isotropy. Elastic subsoil or medium with the three constants similar as in the NE-01 program and the two friction constants C3X, C3Y, together with five physical constants.

NE-07: (a) Triangular 3D-shell element as in the program NE-06, (b) A Mindlin's beam 3D-element, derived by the authors following the Mindlin's idea of the rotation independency and shear effect, fully compatible with the NE-06 element. Fifteen deformation parameters of the same type as on one side of a NE-06 element. Elastic subsoil or medium can be expressed by seven constants: C1X, C2X (friction in the axis direction), C1Y, C1Z (Winklerian constants or subbase moduli), C2Y, C2Z (Pasternakian constants of soil shear effect), CFX (torsion reaction constant). From the geometrical point of view the beam element represents the one-dimensional Cosserat's continuum with all six degrees of freedom of any point.

Together with the NE-06 element the program NE-07 represents the solution of any arbitrary Cosserat's continuum composed by plane shell and axial beam parts. The program can solve e.g. box structure on plate with pile foundation, bored piles systems in interaction with the building, many complex machine structures etc.

NE-08: 2D-element with Saint Venant torsion hypothesis, triangular, cubic course of the torsional function T, 9 parameters of the type T, T_x, T_y.

NE-09: 2D-element in steady state heat conduction condition, similar to the NE-08 element, 9 parameters of the type t (temperature), t_x, t_y.

NE-10: 2D-thick plate isoparametric element of Mindlin's type with quadrilateral shape. Bilinear shape and base functions for independent deflection w and rotations ψ_x, ψ_y, 12 deformation parameters. The element can be supported by a generally layered subsoil of various physical properties which can express any arbitrary geological and geotechnical conditions. Subelement of a layer is a 3D-brick with trilinear course of settlements. Its deformation and shear moduli EO, GXZ, GYZ can vary linear with the depth. In a nodal point of the solved plate (3 + n) deformation parameters are defined, when the subsoil is divided in n layers.

APPENDIX: SN-01 element is a problem-oriented element similar to the NE-01 element. It is destined to the prismatic (one-dimensional) problems of the highway and traffic structures. PTP-element is a simple beam element destined to physically nonlinear analysis of axially loaded beam systems in complicated physically nonlinear support conditions. The solution is performed by the incremental method.

PROGRAM STRUCTURE AND USER COMFORT

Data can be entered through traditional modes as card deck or disk etc. or interactively through a terminal console in DIALOG without the use of manual. Both input and output data can be visualized graphically on screen (colour), hard copy or plotter drawing. The preprocessor arranges the mesh generation in all regular cases and in all cases the SEQ- input (in sequences or series) of all data is possible. Node and element numbering of the mesh generation is performed automatically, the bandwidth is minimized by the internal node numbering, the results are printed in the user's numbering following the technical drawing. The postprocessor can arrange the form, content and extent of results according to the user's demands including the colour drawings etc.

In any case the nodal checking forces and moments can be printed, calculated independent from the deformation parameters, which represents a test of the numerical stability of solution and informs the user about the necessity of refining the division, if needed (residual test forces). The language of the program is FORTRAN.

HARDWARE COMPATIBILITIES

Minimum configuration: 32 KB internal and 2 MB disk memory, a terminal console. An optimum configuration (capacity about 1000 nodes): 256 KB and 28 MB.

Type of computers and other hardware· IBM, DEC, HP.

Peripherals: Plotter, digitizer, graphic colour display.

Operating system: OS, RXS 11 MS DOS

Media: Source code complete listing and cards, complete program on disk, user's and theoretical manuals.

Services: Regular updates, new capabilities, free consultations, technical help in implementation, user's training etc.

EXAMPLES OF APPLICATION

Test Cases

In reference 4 many test examples are presented. Other test examples were presented by many other authors using the same finite element method (O. C. Zienkiewicz, R. H. Gallagher, J. H. Argyris, R. W. Clough, B. M. Irons, E. L. Wilson etc.). The practice of the years 1956 to 1984 shows the full reliability of the method.

NE-XX: A Finite Element Program System

Practical Examples in Industry

Motorway bridge (Figs. 1 and 2). Five span continuous orthotropic plate (one span about 30 m), prestressed concrete structure, live load elastic supports.

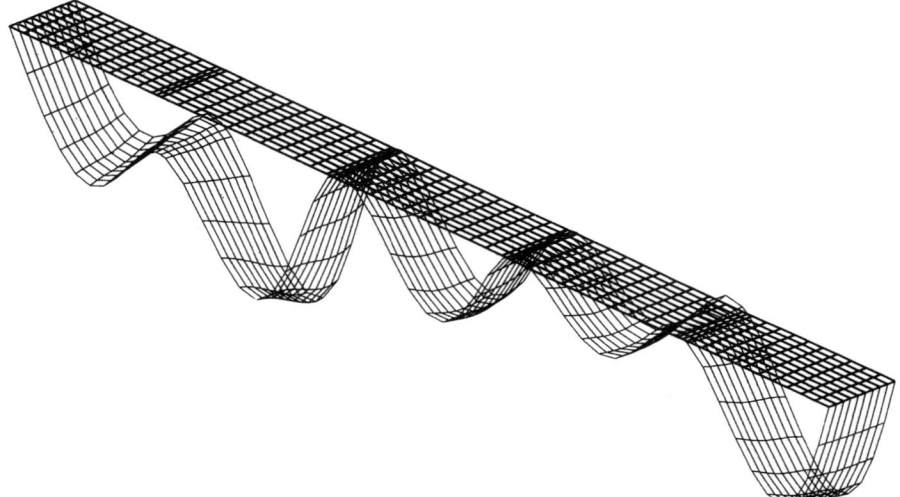

Fig. 1. Five span continuous orthotropic bridge plate 5 x 30 = 150 m with the live load. Division into 522 finite elements of the program NE-10. Deflections 200 : 1 pertaining to the most effective load case.

Application area: civil engineering, traffic structures.

Type of problem: Static.

Drawing of the structure see Fig. 1 (thick lines).

Discretization see Fig. 1.

NE-10 quadrilateral isoparametric finite elements of Mindlin's type with shear effect and physical orthotropy.

Number of elements: 522 (5 different types), number of nodes: 590, number of degrees of freedom: 1744, bandwidth 36, subregions not used, part of program: NE-10, computer and peripherats used: PDP 11, plotter Calcomp 836.

Input data: Interactively in DIALOG.

Output data: Deformation and internal forces. Example: Deflection 200 : 1 (Fig. 1), bending moment $M_{x,dim}$ (Fig. 2).

Computation time: 30 min. Costs: 225 Czech. Crones.

Another motorway bridge (Fig. 3). A precast prestressed bridge structure (concrete), modelled as a box structure by NE-07 program elements. Application area: civil engineering. Type of problem: static. Drawing see Fig. 3, (thick lines), view in the longitudinal direction. Mindlin's thick shell elements (see NE-06). 302 elements, 661 nodes, 6 load cases, 2478 degrees of freedom, bandwidth 507 (before) and 345 (after optimization). Subregions not used. Program part NE-07. Computer PDP 11, plotter Calcomp 836. Computation time 10 hours in multiuser regime. Costs: 3000 Czech. Crones. Input: Cards. Output: Prints and drawings, example on Fig. 3 (deflections 200 : 1).

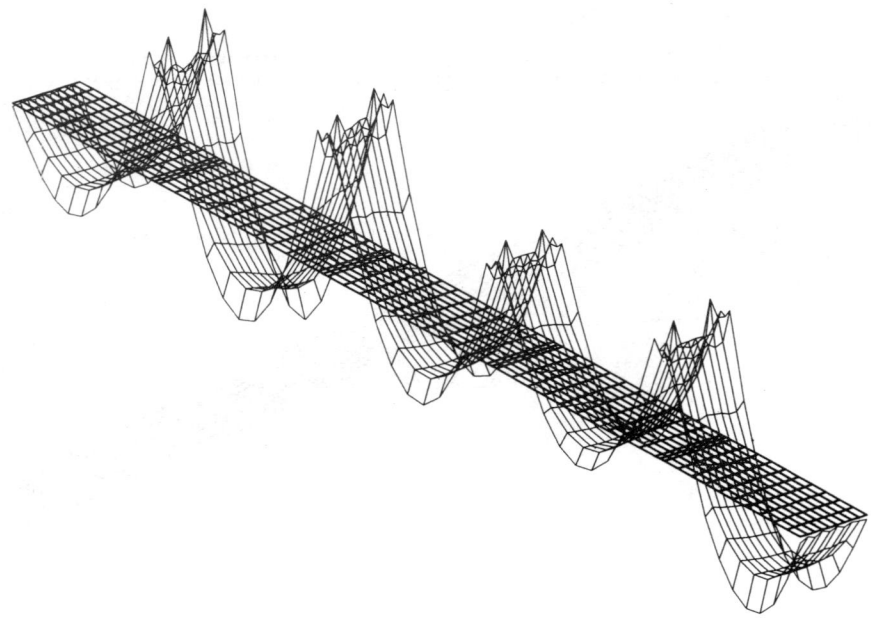

Fig. 2. Longitudinal bending moments $M_{x,dim}$ of the plate on Fig. 1.

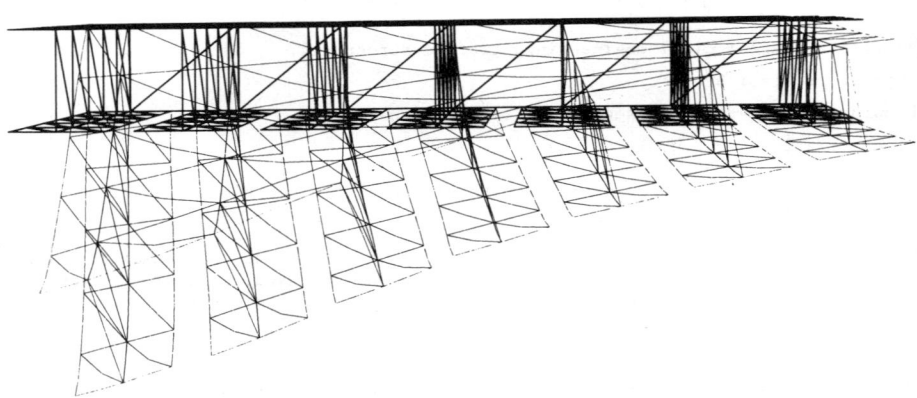

Fig. 3. Cosserat's model (program NE-07) of a precast prestressed bridge box structure 30 x 20 x 2 m, divided into 302 finite elements. Deflections 200 : 1 pertaining to a nonsymmetrical load case.

Mining excavator (Fig. 4). A steel shell with ribs and stiffeners, fore part of the rear structure of an automatic machine. Area: Machines. Problem: Static. Drawing see Fig. 4 with discretization into NE-07 shell and beam elements. 355 shell elements, 23 beam elements, 758 nodes, 2766 degrees of freedom, bandwidth 379, subregions not used, part of program NE-07, computer PDP 11, peripherals: plotter Calcomp 836, input data on cards, output data: prints and drawings. Computation time 6 hours, costs 2700 Czech. Crones.

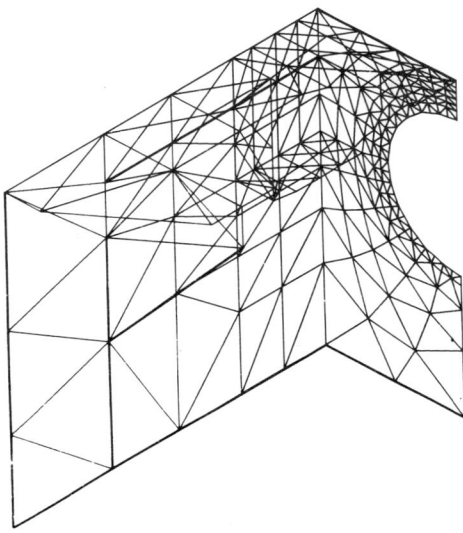

Fig. 4. A half of the steel shell with ribs of an automatic mining excavator about 1.5 m high, divided into 355 shell- and 23 beam- finite elements, solved by the NE-XX program system.

Plate with circular holes (Figs. 5 and 6). A prestressed concrete plate with five circular holes (a half see Fig. 5 with the discretization into NE-08 elements), physically orthotropic, analysis of the torsion of cross section. Application area: Civil engineering. Problem: Static. Type of element: NE-08, triangular, cubic torsional function. 1000 elements (275 various types), 563 nodes, 1530 unknown values, bandwidth 292, subregions not used, part of program NE-08, computer PDP 11, plotter Calcomp 836, input data on cards, output: Prandtl's torsion function T(x,y) in nodes and its first derivatives (shear stress components τ_{xy}, τ_{xz}) printed and visualized, e.g.: The function T (x,y) on Fig. 6.

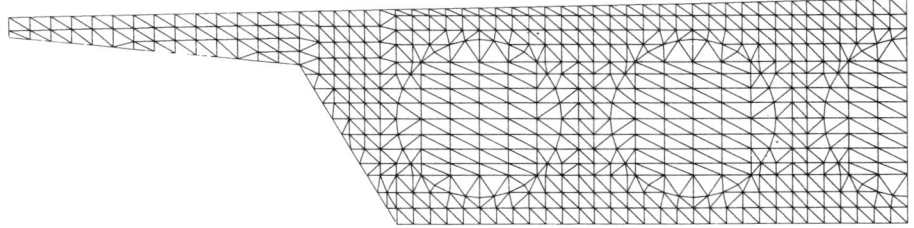

Fig. 5. The right half of a cross section (18 x 2 m) with five circular holes. Division into 1000 finite elements for solving the Saint Venant torsion by the NE-08 program.

Hydraulic clutch box (Fig. 7). Cast iron hydraulic clutch box (0.8 x 1.60 x 1.40 m) of a big engine. Bearing hole diameter 0.28 m, internal box diameter 1.20 m, shell thickness 0.02 m. Power station, static and fatigue problem.

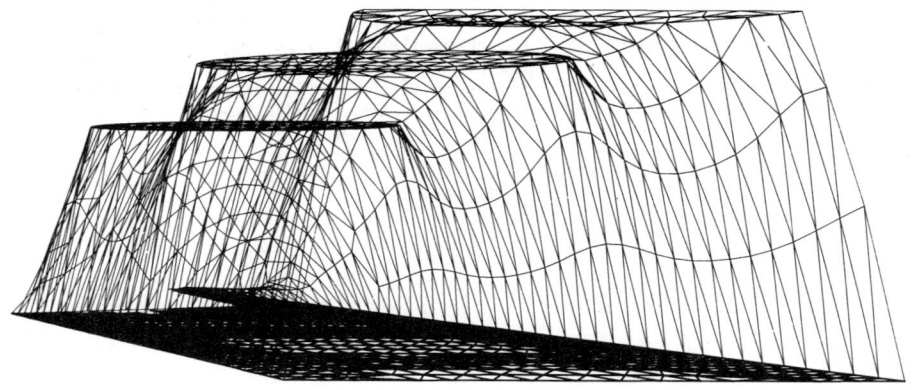

Fig. 6. The Prandtl's torsion function T(x,y) of the cross section on Fig. 5, computed by the NE-08 program and visualized by the plotter CALCOMP 836.

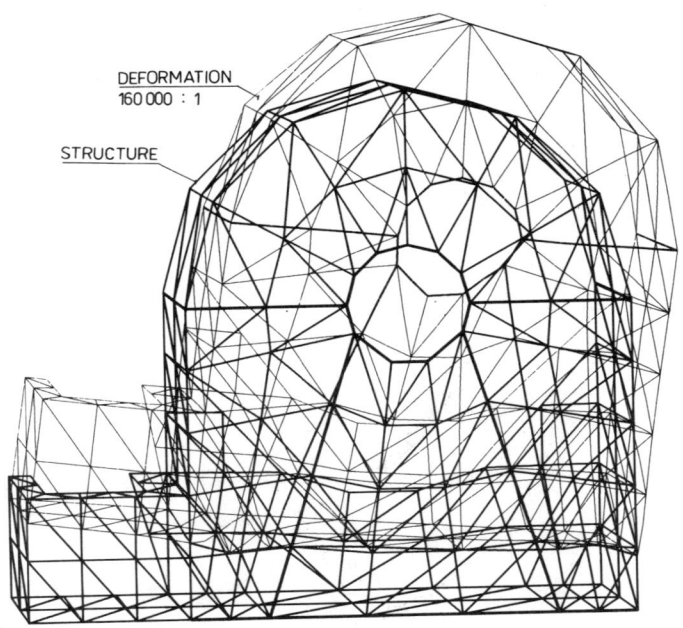

Fig. 7. The fore half of the Mindlin's shell structure of a large hydraulic clutch box (0.8 x 1.60 x 1.40 m). Division into 257 shell- and 34 rib-elements. Deformation caused by four symmetrical 500 kN forces, solved by the NE-07 program. Published with the permission of SKODA Corporation, CKD BLANSKO, Czechoslovakia.

NE-XX: A Finite Element Program System

Division into 257 Mindlin's shell elements and 34 Mindlin's rib elements. 554 nodes, 2400 unknown deformation parameters of the Cosserat's model u, v, w, ψ_x, ψ_y, ψ_z, bandwidth 260. Part of the program: NE-07. Computer: PDP 11, peripherals: plotter Calcomp 836. Input data on cards. Output data: prints and drawings for more load cases. Deformation caused by four symmetrical forces (500 kN each force) see Fig. 7. Computation time 4 hours, user costs 1800 Czech. Crones.

Turbine shaft bearing (Fig. 8). Steel shell with ribs, shell thickness 17 mm and 19 mm, diameter: external 3 m, internal 2 m, cross thickness 0.8 m. Water power station, 500 m water level difference, 5 MPa pressure. Static and fatigue problem. Division into 240 Mindlin's shell elements and 14 Mindlin's rib elements. 546 nodes, 2200 unknown deformation parameters of the Cosserat's model u, v, w, ψ_x, ψ_y, ψ_z, bandwidth 200. Part of the program: NE-07. Computer: PDP 11, peripherals: plotter Calcomp 836. Input data on cards. Output data: prints and drawings for more load cases. Deformation caused by the load acting on the lowest nodes of the internal ring see Fig. 8 (the left half of the structure). Computation time two hours. User costs 900 Czech. Crones.

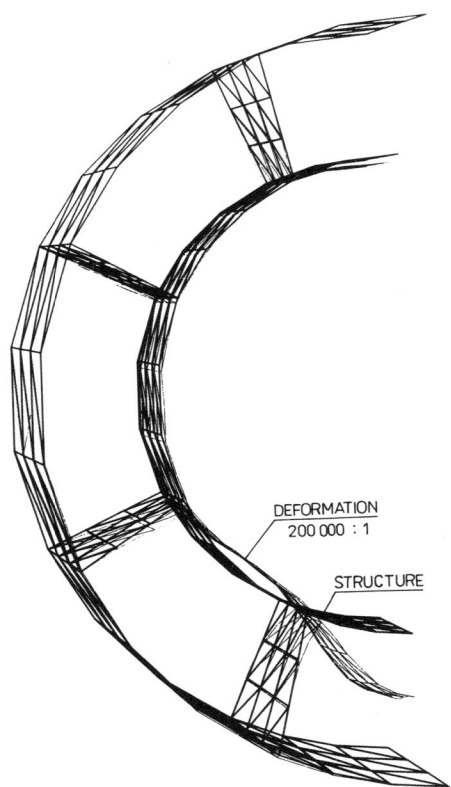

Fig. 8. The left half of the Mindlin's shell structure of a large turbine shaft bearing (exgernal diameter 3 m) Division into 240 shell- and 14 rib-elements. Solution by the NE-07 program, deformation caused by the internal ring load acting on the lowest nodes. Published with the permission of SKODA-Corporation, CKD BLANSKO, Czechoslovakia.

REFERENCES

1. Kolář, V. et al. (1975). *Design of 2D- and 3D- Structures by the Finite Element Method.* Springer, New York. (In German), 425 pp.

2. Kolář, V. and I. Němec (1985). Finite element analysis of structures. *United Nations T.E.M. Workshop CAD,* Prague-Laussane, 245 pp.

3. *5th Danube Conf. Soil Mechanics and Foundations,* September 1977, Bratislava.

4. Kolář, V. and I. Němec (198). An efficient model of soil-structure interaction. Academia, Prague.and Elsevier, Amsterdam - New York, 320 pp.

PAFEC: THE PAFEC FINITE ELEMENT ANALYSIS SYSTEM

P. M. Wheeler

PAFEC Ltd., Strelley Hall, Nottingham, U.K.

ABSTRACT

The PAFEC Finite Element Analysis System is a large, widely used program designed to solve a very wide range of static, dynamic, heat transfer, acoustic, elastohydrodynamic and nonlinear problems. Under continuous development, PAFEC has developed a highly advanced and flexible database which allows full upwards compatibility for the user between new versions. The success of PAFEC has centred on ease of data input, which is discussed, and the range of graphics available.

INTRODUCTION

PAFEC is a general purpose, 3-dimensional linear and nonlinear finite element analysis program where data preparation is the simplest and most convenient yet devised for a large scale finite element system. It employs free format input with engineering keywords, automatic mesh generation and extensive off-line and interactive graphics facilities. The basic package includes a full statics, dynamics and heat transfer capability with a basic nonlinear facility for large displacement, creep and plasticity calculations.

A comprehensive library of elements caters for all types of engineering structure utilizing simple springs masses and beams through to membranes, solid and doubly curved thin shell semi-loof and other thin and thick shell elements. A sophisticated automatic driver system allows the user as much or as little control over the analysis procedure as he wishes and is helped by a large range of restart capabilities which combine both versatility and economy.

To complement the basic package several optional enhancements are available which include advanced nonlinear facilities, full colour graphics, interactive pre- and postprocessing, substructures, stiffness and mass matrix manipulation, interactive data preparation, boundary elements allowing mixed finite element and boundary element models and a highly developed offshore analysis facility.

THEORETICAL BACKGROUND AND PROGRAM STRUCTURE

PAFEC consists primarily of two systems. The first does nothing but perform the actual finite element analysis and the second handles the interaction of the program with the computer on which it is running. This second system is referred to as the PAFEC driver. Because they are kept separate means, in practice, that no interaction between the analytical part and the host computer operating system is needed. This has allowed PAFEC to remain completely machine independent, the machine dependency parts existing solely with the driver.

PAFEC operates with just two databases, one memory active and the second disk resident. Both are one-dimensional and utilize a family of tailored I/O routines to ensure the appropriate data is available for the relevant computations to take place. PAFEC is a highly modular system, all internal and external data being handled in simple packets allowing its operation to be easily understood by the user.

The user's data is processed through up to 10 individual programs which is extended to 14 if an offshore analysis is being undertaken. The link between each program is usually handled automatically but manual override by the expert user is possible. The method by which way the user controls the operation of the analysis is handled through the data and, as a result, requires no knowledge of the operating system of the computer being used. The PAFEC driver analyses the user's data and is able to detect what kind of analysis is to be run, together with any special requirements that the user may have. For example the driver detects whether restarts are to be used or generated displacements or temperatures are to be written to a particular file, etc.

The advanced user may, if he wishes, submit his own Fortran as a single program or as an unlimited number of subroutines which the PAFEC driver will incorporate at run-time. This means that PAFEC can be used in an absolutely non-standard or standard fashion on the same computer at the same time without any duplication of the system or any part of the system.

FIELDS OF APPLICATION

Geometric Modelling Facilities

PAFEC has over one hundred types of finite element which may be divided into families as follows:

- Simple spring with 6 degrees of freedom at each node. Is able to join any two nodes or connect a node to earth.
- Mass element, which may be offset or lumped at any node.
- Beam elements ranging from simple general beam to beams with general offset, shear deformation and rotary inertia, curved beam and tension bar. A semi-loof beam is available for use with doubly curved semi-loof shell elements.
- Axisymmetric element for use with Fourier loading, i.e. axisymmetric structure carrying non axisymmetric loads.
- Isoparametric membrane elements for plane stress, plane strain and axisymmetric problems with axisymmetric loading. Elements may be rectangular with 0,1,2 or 3 midside nodes and may also be triangular with 0,1 or 2 midside nodes.
- Isoparametric brick elements for full 3-dimensional analyses. Elements may be cuboid with 0,1, or 2 midside nodes some having a different number of facial midside nodes than through the thickness. Brick elements may also be triangular again with a varible number of midside nodes.
- Shell elements consisting of simple facets of triangular or quadrilateral shapes together with a thin shell of revolution with up to two midside nodes. Faceted shells may have up to two midside nodes per side. Thick facet shell elements are also available.
- Boundary elements for both 2- and 3-dimensional problems particularly useful

for structures containing discontinuities and for problems where the solution domain extends to infinity. PAFEC boundary and finite elements may be mixed if required.
- Doubly curved thin shell using the semi-loof formulation available as a quadrilateral or triangular element with compatible semi-loof beam. The Ahmad formulation provides a thick shell version.
- 2- and 3-dimensional elements for temperature calculations together with 2- and 3-dimensional heat transfer elements.

Many of the above elements are available with isotropic and orthotropic properties and may be mixed together in a single finite element model.

Materials

PAFEC is able to model a wide range of material types from simple isotropic to multilayer laminates, most physical properties being allowed to vary with temperature. In the case of a temperature analysis material properties are allowed to vary with both time and temperature, which also applies to heat transfer coefficients. Both elastic and plastic moduli can be input. Anisotropic materials can also be catered for.

Analysis Capabilities

The following analyses may be undertaken with PAFEC.

- Linear and nonlinear statics
- Buckling
- Natural frequency calculations of both loaded and unloaded structures
- Frequency response
- Large deflections
- General transient response and nonlinear transient response
- Seismic analyses
- Steady state and transient temperature calculations with automatic temperature feedback for thermal stress analysis
- Fracture mechanics using J-integrals
- Elastohydrodynamic lubrication
- API postprocessing for offshore structures

Loading Conditions

In general any number of load cases can be supplied with any particular analysis as well as the ability to be able to combine any factor loads from different analyses.

Standard loadings permitted are:

- Pressure
- Multiaxis acceleration
- Gravity
- Centrifugal
- Non axisymmetric Fourier loads

- Point loads
- Member loads
- Time and frequency dependent forces
- Temperature
- Inertia relief

Boundary and Support Conditions

- Fixed and simple supports
- Prescribed displacements
- Gaps
- Sliding boundaries
- Friction
- Coupled nodes
- Generalized constraints
- Time dependent temperature and heat flux
- Time dependent displacements, velocities and accelerations

PROGRAM DESCRIPTION

Basic Methods Employed

PAFEC primarily adopts the finite element method but is also able to use boundary element techniques. Boundary elements can greatly reduce the amount of data preparation required for a large number of problems and, indeed, with PAFEC the user can mix finite elements and boundary elements at will in any structure.

PAFEC utilizes almost exclusively Gaussian elimination as a solution technique implemented as the blocked front solution method. Optimizers are available to calculate a suitable front order automatically.

User interface and facilities. The PAFEC Finite Element System combines enormous power for the analyst with extreme ease of use. Data may be prepared interactively using the PAFEC Interactive Graphics System, PIGS or passively, the user constructing a datafile which is submitted to the computer in the normal fashion.

The PAFEC interactive graphics system, PIGS. PIGS is a powerful, full colour interactive pre- and postprocessor for PAFEC. It can be used for developing finite element models from scratch using all the features the modern analyst would expect to find in such a system, as well as being able to correct existing meshes previously submitted for analysis.

Load, restraint and editing facilities are available within PIGS as well as a powerful postprocessor which is able to produce colour or monochrome plots of any structure at any view angle. A full graphical facility is available and the system is operated through a comprehensive menu system in exactly the same way modern computer aided design systems operate. PIGS interacts directly with the generated PAFEC databases and hence requires no intermediate processing of data before the system can be used.

Passive data preparation. If an interactive facility is not available, PAFEC data can be created using data layouts as described in the PAFEC data preparation

PAFEC: The PAFEC Finite Element Analysis System

manuals. Data preparation is made extremely easy with the user being allowed to submit data in any order in a totally free-format environment. Data may be liberally commented as shown in the examples.

Users without suitable graphics terminals still have access to full postprocessing supported by a very wide range of graphics facilities permitting off-line plotting of results. Automatic mesh generation and automatic front or wave ordering is available entirely under user control and the system is available with a very large number of system defaults minimizing the data required. All keywords used in a data file may be shortened to the first four characters and, by implication, may be lengthened to further amplify the purpose of the data. For example the keyword MATERIAL may be shortened to MATE or lengthed to MATERIAL.PROPERTIES.OF.THE.REAR.SUBFRAME. All three versions are perfectly valid.

Research and Development

The PAFEC system together with PIGS and the other subsystems available are all written to exacting Fortran standards. Rigorous quality control ensures the code's computer independency. In some cases the Fortran source is available to the user, the PAFEC driver ensuring that any modifications to the system do not disturb the standard implementation.

PAFEC has always been subjected to a rigorous and wide ranging development plan, the users of the system (numbering some 230 with in-house systems in 1984), having the greatest influence on the developments that are included in new versions released annually.

HARDWARE REQUIREMENTS FOR OPERATING PAFEC

Basic Computer Requirements

PAFEC and to a large extent the PIGS interactive graphics system, is available to operate on most mainframe and supermini computers. The minimum requirement is that a processor must have a minimum word length of 32 bits and, depending upon the computer model, utilize a virtual memory operating system. A minimum of 2 Mbytes of main memory is also required with a working disk space of at least 30 Mb.

Other peripherals which must be available are a plotter, printer and, if PIGS is installed, a suitable graphics display.

Type of Computer

The following computers and operating systems are available to operate PAFEC and PIGS.

Computer	Operating System
Apollo Domain	AEGIS
CDC	NOS/VE
Cray	COS
Data General	AOS/VS
DEC VAX	VMS
DEC mainframe	TOPS
ELXSI	ENBOS

Floating Point Systems	PDS
Harris	VOS II
Hewlett Packard 9000/500	HP UNIX
Honeywell	CP6, Multics
ICL	VME
IBM	MVS,VM/CMS
Norsk Data	SINTRAN
Perkin Elmer	OS32
Prime	PRIMOS
Sperry Univac	OS1100

Peripherals

PAFEC and PIGS incorporate a very large number of drivers for graphics peripheral devices which plotters include Calcomp, Benson and Hewlett Packard plus others and also an interface to GINO-F.

Graphics displays for the PIGS program include Tektronix both colour and monochrome, Sigma, Envision, D-SCAN, Westward, Counting House, DACOLL, Autograph, Lexidata and Datapath. New drivers are constantly being written as hardware becomes available.

Media

PAFEC, together with its subsystems, is supplied usually on 9 track magnetic tape and in almost all cases is primarily installed by qualified PAFEC Implementations Staff. This service is available worldwide. Full "Action Desk" and support and training facilities are available in the UK and through Agencies and PAFEC Limited Subsidiaries throughout the world.

EXAMPLES OF APPLICATION

Standard Test Case

PAFEC has available a large number of standard basic test analyses one of which is described here. This analysis consists of using a semi-loof element to describe a cylindrical segment according to Roark, "Formulaes for Stress and Strain", chapter 12, table 30. A combination of conditions 14 and 15 is used. The stress calculated at the restraining plane should be 1.54 times the nominal axial stress. The data required to describe the mesh shown is given. Results are as follows:

Number of elements (N)	Stress ratio
6	1.46
7	1.52
8	1.54

The PAFEC data which produce these results is as follows:

CONTROL
STRESS
PHASE=6
REACTIONS

```
CONTROL.END
C
C CREATE NODES MODULE USING AXIS TYPE 2
C A CYLINDRICAL POLAR AXIS SET WITH A CONSTANT
C RADIUS OF 1.94 M
C
NODES
AXIS=2
Y=1.94
NODE        X           Z
 1         0.0         0.0
R2   1     0.0        15.0
C
C THE ABOVE ENTRY USES THE REPEAT FACILITY. HERE
C WE ARE ASKING FOR 2 REPEATS OF THE PRECEDING LINE
C ADDING 1 TO THE FIRST ENTRY, ZERO TO THE SECOND
C AND 15 TO THE THIRD WITH EACH REPEAT
C
     4      7.278       0.0
R2   1      0.0        15.0
C
C CREATE A PAFBLOCK WITH THE REQUIRED MESH SPACING
C USING THE MESH MODULE, HERE 8 ELEMENTS ARE DESCRIBED.
C
PAFBLOCKS
BLOCK   ELEMENT.TYPE   PROPERTIES   N1 N2     TOPOLOGY
  1        43210           4         1  2   3,6,1,4,0,2,5,0
MESH
1      183,184,492,659,882,1181,1581,2116
2       1
C
C CREATE A PLATES MODULE ASSIGNING PROPERTIES 4 MATERIAL
C TYPE 1 (MILD STEEL) AND A THICKNESS OF 0.12M
C
PLATES
PLATE.NUMBER    MATERIAL.TYPE    THICKNESS
     4               1              0.12
C
C CREATE A LOCAL AXIS SET SUCH THAT TANGENTIAL
C RESTRAINTS MAY BE APPLIED IN A LOCAL DIRECTION
C CORRECTLY ON THE 30 DEG END OF THE SEGMENT
C
AXES
AXIS    RELATIVE.TO    ANG3
 4           1          30
LOCAL.DIRECTIONS
LOCAL=4
NODE.NUMBER     PLANE
     3            4
RESTRAINTS
NODE    PLANE    DIRECTION
 1        1          0
 1        4         345
 3        4         345
C
C NOW APPLY THE LOADING INFORMATION
C
LOADS
DIRECTION=1
NODE.NUMBER     VALUE.OF.LOAD
     4            -1388888.9
     5            -5555555.6
```

```
     6            -1388888.9
END.OF.DATA
```

Practical Example

Figure 1 shows the PAFEC finite element model of shearer arm (symmetric half) as used on an AMBROAS deep mining drum shearer.

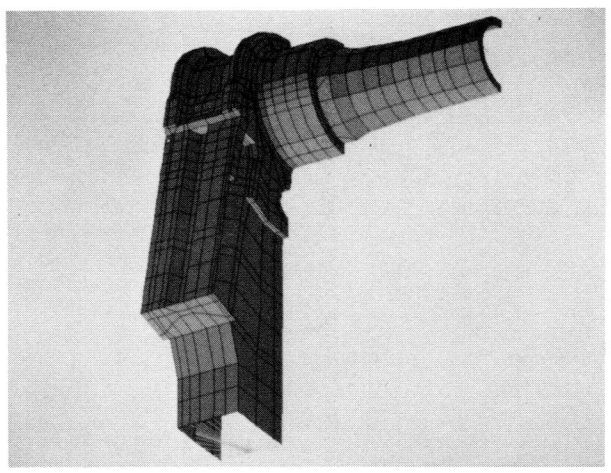

Fig. 1.

The statistics of the analysis are as follows:

```
Type of problem:     Static stress analysis, 3 load cases
Discretization :     1010,20 noded 3-D isoparametric brick elements
                     290, 8 noded plate elements
Number of nodes                 :  14730
Number of degrees of freedom:      22185
Front size (wavefront)          :  722
Computer resources used         :  Prime 750 with 8 Mb of main memory
                                   160 Mb of disk space
Computation time to stressing:     11.5 cpu hours
```

Details about PAFEC

Full details about any part of the PAFEC finite element system can be obtained from:

Finite Element Sales Division, PAFEC Limited, Strelley Hall, Strelley, Nottingham NG8 6PE, England.

Details concerning local subsidiaries or Agencies will be supplied.

PAID: PIPING ANALYSIS AND INTERACTIVE DESIGN WITH THE STAND-ALONE GRAPHICS PACKAGE

Zs. Révész

Electrowatt Engineering Services Ltd., Zurich, Switzerland

ABSTRACT

The kind of projects managed by Electrowatt Engineering Services Ltd., involves integral design of piping systems with supports and the qualification of such for various postulated events. The large number of systems, the growing complexity of the load cases and structure models require human assimilation of large amounts of data. An effort has been made to enlighten evaluation of numerical data and visualize as much of it as possible, thus eliminating a source of error and accelerating analysis/reporting. The product of this effort is PAID, the Piping Analysis and Interactive Design software.

While developing PAID interest has been focused on the acceleration of the work done mainly by PIPESTRESS. This goal is achieved by installing a wide variety of graphic output options. Some installed and tested capabilities of PAID are presented in this paper. Examples are given from the graphic output in report form and the conversation necessary to get such is demonstrated.

BACKGROUND

Humans have an overpowering urge to see as a picture conveys a vast amount of information at a glance. Computer graphics is the best communication method between man and computer since man has an amazing ability to absorb and process graphic data. With the aid of interactive graphics a finite element analyst can, for the first time, climb out from beneath the mountains of output and ascend to a "higher plane" of analysis. Having a wide variety of graphic output options at his/her fingertips means the analyst will have high confidence that the analysis job will be successfully executed and the graphical presentation of the results will help him with insight and more productive use of his time. The development makes it more likely that engineering continues to rely heavily on human intelligence rather than becoming computer dominated.

It is well understood that the potential of engineering misapplication is greatly reduced by graphic interpretation. Still, as experience shows, practical implementation of such simple principles does not follow unless the user has at its disposal a powerful software which he can use conveniently and with minimum effort. Our flexible and interactive graphic system proved to be of immense help to the user permitting examination of numerous details of the model before the solution is effected and the interpretation of the results. An optical control is supported by PAID at all stages of a piping project, from the beginning of general planning, through mesh generation, load input and results representation, until the preparation of adequate documentation.

PAID has been developed to facilitate piping structural engineering work performed for various power plants. The analysis program package which is extensively used by the company is EBASCO/PIPESTRESS[1]. This package is known as one of the best suited software products for the work of this kind. Its theory is well documented[2,3], it is widely applied and is suitable for economic analysis of special problems as reported by Gordis[4] and Révész[5,6].

An engineering company with international clientele cannot be fixed to one specific software product. Therefore communication between PIPESTRESS and PAID has been channeled through a neutral interface file. This development also enables linking of PAID with piping software other than PIPESTRESS (e.g. NUPIPE, KWUROHR, ADLPIPE) as well as transferring data from one program/computer to another.

FIELD OF APPLICATION

The danger of inappropriate use of a structural analysis program can be lessened by extensive use of computer graphics. Engineering misapplications cannot be prevented but their potential is greatly reduced by the use of graphic interpretation. This optical control is possible and necessary at all stages of a project and a flexible and interactive graphic system can be of immense help. With the use of the tools available today such as the PAID software a finite element analyst can look through, inside and around his model as never dreamed before, can study the deformed geometry or the vibration modes and enhance the presentation of his analysis results.

It is seldom that a larger system can be analysed in a single run. For example spectral analysis or static analysis may lead to the change of the system before proceeding to the next step. The interactive graphic capabilities of PAID enable a rational procedure and economic analysis of advanced piping problems.

Four major segments of PAID are released by now, dealing with structure layout, loads, spectral analysis and results. A fifth segment is about to be released to enlighten the handling of thermal transients in piping systems. The program is sufficiently documented and its write-up has also been published by Révész[7,8]. Here a few examples are presented to give an impression of the kind of assistance the user has at its disposal for easier assimilation and better understanding of the model.

PROGRAM DESCRIPTION

The particular features of the software product can be summarized as follows:

- PAID is well tested.
- The strategy of development has been straightforward, dictated by the needs.
- The output is in report form, plots are to scale and properly spaced, texts are legible, sheets dated and identified, ready to be included in reports.
- Interactive execution through conversation with the program in plain English language.
- Minimal written documentation with maximum comfort and user-support during execution.
- Use of procedure file for repetition of actions and for batch execution.

PAID: Piping Analysis and Interactive Design 163

- Quick engineering interpretation and understanding of the elaborated results through manifold graphic capabilities.

Since analysis work is increasingly performed by engineers rather than computer specialists, user-friendly execution with simple interactive program flow has been emphasized. Much assistance is provided for the user when carrying out piping analysis. This assistance consists of internal checks, error messages and explanations if an answer were illegible for the program. Consequently, the program documentation is mostly built-in. Instructions are given and questions are asked during interactive execution. The hard copy documentation is minimal and is basically limited to the correct call sequence, sample results and sample conversations.

HARDWARE COMPATIBILITIES

The full version of PAID is operationable on a CDC computer with BENSON, TEKTRONIX and HEWLETT-PACKARD plotters A version for full graphic output without interactive capabilities is about being released for a PRIME computer with BENSON and CAPTRONIX plotters.

EXAMPLES OF APPLICATION

Several tested capabilities of PAID are presented below. All examples are from the nuclear industry, specifically from piping structural analysis. The presentation of capabilities of a graphic interactive software cannot be exhaustive in the frame given here. Therefore it has been tried to give an impression of the power and sophistication the users get to their fingertips by PAID

Documentation of Isometry

Using the different plotting capabilities of PAID the model of a piping system can be comprehensively documented by means of plots. This prerequisites plotting options as isometric/plane views from an arbitrarily chosen direction and plotting of substructures. Such output can be easily and conveniently generated using PAID.

Processing Earthquake Response Spectra

Response of piping systems on the load of a postulated earthquake is usually calculated by the response spectrum method. The response spectrums for the piping analysis are derived from the seismic input using a structure model of the building. This model is often a bar-type model and the floors of a building are represented by a single mass point thus spectra for piping is valid for a floor of the building. The use of an envelope spectrum of all floor response spectra may be too conservative - still conservativity cannot be ensured if piping supports are at a distance from the lumped mass of the building model.

PAID has been prepared to compute individual response spectra for differential support excitation analysis[9]. Input for PAID here are the response spectra for the three translational and three rotational degree of freedom of a node in the building model. These have to be input individually for the X,Y and Z earthquake. After the definition of the location of a piping restraint PAID can compute the three translational response spectra valid at the restraint given. The superposition of the individual spectra at a point is performed by absolute addition between components of the same earthquake input and then results from the X, Y and Z earthquake are superposed by SRSS method. The conversation of the user with PAID for such an analysis is given in Table 1. As the result of such a

TABLE 1   Sample Conversation with PAID: Combination of Response Spectra

```
     HELLO PAID-USER
     WHAT'S THE SUBJECT OF YOUR ACTIVITY?
LOADS
     DO YOU WORK WITH A PIPESTRESS RESTART FILE?
NO
     ARE YOUR LOADS OF RESPONSE SPECTRUM TYPE?
YES
     WHAT DO YOU WANT TO DO WITH SPECTRA?
COMBINE
     TO GET THE RESPONSE SPECTRUM IN A GIVEN POINT
     THE FLOOR RESPONSE SPECTRA HAVE TO BE COMBINED.

     INPUT SIX FLAGS FOR EACH EQ. COMPONENTS
        (SEPARATED BY COMMAS).  THEY WILL INSTRUCT
     THE PROGRAM IF THE CORRESPONDING COMPONENT
        (X,Y,Z,XX,YY,ZZ) WERE RELEVANT FOR THAT
     EARTHQUAKE COMPONENT (1) OR NOT (0)

     FLAGS FOR FLOOR RESPONSE SPECTRA
       FROM X EARTHQUAKE EXCITATION?
1,0,0,0,0,1

     FLAGS FOR FLOOR RESPONSE SPECTRA
       FROM Y EARTHQUAKE EXCITATION?
0,1,0,0,0,0

     FLAGS FOR FLOOR RESPONSE SPECTRA
       FROM Z EARTHQUAKE EXCITATION?
0,0,1,1,0,0

     ARE THE COMPONENT SPECTRA STORED ON THE INPUT FILE
     IN THE SEQUENCE AS THE NON-ZERO FLAGS HAVE BEEN DEFINED?
YES
     MAGNITUDE OF THE FACTOR WITH WHICH
     THE X COMPONENT FROM THE X EARTHQUAKE
     IS TO BE MULTIPLIED?
0.00
     MAGNITUDE OF THE FACTOR WITH WHICH
     THE ZZ COMPONENT FROM THE X EARTHQUAKE
     IS TO BE MULTIPLIED?
0.82
     MAGNITUDE OF THE FACTOR WITH WHICH
     THE Y COMPONENT FROM THE Y EARTHQUAKE
     IS TO BE MULTIPLIED?
1.00
     MAGNITUDE OF THE FACTOR WITH WHICH
     THE Z COMPONENT FROM THE Z EARTHQUAKE
     IS TO BE MULTIPLIED?
0.00
     MAGNITUDE OF THE FACTOR WITH WHICH
     THE XX COMPONENT FROM THE Z EARTHQUAKE
     IS TO BE MULTIPLIED?
0.84

     COMBINATION HAS BEEN PERFORMED
     ENTER SPECTRUM SPECIFICATION FOR THE RESULT
```

Table 1 (cont'd)

```
          EVENT NO.?
     1
          LEVEL NO.?
     1
          DIRECTION?
     X
          CARD DECK FOR SPECTRUM HAS BEEN PREPARED
          BYE.
```
---

conversation PAID outputs the translational response spectra valid for the given restraint, optionally in form of a listing, a plot or PIPESTRESS input desk.

The installed and presented program flow also enables easy interpolation/ extrapolation of spectral data. PAID is also enabled for graphic/numeric input of spectra.

Modal Time-history Analysis

In the response spectrum method the system response will be evaluated for the support structure (usually a building model) separately, after having the damping introduced. The results are then to be input into the piping model in form of the already mentioned response spectra. When generalized response analysis is performed the load cannot be considered indpendently from the piping structure and the frequency range cannot be assessed from the response spectra of the individual load components. The response of a system on such a load case is composed of the scalar modal amplitude defined by:

$$y_j(t) = \frac{1}{\omega_j} \int_0^t \mathbf{e}_j^T \mathbf{f}(\tau) \sin \omega_j (t-\tau) \, d\tau \tag{1}$$

where $\mathbf{e}_j$ the mode shape vector corresponding to the circular frequency ω_j, $\mathbf{f}(t)$ a vector containing all forcing function components and the damping is still to be introduced.

Efficient data handling performed by PAID enables the user easy solution/ documentation of such problems. Among the most advanced problems solved by this method was the equation[10]

$$M_p \ddot{\mathbf{x}}^{(n+1)} + K \mathbf{x} = \mathbf{f}(t) + M_c \ddot{\mathbf{x}}^{(n)} \tag{2}$$

where M_p stands for the mass matrix of a piping system with scalar mass model and M_c is the correcting mass matrix containing all those added effective mass components of a weak fluid-structure interaction problem which are not included in M_p. Equation 2 can only be solved iteratively. This application (and also less sophisticated ones) are published in the papers listed in the References. All these analyses probably could not have been carried out with this efficiency and accuracy[11,12] without the aid of PAID.

Evaluation of Results from Structural Analysis

PAID assists the user in the documentation of the results too. The available options enable the analyst both to have an overall view of the entire piping system as a single entity and to plot specified segments for documentation purposes. The point of view can be chosen arbitrarily, the plots are to scale and computed results such as forces and displacements can be blended with the structure layout. This enables the analyst to depict realistic geometrical configurations of the system and to place restraints at structurally convenient locations.

The physically meaningful results of a complex dynamic analysis can be presented on a few plots whereas a full listing may cover several hundred printed pages. To lighten the work of the analyst and to reduce the work connected with documentation PAID has been prepared to visualize as much from the tabulated output as possible.

Options available are

- plot of displacements in static load cases,
- plot of mode shapes,
- plot of bound solutions for dynamic load cases,
- plot of reaction forces in individual load cases and
- plot of internal moments.

These plots for different segments of a piping system for different load cases in the requested isometric or plane view are available using PAID upon completion of the analysis.

Extensive graphic interpretation of results can serve different purposes:

- It can be part of a special investigation, e.g. on the coupling of substructures, or

- a part of an investigation on a special vibration problem, both requiring display of certain resonance mode shapes.

- Further it can serve representation of computed data in compressed form for shorter documentation, better understanding, and lastly

- checks of the computation right after completion before listings could be printed.

The attached plots on Figs 1 - 3 advocate themselves. Such plots can be requested using PAID one-by-one in an interactive manner as soon as the analysis program has performed the computation, or they can be requested in a batch job connected with the analysis run.

Figure 1 shows the second mode shape of an auxiliary cooling water pipeline in a nuclear power plant in isometric view. Graphic interpretation of such mode shapes is not only to help understanding of the vibration phenomenon but, as mentioned before, it may be essential in investigations on specific vibration problems.

The bound solution obtained for an earthquake analysis of the same piping is portrayed on Fig. 2. A first glance at Fig. 2 perhaps does not convince all engineers to do away with printed output. However, interrogation of such a picture of computed displacements can evidently give more information than study of lengthy listings of nodes and modal displacements. In Fig. 2 the maximum displacements are plotted to each point of computation along each coordinate axis. Note that the shown displacements are not real physical displacements but the results of modal superposition in global coordinates. The real position of a node during the transient process can be somewhere inside the surface of that rectangular prism, which has side lengths equal to the shown displacements. Such a plot informs the engineer at a glance about the maximal displacements and

PAID: Piping Analysis and Interactive Design

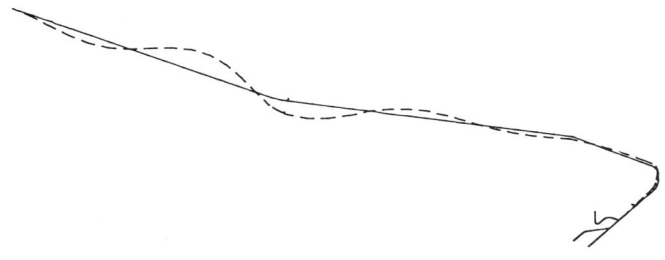

Fig. 1. Isometric view of a typical vibration mode shape for a section of power piping.

critical sections. In case of difficulties selected substructures can be plotted with user-defined view-angle.

Results from the bound solution for generalized response analysis during a transient initiated by loss of power to operating pumps is given in Fig. 3. The pipeline analysed here is identical with that presented in the previous two plots. The form of representation is a plane view. Geometry and reaction forces are to scale, thus this plot informs the analyst at a glance about the magnitude of the reaction forces the support system experiences in this computationally rather complex load case. Again, different views of specific substructures can be displayed with similar ease.

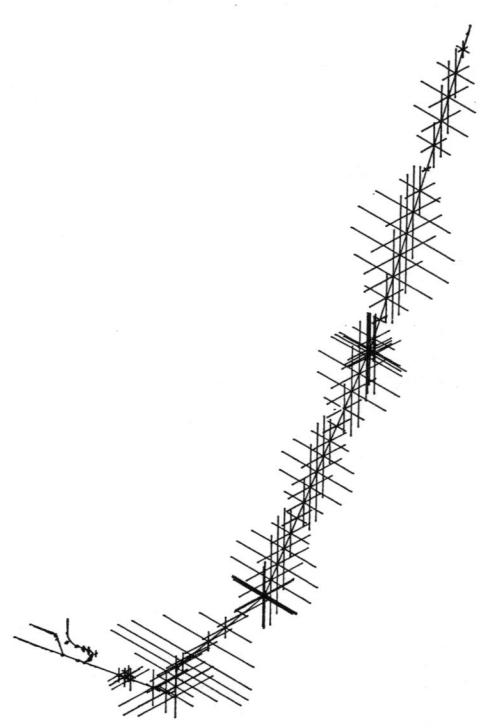

Fig. 2. Graphic representation of the computed displacements for the bound solution for earthquake response analysis – an isometric view.

Using PAID it is equally easy to get both full graphic output and selected plotting. The graphic output can be requested in an interactive manner or in batch processing. In the latter case a procedure file of a previous interactive execution with the same program flow is to be made available for the program.

A default procedure file is in preparation for the full graphic output. Full graphic output may cut down computer resources necessary for graphic postprocessing by magnitudes through better organization of the program flow and using parts of previous plots. Manual interactive plotting still remains necessary and reasonable to display hidden details which cannot be discovered without looking closer at the plots.

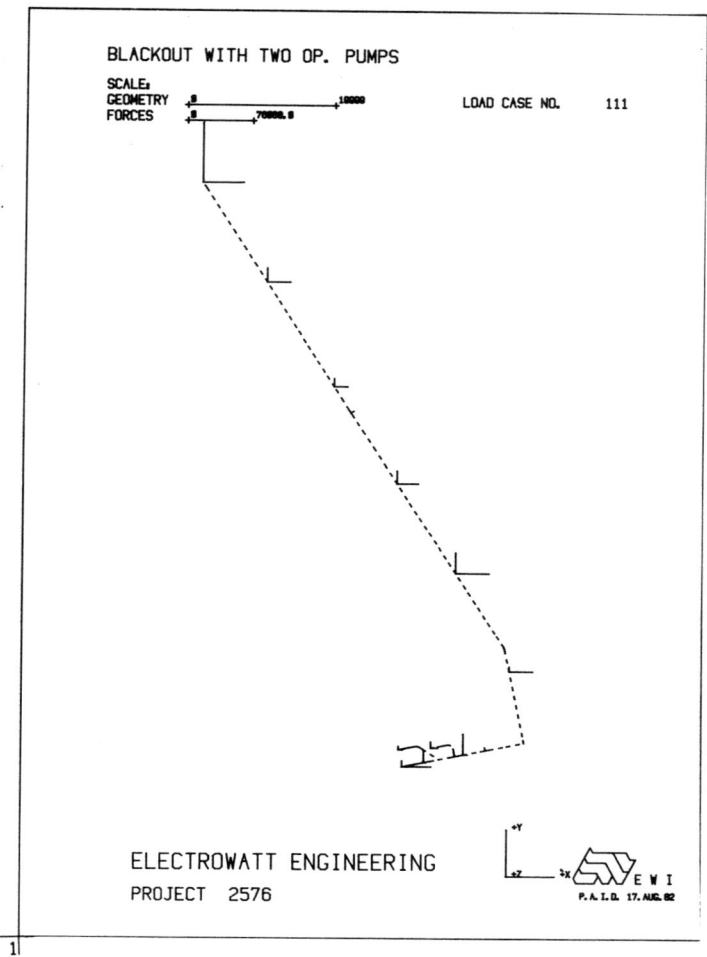

Fig. 3. Graphic representation of the computed reaction forces for the bound solution for generalized response (modal time-history) analysis of hydrodynamic loading - a plane view.

SUMMARY

Extensive use of computer graphics enables competitive engineering services. Economically it offers very good prospects because the hardware prices are continuously decreasing and because graphic representation enables a radical reduction of the hard copy documentation, i.e. no longer large printouts be stored and distributed for control. Other advantages are the very fast result representation, the ease of evaluation of large systems and the support of analysts at solving detail problems.

The result of our development, PAID, has made interactive graphic data processing and so finite element analysis of piping structures more convenient and more feasible. It makes fast, efficient and condensed documentation possible.

The PAID software is sufficiently documented and is installed in a commercial computing centre with international network and also on a mini computer.

The program is available to users on basis of bilateral agreements.

REFERENCES

1. Gordis, K. (1981). *EBASCO/PIPESTRESS*. *Version 3.2 User's Guide*. DST Computer Services, Geneva.

2. Gordis, K. (1979). Outline of dynamic analysis for piping systems. *Nuclear Engineering and Design*, 52 (1)

3. Gordis, K. (1981). Mathematical basis of dynamic structural analysis methods for piping systems *Denver ASME Joint Conference Proceedings*.

4. Gordis, K. (1981). Generalized response analysis: A method for calculating bound solutions for elastic structures subject to dynamic applied force loads. *Trans. 6th Conf. SMIRT*. North-Holland, Amsterdam.

5. Révész, Zs. (1982). Analysis of relief/safety discharge loads by modal approach. *International Modal Analysis Conference Proceedings*, Union College, Schenectady.

6. Révész, Zs. (1983). Comprehensive stress analysis of BWR discharge lines subjected to high frequency fluid dynamic excitation. *Int. J. Pressure Vessels and Piping*, 12 (4).

7. Révész, Zs. (1982). *Piping Analysis and Interactive Design. Version A/2, User's Manual*, Electrowatt Engineering Services, Zurich.

8. Révész, Zs. (1983). PAID: An interactive graphics package to support piping analysis. R. A. Adey, (ed.). In *Engineering Software III*, Springer Verlag, Berlin.

9. Clough, R. W. and J. Penzien, (1975). *Dynamics of Structures*. McGraw-Hill Kogakusha, Tokyo.

10. Révész, Zs. (1983). Analysis of hydrodynamic transients in thin-walled piping applying mass-correcting forces, *Trans. 7th Conf. SMIRT*. North-Holland, Amsterdam

11. Révész, Zs., F. Ferroni and L. Bollok, (1984). Verifying computations for advanced piping problems with in situ measurements. G. A. Keramidas and C. A. Brebbia (eds.). In *Computational Methods and Experimental Measurements*. Springer, Berlin.

12. Révész, Zs. and F. Ferroni, (1984). Experimental validation of a model for weak fluid-structure coupling in a pipeline. In J. Robinson (ed.) ART in FEM Technology, R & A, Dorset.

PANDA: INTERACTIVE PROGRAM FOR MINIMUM WEIGHT DESIGN OF COMPOSITE AND ELASTIC-PLASTIC STIFFENED CYLINDRICAL PANELS AND SHELLS

D. Bushnell

Lockheed Applied Mechanics Laboratory, Department 93-30, Building 255, 3251 Hanover Street, Palo Alto, California 94304, U.S.A.

ABSTRACT

PANDA is an interactive program through which minimum weight designs of composite, elastic-plastic, stiffened, cylindrical panels and complete cylindrical shells can be obtained subject to general and local buckling constraints and stress and strain constraints. The panels or shells are subjected to arbitrary combinations of uniform in-plane axial, circumferential, and shear loads. Nonlinear material effects are included if the material is isotropic or has stiffness in only one direction (as does a discrete or smeared stiffener). The panel can be stiffened with both stringers and rings. Several types of general and local buckling modes are introduced as constraints in the optimization process, including general instability, panel instability with either stringers or rings smeared out, local skin buckling, local crippling of stiffener segments, and general, panel, and local skin buckling including the effects of stiffener rolling. Certain stiffener rolling modes in which the panel skin does not deform but the cross section of the stiffener does deform are also accounted for. The interactive PANDA system consists of three independently executed modules, BEGIN, DECIDE, PANCON, that share the same data base. In the first module, BEGIN, an initial design concept and rough (not necessarily feasible or accurate) dimensions are provided by the user in a "conversational" mode. In the second module, DECIDE, the user decides which of the design parameters of the concept are to be treated by PANDA as decision variables in the optimization phase. In the third module, PANCON, the optimization calculations are carried out with the help of the well-known optimizer CONMIN.

THEORETICAL BACKGROUND AND PROGRAM OVERVIEW

The PANDA computer program was developed in response to the need for a very easy-to-use and rapid preliminary design capability for flat panels and cylindrical shells with stiffeners running both axially and transversely or circumferentially. The panels or shells may be made of layers of any type of material, including fibre-reinforced lamina, and the stiffeners are assumed to consist of assemblages of rectangular segments that constitute typical cross sections, such as rectangular, L-shaped, T-shaped, I-shaped, etc. Design variables include the thickness and material orientation (winding angle) of each layer of the panel skin, spacings of the stiffeners, and cross section dimensions of each part of each stiffener. With so many possible decision variables, it is essential to be able speedily to calculate critical loads corresponding to all types of buckling (general, panel, local, crippling, rolling).

The panel or shell is loaded uniformly by any combination of axial resultant N_x, transverse or circumferential resultant N_y, and shear resultant N_{xy}. These loads (units of N/mm or lb/in, for examples) act in the plane of the panel skin. The loading is assumed in most cases to result in uniform membrane strain in both skin and in the stiffeners. Prebuckling inter-ring bending is accounted for in multilayered shells with rings only. Shear is carried by the skin only.

PANDA uses approximate but reasonably accurate estimates for buckling and stress, so that preliminary designs for rather complex geometries can be obtained interactively on a minicomputer such as the VAX. The goal here is *preliminary design*. Once a reasonable preliminary design has been obtained, the user can go to other more general computer programs, such as BOSOR4 and BOSOR5, for confirmation of the adequacy of the PANDA design and for refinement of the design.

Table 1 lists the characteristics and status of PANDA as of November 1, 1984. The program is currently in widespread use and is maintained by the developer. Notices of bugs found are distributed to all known users. PANDA has been thoroughly checked out by comparisons with other known solutions, tests, and by extensive use at several institutions the world over for about 3 years.

TABLE 1 PANDA at a Glance

KEYWORDS: panels, cylinders, buckling, optimization, composite, stiffened, elastic-plastic

PURPOSE: To optimize, with respect to weight, cylindrical panels or complete cylindrical shells with discrete or smeared stringers and/or rings, subjected to simultaneous uniform in-plane axial compression, hoop compression and shear. Shell wall can be layered composite material. Plasticity can be included in isotropic layers and in stiffeners. General and local buckling loads as well as maximum effective stress or maximum strain components act as constraints on the design. Local buckling modes include skin buckling between stiffeners and local buckling of parts of the stiffeners, such as web crippling. Stiffener rolling modes are included also. PANDA can be used to calculate buckling loads and interaction curves for known designs. Empirically derived knockdown factors (ASME Code Case N-284, modified) are included in this branch.

DATA: 1982, Latest update, 1984

DEVELOPER: Dr. David Bushnell, 93-30/255, Lockheed Applied Mechanics, 3251 Hanover St., Palo Alto, California 94304, (415) 858-4037

METHOD: Buckling modes are assumed to follow simple $sinmx\ sinny$ pattern for both general and local buckling. For buckling under shear loading and unbalanced laminates the pattern $sinny\ sinm(x - cy)$ is used. Other simple analytical expressions are used to describe buckling modes for parts of the stiffeners or rolling, crippling, and wide column buckling of stringers and rings. Optimization is by method of feasible directions. The optimizer (CONMIN) was written by Vanderplaats. Donnell equations are used, with suitable correction factors applied for low circumferential wave numbers in the case of deep cylindrical panels and complete cylindrical shells. Complete (360 deg.) cylindrical shells are modelled as panels that subtend 180 degrees of circumference.

RESTRICTIONS: Panel is simply supported; prebuckling stress resultants uniform in the panel; thicknesses and stringer and ring spacings uniform; all stringers are identical; all rings are identical; stringers may be different from rings;

PANDA: Interactive Program for Minimum Weight Design

panel must be cylindrical or flat; plasticity is allowed in isotropic material only; no postbuckled skin is allowed. Imperfections are not accounted for in optimization analysis: User must design panel to increased loads in order to compensate for imperfections.

LANGUAGE: FORTRAN 77

DOCUMENTATION: References 1, 2 and 3 and the program itself. A file, HELPER.PAN, tells about how to use VAX-PANDA.

INPUT: The interactive PANDA system consists of three independently executed modules that share the same data base. In the first module an initial design concept and rough (not necessarily feasible or accurate) dimensions are provided by the user in a "conversational" mode. In the second module the user decides which of the design parameters of the concept are to be treated by PANDA as decision variables in the optimization phase. In the third module, in which the optimization calculations are carried out, the user decides whether or not to continue design iterations, whether or not to display design information and buckling margins and modes, and whether or not to save output on a permanent file.

OUTPUT: panel weight, in-plane loads, dimensions, buckling modes, buckling margins, stress margins, combined load interaction curves for perfect and imperfect shells, knockdown factors for geometric imperfections, knockdown factors for plasticity.

HARDWARE: VAX11/780. Should be easy to convert to other computers.

SIZE: 524 blocks required for storage of PANDA absolute elements; 1052 blocks required for source files; 848 blocks required for relocatables; about 300 blocks required for I/O for a typical case.

USAGE: About 25 institutions are using PANDA.

RUN TIME: Typically an optimum design can be obtained with an hour or so at the terminal and about half a minute of computer time. Cost for the computer time (usually seconds or a couple of minutes at the most) is negligible.

AVAILABILITY: VAX version available from developer (address above); Price: $1000.00 includes all documentation, magnetic tape with source, relocatables, absolutes, test cases, and further documentation. One time purchase price.

MAINTENANCE: Developer sends out notices of bugs and other news from time to time.

FIELD OF APPLICATION

Figures 1 and 2 show the geometry and loading. PANDA does three types of analysis:

1. Buckling and stress analysis of a fixed design (no optimization). This branch is used to evaluate a given design under a single set of in-plane uniform resultants, N_x, N_y, N_{xy}.

2. Optimization analysis, in which the design is constrained by buckling and stress or strain allowables. The user provides a starting design and tells PANDA which variables are allowed to change, what are the upper and lower bounds of these variables, and which variables are linked to those that are allowed to change.

3 Calculation of critical load interaction curves for perfect and imperfect panels. These curves identify combinations (N_x, N_y), (N_x, N_{xy}), and (N_y, N_{xy}) that cause general instability, local buckling, panel buckling (smeared rings), panel buckling (smeared stringers), and maximum allowable (von Mises) stress. The dimensions of the panels are fixed in this branch. Knockdown factors for geometric imperfections and for plasticity are generated via a modified version of ASME Code Case N-284 (1980).

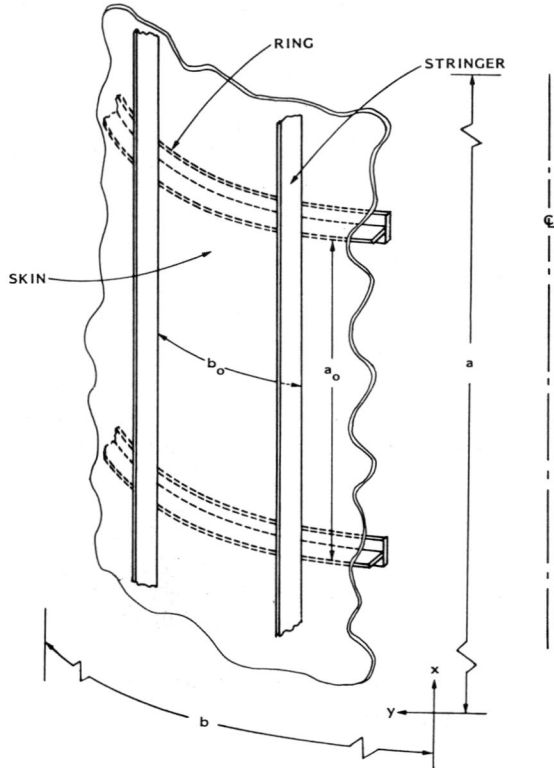

Fig. 1. Stiffened cylindrical panel with overall dimensions (a,b), ring spacing a_0, and stringer spacing, b_0.

The various types of buckling checked for by PANDA are listed in Table 2. Buckling in any of these modes is considered to constitute failure of the panel. Therefore the ratios of these buckling loads to the given applied loads, N_x, N_y, N_{xy}, are introduced as constraints in the optimization analysis. Any constraint less than unity signifies that the design is in the unfeasible region.

GOVERNING ASSUMPTIONS IN THE PANDA THEORY

The assumptions upon which PANDA is based are:

1. If the material is orthotropic it is linear elastic. If it is isotropic, it

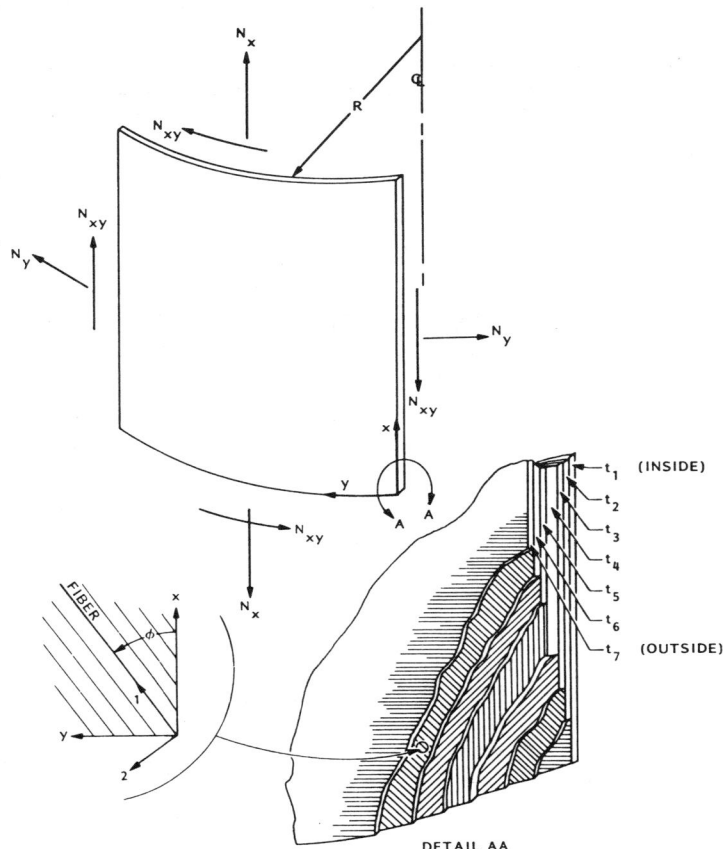

Fig. 2. Coordinates, loading, and wall construction.

can be elastic-plastic if the user so specifies. In this case the stress-strain curve is monotonically increasing, the material hardens isotropically, and deformation theory plasticity is used.

2. Thin shell theory holds: normals to the undeformed surface remain normal.

3. The panel is cylindrical or flat. It is simply supported on all four edges. All stringers are identical and are uniformly spaced. All rings are identical and are uniformly spaced.

4. The rectangular segments from which the stiffeners are built up are slender: that is, their widths (heights) are large compared to their thicknesses and their lengths are large compared to their widths (heights).

5. The nodal lines in any buckling pattern are straight. The normal deflection $w(x,y)$ in its most general form is given by

$$w(x,y) = C\{\cos[(n + mc)y - (m + nd)x] - \cos[(n - mc)y + (m - nd)x]\}$$

TABLE 2   Types of Buckling Included in PANDA Analysis

GENERAL INSTABILITY: Buckling of skin and stiffeners together with smeared rings and stringers. Panel is simply supported along the edges $x = y = 0, x = a, y = b$.

LOCAL INSTABILITY: Buckling of skin between adjacent rings and adjacent stringers. Portion of panel bounded by adjacent stiffeners is simply supported. Stiffeners take their share of the load in the prebuckling analysis, but are disregarded in the stability analysis.

PANEL INSTABILITY (between rings with smeared stringers): Buckling of skin and stringers between adjacent rings. Portion of panel bounded by adjacent rings is simply supported. Stringers are smeared. Simple support conditions are imposed at $y = 0$ and at $y = b$. Rings take their share of the load in the prebuckling analysis, but are disregarded in the stability analysis.

PANEL INSTABILITY (between stringers with smeared rings): Buckling of skin and rings between adjacent stringers. Portion of panel between adjacent stringers is simply supported. Rings are smeared. Simple support conditions are imposed at $x = 0$ and at $x = a$. Stringers take their share of the load in the prebuckling analysis, but are disregarded in the stability analysis.

LOCAL CRIPPLING OF STIFFENER SEGMENTS ("internal" segments): Individual stiffener segment buckles as if it were a long flat strip simply supported along its two long edges. Loading is uniform compression along the stiffener axis. Curvature of ring segments is ignored.

LOCAL CRIPPLING OF STIFFENER SEGMENTS ("end" segments): Individual stiffener segment buckles as if it were a long flat strip simply supported along the long edge at which it is attached to its neighbouring segment or to the panel skin and free along the opposite edge. Loading is compression along the stiffener axis. Number of half waves along the stiffener axis is assumed to be the same as that of the part of the structure to which the "end" is attached. Curvature of ring segments is ignored.

LOCAL ROLLING WITH SKIN BUCKLING BETWEEN STIFFENERS: Same as "LOCAL INSTABILITY", except that the strain energy in the stiffeners and the work done by prebuckling compression in the stiffeners are included in the buckling formula. Stiffener cross sections do not deform as stiffeners twist about their lines of attachment to the panel skin.

ROLLING INSTABILITY (with smeared stringers): Same as "PANEL INSTABILITY (between rings with smeared stringers)", except that the strain energy of the rings and work done by prebuckling compression along the ring centroidal axis are included in the buckling formula. The ring cross section does not deform as the ring twists about its line of attachment to the panel skin.

ROLLING INSTABILITY (with smeared rings): Same as "PANEL INSTABILITY (between stringers with smeared rings)", except that the strain energy of the stringers and work done by prebuckling compression along the stringer centroidal axis are included in the buckling formula. The stringer cross section does not deform as the stringer twists about its line of attachment to the panel skin.

ROLLING OF STRINGERS, NO BUCKLING OF SKIN: Stringer web cross section deforms but the flange cross section does not. The buckling mode has waves along the stiffener axis.

ROLLING OF RINGS, NO BUCKLING OF SKIN: Ring web cross section deforms, but the flange cross section does not. The buckling mode has waves along the ring axis. This mode is sometimes called "frame tripping" by those interested in submarine structures.

AXISYMMETRIC ROLLING OF RINGS, NO SKIN BUCKLING: Same as "ROLLING OF RINGS...", except that the buckling mode has zero waves around the circumference of the panel.

PANDA: Interactive Program for Minimum Weight Design

in which either c or d are zero, depending on the aspect ratio and stiffness components of the entire panel or whatever portion of the panel is being treated by PANDA at the moment.

6. For cylindrical panels or shells, a typical cross section dimension of a stiffener is small compared to the radius of the panel.

7. In rolling modes with skin participation (see Table 2) the cross section of a stiffener remains undeformed as the structure deforms, and rotations about a stiffener centroid are equal to the rotation of the panel skin at the attachment point of the stiffener to the skin.

8. The stiffener centroids coincide with their shear centres.

9. General instability is calculated by smearing both sets of stiffeners; panel instability by smearing one of the sets of stiffeners. It is assumed that the stiffeners are numerous enough to do this and that there is no interaction between local stiffener deformation and overall buckling in the general and panel modes of buckling. Also, there are no local disturbances in the panel skin buckling in the general instability mode due to discrete attachment of stiffeners.

10. The overall dimensions of the panel are large compared to the spacings between stiffeners. Hence, PANDA gives more accurate results when there are several stiffeners along either or both spans a and/or b than when there are only one or two stiffeners along a and/or b. PANDA does not distinguish whether or not there are stiffeners right at the edges or whether the edges are halfway between stiffeners.

METHOD AND DISCRETIZATION

Details of the method are given in references 1, 2 and 3. It is emphasized that there is no discretization in PANDA. The analysis is based on relatively simple formulas for buckling as a function of the number of half waves in each of two coordinate directions. An elaborate search is conducted to ensure that the smallest buckling load has been obtained as a function of the number of axial half waves m, the number of circumferential half waves n, and the slope of the buckling nodal lines c or d. The analysis is based on the principle of minimum potential energy. Plasticity is modelled as if the material were hypoelastic.

The optimizer CONMIN, written by G. N. Vanderplaats, is used in PANDA for finding minimum weight designs. This subroutine is based on a nonlinear constrained search algorithm due to Zoutendijk.

USER-FRIENDLY FEATURES OF PANDA

PANDA runs entirely interactively The user can choose short or long prompts, depending on his or her familiarity with the program. A file containing the user's responses to questions is saved, so that if the user inadvertantly hits the wrong key, he or she can leave PANDA, exit this file, and start again. PANDA then reads responses from the file until the end of the file, then returns control to the user so that he or she can complete the case in the interactive mode.

The PANDA program set contains a file, HELPER.PAN that instructs the user what PANDA does and how to use it.

EXAMPLES OF APPLICATION

DETAILED EXAMPLE CASE: Table 3 (see Appendix) contains a sample runstream for interactive input of data and optimization of a steel containment shell. The starting design is displayed in Fig. 3.

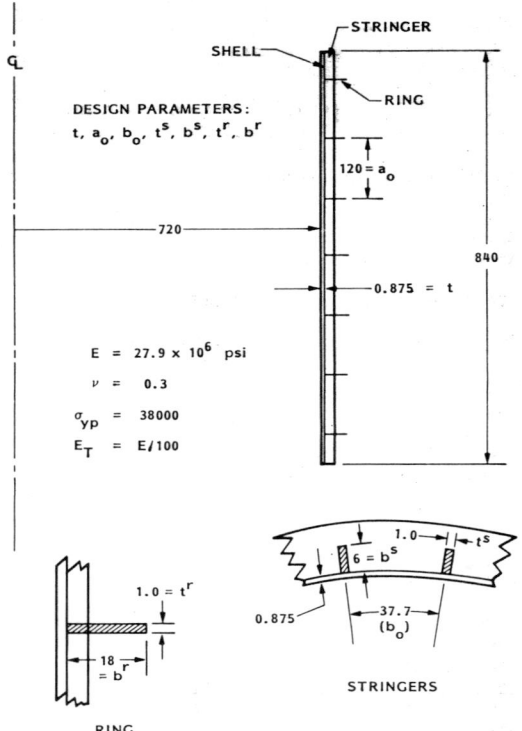

Fig. 3. Ring and stringer stiffened cylindrical shell with dimensions typical of a large containment vessel for a nuclear reactor. This structure is to be optimized with PANDA. Dimensions $t, a_O, b_O, t^s, b^s, t^r, b^r$ are allowed to vary during design iterations.

The purposes of the rather long PANDA runstream listed in Table 3 are:

1. to provide PANDA with a starting design. This is done by the user's executing BEGIN and providing starting dimensions and material properties.

2. to tell PANDA what kind of analysis to do, and if it is an optimization analysis, to tell PANDA which design parameters are to vary during optimization and what are their lower and upper bounds. This is accomplished by execution of DECIDE.

3. To perform the optimization. This is done by execution of PANCON. Three sets of iterations are performed, after which the final design and buckling and stress margins are listed.

4. To check the adequacy of the optimized design in the presence of geometrical imperfections and prebuckling plasticity. How this is done is described in

PANDA: Interactive Program for Minimum Weight Design

Table 3 immediately following the list of the optimized design and establishment of a permanent file for the output, CYLOUT.DAT.

5. To plot interaction curves for the perfect and imperfect optimized design. This is done by execution of PLOTIT. The plots are displayed in Figs. 4, 5, and 6.

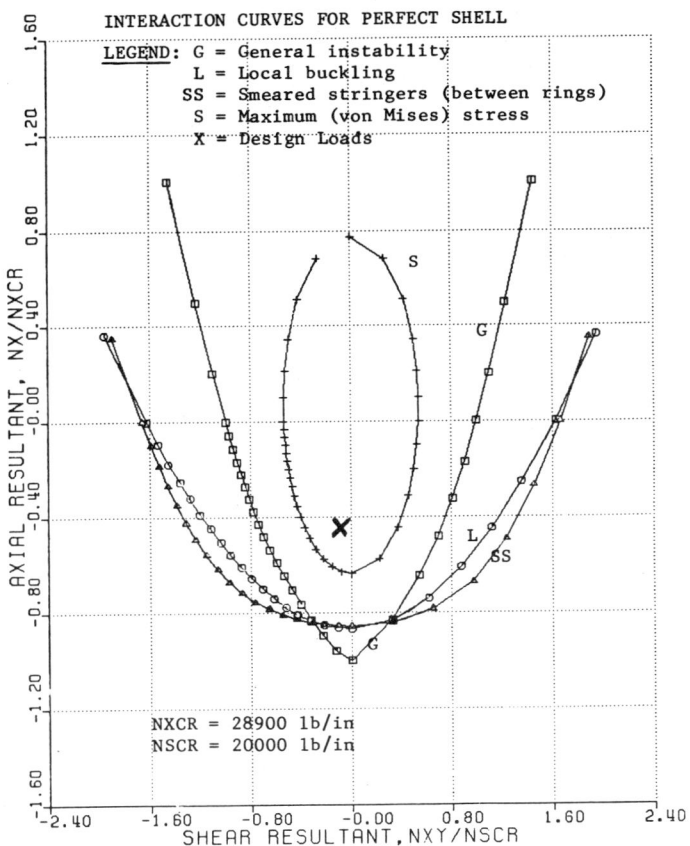

Fig. 4. Interaction curves (N_x, N_{xy}) for maximum (von Mises) stress and various modes of buckling for perfect ring and stringer stiffened steel containment shell optimized under axial compression N_x = 24000 lb/in, circumferential tension, N_y = 4000 lb/in, and in-plane shear N_{xy} = 6300 lb/in. The points where the three curves (G, L, SS) cross correspond to this loading state (sign of N_{xy} unimportant in this problem.) Yield stress = 38000 psi. The Design Point X corresponds to N_x = -12000 lb/in, N_y = 4000 lb/in, and N_{xy} = 4400 lb/in. The interaction curves were obtained with N_y held constant at 4000 psi.

Several aspects of this sample case should be emphasized:

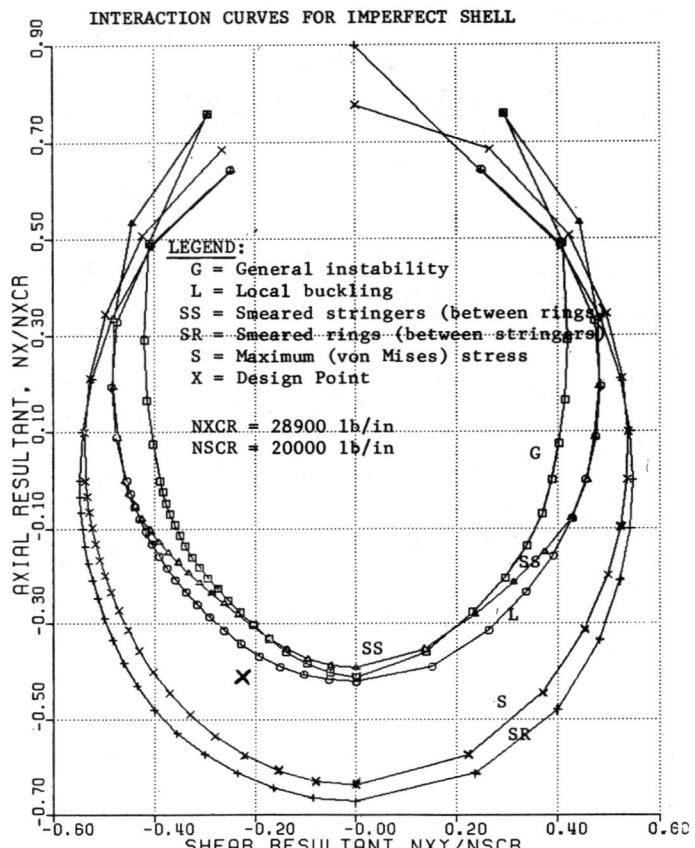

Fig. 5. Interaction curves for same optimized containment shell, but including knockdown factors for geometric imperfections and local plastification. The Design Point X lies outside the three curves corresponding to General Instability, Local skin buckling, and Panel buckling (between rings with smeared stringers). Therefore, the optimized design is unsafe. The user should design the structure using somewhat higher loads and somewhat lower yield stress.

1. The starting design represents a realistic one for operating loads that are expected to generate stress resultants N_x = 6000 lb/in (axial compression), N_y = 2000 lb/in (circumferential tension), and N_{xy} = 2200 lb/in (in-plane shear).

In cases such as this it is normal to apply a factor of safety of 2.0 to allow for uncertainty in the load, so that the structure should be designed to withstand N_x = 12000 lb/in compression, N_y = 4000 lb/in tension, and N_{xy} = 4400 lb/in shear.

2. In the optimization branch, PANDA does not account for initial imperfections or nonuniform prebuckling behaviour such as local bending around stiffeners.

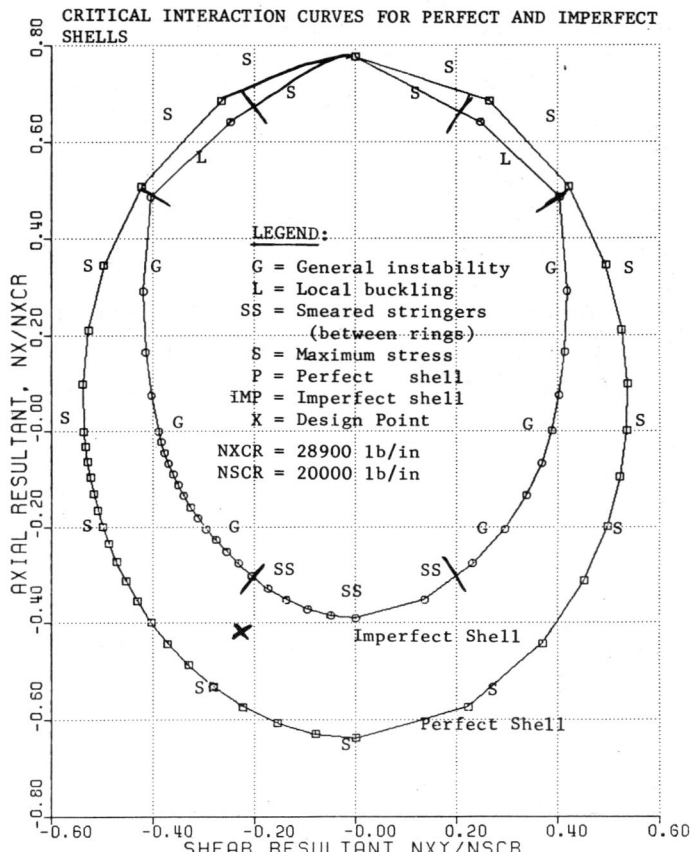

Fig. 6. Same as Figs. 4 and 5, except that only the curves nearest the origin for the perfect and the imperfect structure have been plotted. The type of buckling predicted for the imperfect shell depends on the location in loading space (N_x, N_{xy}).

Nor does PANDA account for residual stresses caused by cold bending sheet into cylindrical sections, welding the sections together, and then welding stiffeners to the assembled cylindrical shell. In order to compensate for this lack, the user must obtain an optimum design to a higher set N_x, N_y, N_{xy}, than $N_x = -12000$, $N_y = 4000$, $N_{xy} = 4400$. Based on previous experience, the user in this case estimated that he should double N_x, leave N_y alone, and increase N_{xy} from 4400 lb/in to 6300 lb/in. The optimum design, were it perfect, would withstand these higher loads. Therefore, it seems likely that the actual structure when constructed will withstand the lower load set, $N_x = -12000$, $N_y = 4000$, $N_{xy} = 4400$, which is identified in Figs. 4, 5, and 6 as a large X called the "Design Point".

3. While buckling loads are often quite sensitive to initial imperfections, the maximum stresses experienced by the structure are less so. Therefore, corresponding to the increases in N_x and N_{xy} just specified, one should increase the yield strength of the material from its actual value to a somewhat exaggerated value. If this is not done the structure will be unnecessarily heavy. In this

case the user doubled the yield stress, increasing it from 38000 psi to 76000 psi. (As the results show, this was perhaps too generous an increase, leading to an unacceptable design, as we have seen.)

Table 3 contains the PANDA runstream. Note near the beginning of Table 3 that results from the interactive BEGIN session are to be saved on a file called CYLBEG.DAT. Table 4 contains a list of this file. It can be edited and used in future runs of BEGIN, thereby allowing the user to bypass the interactive session in BEGIN, running BEGIN instead in a sort of "batch" mode.

TABLE 4   Input File, CYLBEG.DAT, Created by BEGIN Processor
This file can be modified and used for future runs of BEGIN in which the user wants to avoid having to answer all the questions interactively

```
              STIFFENED CYLINDRICAL SHELL
       N      WANT MORE INFORMATION?
       N      WANT LONG PROMPTS?
       N      WANT CHANCE TO CORRECT DATUM?
       N      IS PANEL FLAT?
    0.240E+05 AXIAL STRESS RESULTANT
       Y      IS NX COMPRESSIVE?
    0.400E+04 CIRCUMFERENTIAL STRESS RESULTANT
       N      IS NY COMPRESSIVE?
    0.630E+04 IN-PLANE SHEAR STRESS RESULTANT
       N      WANT TO INPUT NXFIXED,NYFIXED?
      720.    CYLINDER RADIUS
      840.    AXIAL LENGTH OF PANEL
    0.226E+04 CIRCUMFERENTIAL LENGTH OF PANEL
       1      NUMBER OF LAYERS IN PANEL SKIN
       0.875    TLAYER
    0.000E+00   ALPHA
       1        MATL
       Y      IS THE PANEL STIFFENED?
       Y      ARE THERE STRINGERS?
       37.7   STRINGER SPACING
       N      IS THE STIFFENER INTERNAL?
       R      CROSS SECTION TYPE FOR THE STRINGERS
       N      ANY STR. SEGS OF COMPOSITE MAT?
       1.00     TS
       6.00     BS
       1        MATL
       N      DO STRINGER STUFF AGAIN?
       Y      ARE THERE RINGS?
      120.    RING SPACING
       N      IS THE STIFFENER INTERNAL?
       R      CROSS SECTION TYPE FOR THE    RINGS
       N      ANY RING SEGS OF COMPOSITE MAT?
       1.00     TR
      18.0      BR
       1        MATL
       N      WANT TO DO RING STUFF AGAIN?
    0.279E+08 MODULUS IN FIBER DIRECTION
       Y      IS THIS MATERIAL ISOTROPIC?
       0.300  POISSONS RATIO
       0.300  WEIGHT DENSITY
       N      WANT TO INPUT STRESS-STRAIN?
       Y      WANT TO INPUT EFFECTIVE STRESS?
    0.760E+05 MAXIMUM EFFECTIVE STRESS
       N      DO MATERIAL PROPERTIES AGAIN?
       N      IS LOCAL/GENERAL NOT = UNITY?
```

PANDA: Interactive Program for Minimum Weight Design   183

Following the design iterations, during which "PANEL WEIGHT=..." appears many times on the screen (See page 6 of Table 3), the current loading, design, and load factors for buckling and critical stress are listed. Load factors near unity signify an active constraint. That is, the particular failure mode identified is critical for the current design. In this example, general instability, local skin buckling, and panel buckling (between rings with smeared stringers) are all critical. In addition, buckling of the stringer segment is critical, local rolling with skin participation is critical, and axisymmetric deformation of the very slender ring is critical.

The information listed on pp. 6-9 of Table 3 is stored on a permanent file with the name CYLOUT.DAT.

Once the optimum design has been obtained, it is checked, including the effects of imperfections and local plastification, as described on page 10 of Table 3. Figures 4, 5, and 6 show the results from PLOTIT. Notice that the Design Point X (N_x = -12000 lb/in, N_y = 4000 lb/in, and N_{xy} = 4400 lb/in) falls inside all of the critical curves for the perfect optimized shell (Fig. 4), but that X falls outside the load interaction curves for general instability, local skin buckling, and panel instability (between rings, smeared stringers) for the actual (imperfect and fabricated) shell (Fig. 5). Figure 6 shows the relation of the Design Point X to the most critical curve for the perfect shell (von Mises stress) and the most critical curve for the imperfect shell (a combination of various buckling modes, depending on the ratio of N_x to N_{xy}).

The optimum design is unsafe. The user should perform another optimization analysis, in this case probably using a combination of somewhat higher loads than (N_x = -24000 lb/in, N_y = 4000 lb/in, N_{xy} = 6300 lb/in) and somewhat lower yield stress than 78000 psi.

OPTIMUM DESIGN OF ELASTIC-PLASTIC, RING-STIFFENED CYLINDRICAL SHELLS UNDER HYDROSTATIC COMPRESSION: Tables 5 and 6 and Figs. 7-11 pertain to this investigation. One of the purposes of the study is to compare PANDA buckling predictions for optimum designs and BOSOR5 predictions for the same designs for a range of loading over which the amount of prebuckling plastic flow varies. BOSOR5 is an appropriate standard of comparison for ring-stiffened cylindrical shells stressed under hydrostatic compression beyond the material proportional limit because there exist numerous comparisons with tests. (See the BOSOR5 paper and the reference therein, especially Figs. 13-16 of reference 2.

PANDA RESULTS: The optimum designs and buckling pressure load factors and modes from PANDA are listed in Tables 5 and 6. A typical configuration is displayed in Fig. 7(a). The decision variables in the optimization process are the six dimensions listed as headings in columns 3-8 of Table 5. The results for each design pressure in Table 5 were obtained by first optimizing such that the ring spacing was included as a decision variable. The ring spacing was then set to a new value as near the optimum value as possible consistent with the condition that there be an integral number of rings within the cylinder length of 172 inches. A new optimum was then calculated corresponding to this new value of ring spacing, which was not allowed to vary during this second optimization process. It is seen from Table 6 that for a wide range of design pressures the optimum design is characterized by many nearly simultaneous buckling modes.

BOSOR5 RESULTS: Figures 7 and 8 show the BOSOR5 models. Half the length of the shell is modelled, with symmetry and antisymmetry conditions applied as indicated in Fig. 7. (The midlength of the cylinder is at the top of the figure.) The reference surface of the cylindrical shell is taken to be the inner surface. The web of each ring, treated as a flexible shell branch, is assumed to penetrate the flange to the middle surface of the flange. The material of the ring at the structural plane of symmetry at the bottom of Fig. 7 has half the stiffness of the other rings. All flanges except the two nearest the midlength of the shell

TABLE 5 Optimum Designs of Hydrostatically-Compressed, Ring-Stiffened Cylinders Derived by PANDA. Radius to Shell Middle Surface = 44.625 in.; Length = 172 in.; T-Shaped Internal Rings

Pressure P_o (psi)	Weight (lbs)	Thickness of shell (inches)	Ring Spacing (inches)	WEB		FLANGE	
				Thickness (inches)	Height (inches)	Thickness (inches)	Height (inches)
677	2898	0.289	5.93	0.112	3.44	0.077	1.65
1355	4951	0.493	9.05	0.202	5.03	0.131	2.51
2032	6835	0.688	11.47	0.276	6.21	0.177	3.17
2710	8662	0.807	11.47	0.351	6.95	0.262	4.51
2710*	8724	0.822	11.47	0.346	6.85	0.261	4.46
3388	10694	0.998	13.23	0.460	7.82	0.310	4.97
4066	12682	1.244	15.65	0.560	7.96	0.377	4.75
4743	14667	1.519	19.11	0.651	8.42	0.394	4.60

*Model in which the shell wall is treated as if it consists of five identical layers, in order to account for the variation of midbay prebuckling axial strain through the wall thickness.

TABLE 6 Buckling Pressure Factors and Modes for Various Types of Instability Predicted by PANDA

Design Pressure, Po (psi)	General Instability	Local Skin Buckling	Buckling of Web	Buckling of Flange	Rolling with Skin Buckling	Rolling, No Skin Buckling	Axisymmetric Rolling
677	1.0011(1,2)[a]	1.0003(1,16)	1.0(40)	1.0(40)	1.0(1,10)	1.11(4)	1.21(0)
1355	1.0(1,2)	1.0(1,13)	1.0(29)	1.0(29)	1.0(1,9)	1.05(3)	1.10(0)
2032	1.0(1,2)	1.0(1,12)	1.0(24)	1.0(24)	1.0(1,8)	1.01(2)	1.03(0)
2710	1.0(1,2)	1.0(1,11)	1.0(22)	1.0(22)	1.01(1,6)	1.00(1)	1.00(0)
2710[b]	1.0(1,2)	1.0(1,10)	1.0(22)	1.0(22)	0.999(1,5)	1.01(2)	1.02(0)
3388	1.02(1,2)	1.03(1,10)	1.06(19)	1.04(19)	1.03(1,6)	1.03(2)	1.04(0)
4066	1.0(1,2)	1.07(1,9)	1.10(19)	1.11(19)	1.06(1,7)	1.07(2)	1.08(0)
4743	1.0(1,2)	1.07(1,8)			1.06(1,7)	1.07(3)	1.08(0)

[a] Numbers in parentheses are (axial, circumferential) waves in buckling pattern (axial halfwaves, circ. full waves). For local skin buckling and rolling with skin buckling the axial half-wave-index refers to the number of half-waves between adjacent rings. Where only one number is given, it refers to the number of full circumferential waves.

[b] 5-layered model.

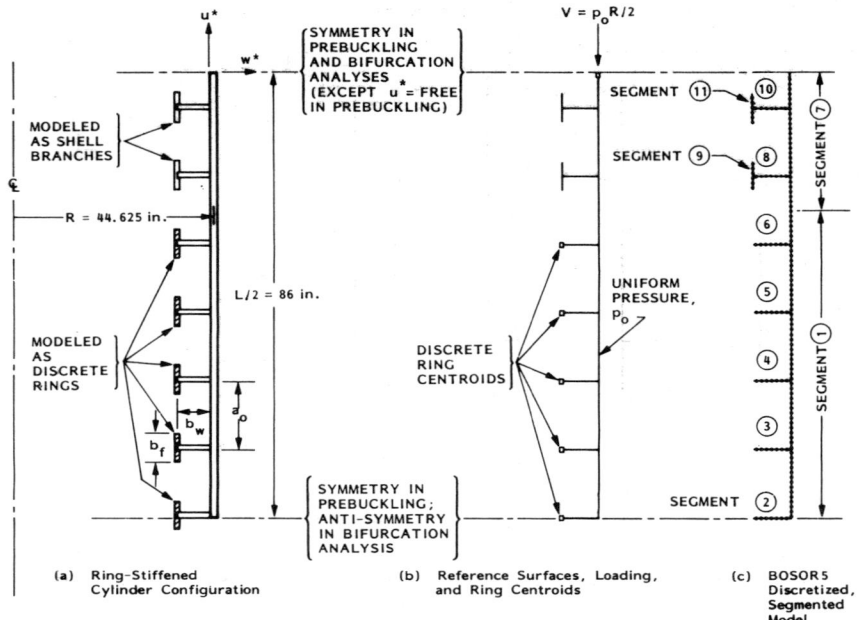

Fig. 7. Hydrostatically compressed, internally ring stiffened cylindrical shells: Modelling strategy for BOSOR5 analyses of the minimum weight designs obtained by PANDA for design pressures p_O ranging from p_O = 677 psi to p_O = 4743 psi. Dimensions are listed in Table 5.

are modelled as discrete rings; the top two flanges are modelled as flexible shell branches. The stress-strain curve for the material appears in Fig. 9.

Figure 8 shows the nodal points in the discretized BOSOR5 models of the optimum designs corresponding to each of the design pressures p_O listed in Tables 5 and 6. Nodal points are concentrated in the portion of the structure nearest the plane of symmetry at the top of Fig. 7(c) in order to obtain converged buckling pressures for local shell, web, and flange buckling modes.

Figure 9 demonstrates that all of the optimum designs are stressed beyond the material proportional limit at design pressures p_O from 677 to 4743 psi. It is interesting that for optimum designs with p_O from 2710 to 4743 psi the effective plastic strains at the midsurface halfway between rings are close to the 0.2 per cent yield strain, a result obtained from a rather rigorous analysis that confirms the appropriateness of earlier engineering design practice.

COMPARISON OF PANDA AND BOSOR5 BUCKLING PRESSURE: As shown in Fig. 10, the PANDA and BOSOR5 results agree with each other within about 10 per cent. Agreement is even better if in PANDA the shell wall is modelled as if it consists of 5 layers. By doing this the user permits simulation of prebuckling meridional bending between rings. Please see reference 3 for further details.

Figure 11 displays four failure modes of the optimum design for the highest applied pressure, p_O = 4743 psi. The buckling modes shown here are more

PANDA: Interactive Program for Minimum Weight Design

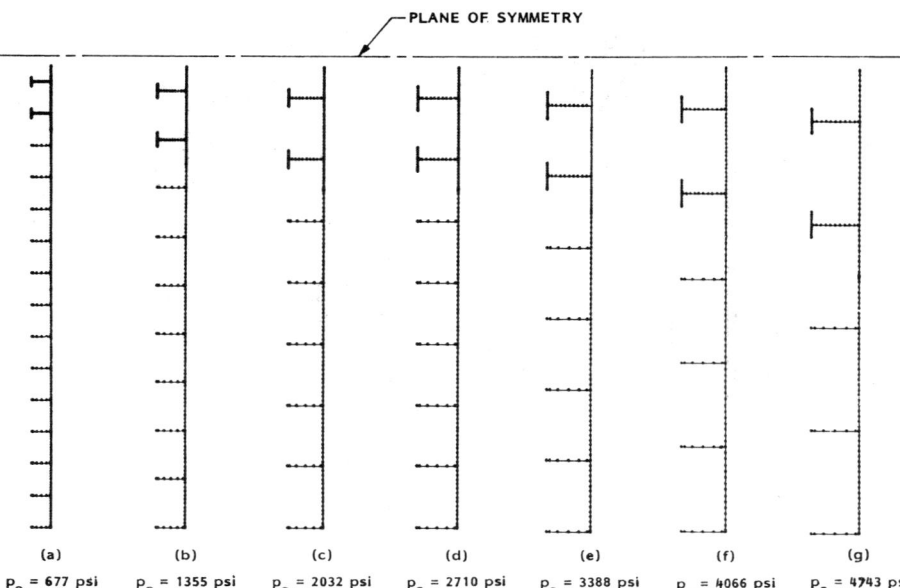

Fig. 8. Hydrostatically compressed, internally ring stiffened cylindrical shells: BOSOR5 discretized models corresponding to minimum weight designs obtained by PANDA for design pressures ranging from $p_O = 677$ psi to $p_O = 4743$ psi. Dimensions are listed in Table 5.

elaborate than those used in the PANDA analysis because there is obvious coupling between local deformation of the stiffeners, local deformation of the cylindrical shell wall, and general buckling of the structure. Even so, it is clear from Fig. 10 that the results obtained with PANDA are in very good agreement with those obtained with the more elaborate and rigorous BOSOR5. Therefore PANDA can profitably be used to obtain preliminary optimum designs, especially when one considers that the adequacy of these preliminary designs can always be checked by use of other more rigorous computer programs.

REFERENCES

1. Bushnell, D., Panel optimization with integrated software (POIS), Volume 1: PANDA-Interactive program for preliminary minimum weight design, Report No. AFWAL-TR-81-3073, Flight Dynamics Laboratory, Air Force Wright Aeronautical Laboratories, Wright Patterson AFB, Ohio (July, 1981).

2. Bushnell, D., PANDA-Interactive program for minimum weight design of stiffened cylindrical panels and shells, COMPUTERS AND STRUCTURES, Vol. 16, pp. 167-185, 1983.

3. Bushnell, D., PANDA-Interactive computer program for preliminary minimum weight design of composite or elastic-plastic, stiffened cylindrical panels and shells under combined in-plane loads, Proc. International Symposium on Optimum Structural Design, Oct. 19-22, 1981, Univ. of Arizona.

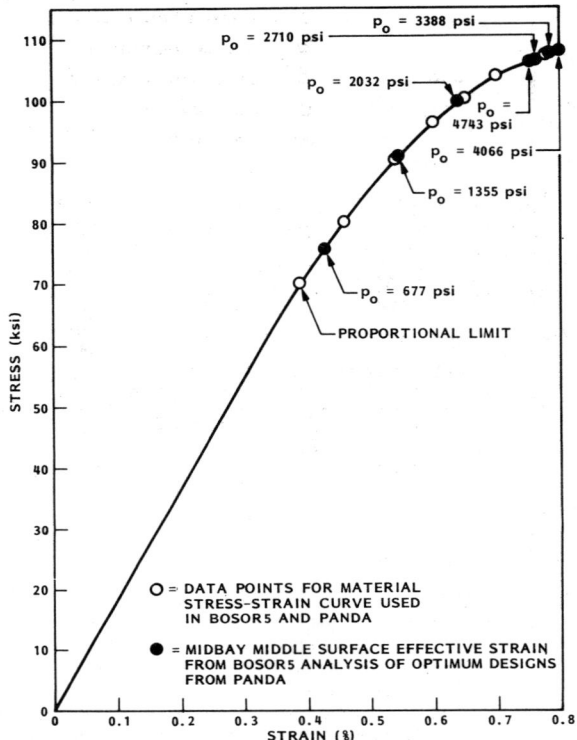

Fig. 9. Hydrostatically compressed, internally ring stiffened cylindrical shells: Midbay effective membrane strains at the design pressures for the optimized configurations displayed in the previous figure.

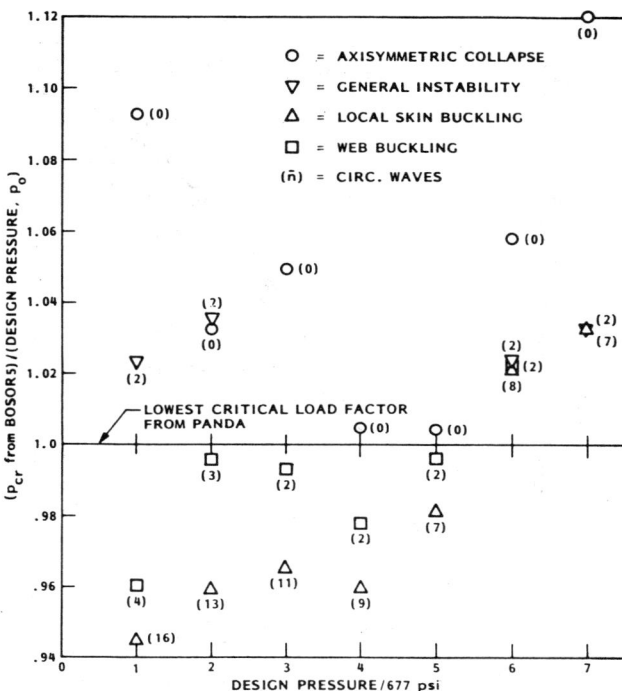

Fig. 10. Hydrostatically compressed, internally ring stiffened cylindrical shells: Comparison of buckling load factors obtained from PANDA and from BOSOR5. This is a direct comparison only because the critical load factors from PANDA are all very nearly unity, as seen from Table 6.

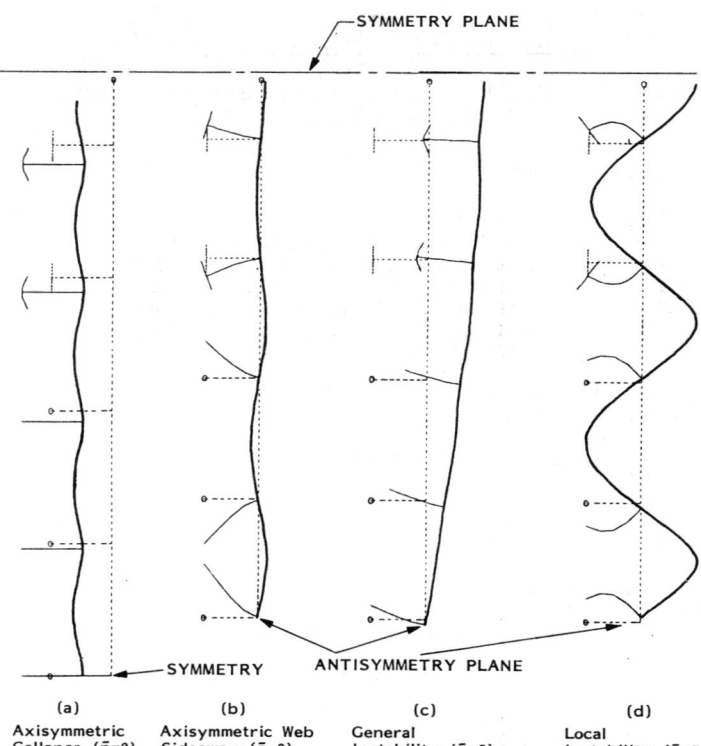

Fig. 11. (a) Prebuckling deformation near collapse, and (b,c,d) buckling modes predicted by BOSOR5 for the optimum design corresponding to the external hydrostatic pressure, p_o = 4743 psi.

PANDA: Interactive Program for Minimum Weight Design 191

APPENDIX

TABLE 3 Sample Case: Optimum Design of Steel Containment Vessel
BEGIN, DECIDE, and PANCON Processors are first run to get an
Optimum Design:
Then load interaction curves for perfect and imperfect optimized shells
are calculated and plotted.

```
$ RUN [BUSHNELL.PANDA]BEGIN

THE INPUT DATA PROVIDED BY YOU DURING EXECUTION
OF THIS PROGRAM ARE SAVED ON A FILE. PLEASE PROVIDE A NAME
FOR THIS FILE (9 CHARACTERS OR LESS ON VAX, 7 OR LESS ON CDC)
THIS FILE CAN BE USED AS INPUT TO "BEGIN" IN FUTURE RUNS OF
THE SAME OR A SIMILAR CASE. RATHER THAN ANSWER THE QUESTIONS
ASKED BY "BEGIN" INTERACTIVELY, YOU CAN EDIT THIS FILE AND
USE IT AS INPUT INSTEAD OF YOUR TERMINAL KEYBOARD.

PERMANENT FILE NAME =  CYLBEG
PERMANENT FILE NAME =CYLBEG    .DAT

IS PART OR ALL OF THE INPUT FOR THIS CASE STORED ON
THIS FILE YET?  N

PLEASE TYPE A SHORT TITLE ON THE NEXT LINE..
   STIFFENED CYLINDRICAL SHELL

DO YOU WANT MORE INFORMATION ABOUT PANDA (Y or N) ?  N
DO YOU WANT LONG PROMPTS?  N
DO YOU WANT A CHANCE TO CORRECT EACH INPUT DATUM
RIGHT AFTER YOU PROVIDE THAT DATUM? (This feature involves an
extra YES or NO question and answer for each datum. May get
somewhat tiresome for multilayered wall construction.)?  N
IS THE PANEL FLAT  (YES OR NO)?  N
AXIAL RESULTANT, NX =  24000
IS THE AXIAL LOAD COMPRESSIVE...(Y or N) ?  Y
NX =-0.240E+05

CIRCUMFERENTIAL RESULTANT, NY = 4000
IS THE CIRCUMFERENTIAL LOAD COMPRESSIVE...(Y or N) ?  N
NY = 0.400E+04

IN-PLANE SHEAR RESULTANT, NXY = 6300
NXY = 0.630E+04

DO YOU WANT TO READ VALUES FOR AXIAL RESULTANT AND
HOOP RESULTANT  (NXFIXED, NYFIXED) WHICH ARE NOT BUCKLING
PARAMETERS BUT REPRESENT A FIXED PRESTRESSED STATE...
(THE USUAL ANSWER IS NO) ?  N

CYLINDRICAL PANEL RADIUS OF CURVATURE= 720
AXIAL LENGTH OF PANEL = 840
CIRCUMFERENTIAL LENGTH OF PANEL =   2262

NUMBER OF LAYERS IN PANEL SKIN=  1
PANEL SKIN LAYER THICKNESSES (INNER LAYER IS FIRST)
TLAYER( 1) =  0.875
PANEL SKIN LAYER WINDING ANGLES (DEGREES)
ALPHA( 1) =  0.000E+00
PANEL SKIN LAYER MATERIAL TYPES (1, 2, 3, ...)
MATL( 1) =    1
```

```
IS THE PANEL STIFFENED... (YES OR NO) ?  Y
DOES THE PANEL HAVE STRINGERS (AXIAL STIFFENERS) ?  Y
NEXT PROVIDE INPUT DATA FOR STRINGERS...
ARC LENGTH, BO, BETWEEN ADJACENT STRINGERS = 37.7
YOU MUST NOW SPECIFY WHETHER THE STRINGERS ARE INTERNAL OR EXTERNAL.
ARE THE STRINGERS INTERNAL?  N
WHAT TYPE OF CROSS SECTION DO THE STRINGERS HAVE?

CHOOSE ONE OF THE FOLLOWING...
RECTANGULAR (please type the letter R)
T-SHAPED (FLANGE AWAY FROM PANEL SKIN) (type the letter T)
INVERTED T (FLANGE NEXT TO PANEL SKIN)  (type T2)
Z-SHAPED    (type the letter Z)
I-SHAPED    (type the letter I)
[-SHAPED (CHANNEL)     (please type C)
L-SHAPED (ANGLE) (FLANGE AWAY FROM SKIN) (please type L)
INVERTED L (ANGLE) (FLANGE NEXT TO SKIN) (please type L2)
OTHER    (please type the word OTHER)

R
STRINGERS CROSS SECTION INDICATOR=R

                                I
                                I
                                I.....STIF. SEG. NO. 1
                                I
            SHELL SKIN-.        I
                                I
                         .      I
                                I
            ------------------------------------------

WHILE PROVIDING THE FOLLOWING INPUT DATA, PLEASE REFER
TO THE SKETCH ABOVE FOR A DIAGRAM OF THE STRINGER CROSS SEC-
TION AND THE STRINGER SEGMENT NUMBERING SCHEME.

ARE ANY OF THE STRINGER SEGMENTS MADE OF
LAYERED OR COMPOSITE MATERIAL?  N
STRINGER SEGMENT THICKNESSES
TS( 1) =    1.00
STRINGER SEGMENT HEIGHTS (NOT THICKNESS)
BS( 1) =    6.00
STRINGER SEGMENT MATERIAL TYPES (1, 2, 3,...)
MATL( 1) =    1

REVIEW OF STRINGER INPUT DATA FOR 1 STRINGER SEGS.
SEG.  BS(I)      ANGLES(I)    THICKNESS    CONNECT.   FREE END   MATL TYPE
 1   6.00E+00    0.00E+00     1.00E+00        0         YES          1

DO YOU WANT ANOTHER CHANCE TO PROVIDE STRINGER INPUT DATA(YES OR NO)?  N

DOES THE PANEL HAVE RINGS ?   Y
NEXT PROVIDE NUMERICAL INPUT DATA FOR RINGS...
SPACING, A, BETWEEN ADJACENT RINGS = 120
YOU MUST NOW SPECIFY WHETHER THE    RINGS    ARE INTERNAL OR EXTERNAL.
ARE THE    RINGS    INTERNAL?  N
WHAT TYPE OF CROSS SECTION DO THE    RINGS    HAVE?  R
ARE ANY OF THE RING SEGMENTS MADE OF
LAYERED OR COMPOSITE MATERIAL?  N
RING SEGMENT THICKNESSES
TR( 1) =    1.00
RING SEGMENT HEIGHT (NOT THICKNESS)
```

PANDA: Interactive Program for Minimum Weight Design   193

```
BR( 1) =    18.0
RING SEGMENT MATERIAL TYPES (1, 2, 3,...)
MATL( 1) =    1


   REVIEW OF RING INPUT DATA FOR 1 RING SEGMENTS
SEG.  BR(I)   ANGLES(I)  THICKNESS   CONNECT.  FREE END  MATL TYPE
 1   1.80E+01  0.00E+00   1.00E+00       0        YES        1

DO YOU WANT ANOTHER CHANCE TO PROVIDE RING INPUT DATA (YES OR NO)? N

NEXT PROVIDE MATERIAL PROPERTIES...
NUMBER OF DIFFERENT MATERIALS SPECIFIED = 1
MATERIAL PROPERTIES FOR MATERIAL TYPE 1
MODULI E1, E2, AND G; POISSONS RATIO NU; AND
DENSITY RHO ARE TO BE PROVIDED NOW FOR MATERIAL TYPE NO.  1

MODULUS E1 IN MATL FIBER DIRECTION OR STIFFENER AXIS= 27900000.
IS THIS MATERIALISOTROPIC?  Y
POISSONS RATIO NU = 0.3
WEIGHT DENSITY (e.g. LB/CUBIC INCH), RHO = 0.3
DO YOU WANT TO PROVIDE A STRESS-STRAIN CURVE FOR THIS MATERIAL?  N
FOR THIS MATERIAL YOU MAY SPECIFY EITHER A SINGLE
MAXIMUM ALLOWABLE EFFECTIVE STRESS   OR  FIVE ULTIMATE
STRAIN COMPONENTS.
DO YOU WANT TO SPECIFY A SINGLE MAXIMUM ALLOWABLE
EFFECTIVE STRESS...(Y or N)?  Y
MAXIMUM ALLOWABLE EFFECTIVE STRESS = 76000.

DO YOU WANT ANOTHER CHANCE TO PROVIDE MATERIAL
PROPERTIES FOR THIS MATERIAL TYPE?  N

IS THE RATIO PHI=(LOCAL/GENERAL) BUCKLING LOAD
DIFFERENT FROM UNITY? (ANSWER IS USUALLY NO)?  N

NEXT, RUN DECIDE

$ RUN [BUSHNELL.PANDA]DECIDE
   STIFFENED CYLINDRICAL SHELL

INPUT DATA SUPPLIED INTERACTIVELY BY YOU IN THIS RUN
ARE STORED ON A FILE.  THIS FILE CAN BE USED AS INPUT FOR
FUTURE RUNS OF "DECIDE".  ITS NAME MUST BE DIFFERENT FROM
THE NAME OF THE FILE USED FOR "BEGIN" INPUT DATA.

WHAT IS THE NAME OF THIS FILE?  CYLDEC
PERMANENT FILE NAME =CYLDEC     .DAT

IS PART OR ALL OF THE INPUT FOR THIS CASE STORED ON
THIS FILE YET?  N

DO YOU WISH TO CHANGE THE TITLE OF THIS CASE?  N
DO YOU WANT MORE INFORMATION ON THIS PROGRAM?  Y

    THE FOLLOWING TERMS ARE USED IN THIS PROGRAM, AND
IT IS VERY IMPORTANT THAT YOU UNDERSTAND THEIR MEANING...

******************************************************
******************************************************
```

(1) DESIGN PARAMETER-- ANY STRUCTURAL DIMENSION OR WINDING ANGLE.

(2) DECISION VARIABLE- A DESIGN PARAMETER WHICH HAS BEEN SELECTED AS A PRIMARY VARIABLE IN THE OPTIMIZATION PROCESS.

(3) LINKED VARIABLE-- A DESIGN PARAMETER WHICH HAS BEEN SELECTED AS A SECONDARY VARIABLE IN THE OPTIMIZATION PROCESS. A LINKED VARIABLE IS PROPORTIONAL TO ONE OF THE DECISION VARIABLES, BUT IS NOT A MEMBER OF THE VECTOR OF DECISION VARIABLES, (X(I),I=1,NDV).

****
****

DO YOU WANT MORE INFORMATION? Y

FIRST YOU DECIDE WHICH OF THE DESIGN PARAMETERS (DIMENSIONS AND WINDING ANGLES) ARE TO BE DECISION VARIABLES IN THE OPTIMIZATION PROCESS. YOU CAN SELECT AS DECISION VARIABLES ANY SUBSET OF DESIGN PARAMETERS YOU WISH. OTHER DESIGN PARAMETERS CAN BE SELECTED AS BEING LINKED TO DECISION VARIABLES. A LINKED VARIABLE Y WILL VARY IN PROPORTION TO A USER-SELECTED DECISION VARIABLE X, AS FOLLOWS...

Y(A LINKED VARIABLE) = C * X(A DECISION VARIABLE)

IN WHICH YOU CHOOSE THE CONSTANT OF PROPORTIONALITY C AND WHICH DECISION VARIABLE X THAT Y IS LINKED TO. THE OPTIMUM DESIGN IS OBTAINED ITERATIVELY BY THE METHOD OF FEASIBLE DIRECTIONS (VANDERPLAATS). SEE THE AFFDL REPORT AFWAL-TR-81-3073, VOL. I AND ITS REFERENCES FOR FURTHER DETAILS.

DO YOU WISH TO CHANGE THE LOADS OR THE FACTOR PHI? N
DO YOU WANT TO MAKE THE PANEL FLAT? N

YOU CAN DO THREE TYPES OF ANALYSIS WITH PANDA:

    1   AN OPTIMIZATION ANALYSIS;
    2   A SIMPLE BUCKLING ANALYSIS (FIXED DESIGN) WITH A SINGLE SET OF IN-PLANE LOADS, Nx, Ny, Nxy;
    3   FIXED DESIGN BUCKLING ANALYSIS TO GET INTERACTION CURVE FOR CRITICAL IN-PLANE LOAD COMBINATIONS.

PLEASE RESPOND WITH 1 or 2 or 3.
1

DO YOU WISH TO CHANGE SOME OF THE STRUCTURAL DIMENSIONS OR WINDING ANGLES? N

YOU CAN DO BUCKLING OR OPTIMIZATION ANALYSES EITHER

    (1) NEGLECTING
        or
    (2) INCLUDING

PANDA: Interactive Program for Minimum Weight Design

THE EFFECT OF TRANSVERSE SHEAR DEFORMATION IN THE PLANES OF
THE STIFFENER WEBS FOR PREDICTION OF GENERAL OR PANEL (SEMI-
GENERAL) INSTABILITY MODES (MODES IN WHICH STIFFENERS ARE
SMEARED OUT). PLEASE NOTE THAT CHOICE (2) INVOLVES MUCH MORE
COMPUTER TIME THAN DOES CHOICE (1). USE (2) SPARINGLY FOR
OPTIMIZATION RUNS OR RUNS FOR CALCULATION OF INTERACTION
CURVES, AND THEN ONLY AFTER YOU HAVE FIRST PERFORMED THESE
TYPES OF ANALYSIS WITH USE OF CHOICE (1).

NOW PLEASE CHOOSE OPTION (1) OR (2) BY TYPING 1 OR 2:   1

NOTE... DEFAULT VALUES FOR LOWER AND UPPER BOUNDS OF A
DECISION VARIABLE X ARE...

       LOWER BOUND = X/100
       UPPER BOUND = X*100

IN WHICH X IS THE CURRENT VALUE OF THE DECISION VARIABLE.

ARE ANY OF THE SHELL WALL LAYER THICKNESSES DECISION VARIABLES?   Y
DO YOU WANT TO USE DEFAULT VALUES FOR LOWER AND UPPER BOUNDS
OF SHELL WALL LAYER THICKNESSES?   N
LOWER BOUND FOR THICKNESS OF LAYER NO. 1 =   .5
UPPER BOUND FOR THICKNESS OF LAYER NO. 1 =   2
ARE ANY OF THE LAYER WINDING ANGLES DECISION VARIABLES?   N
IS THE STRINGER SPACING A DECISION VARIABLE?   Y
DO YOU WANT TO USE DEFAULT VALUES FOR LOWER AND UPPER BOUNDS
OF STRINGER SPACING?   N
LOWER BOUND FOR STRINGER SPACING=   20
UPPER BOUND FOR STRINGER SPACING=   200
ARE ANY OF THE STRINGER SEGMENT THICKNESSES DECISION VARIABLES?   Y
DO YOU WANT TO USE DEFAULT VALUES FOR LOWER AND UPPER BOUNDS
OF STRINGER SEGMENT THICKNESSES?   N
LOWER BOUND FOR THICKNESS OF STRINGER SEGMENT   1=   .2
UPPER BOUND FOR THICKNESS OF STRINGER SEGMENT   1=   2
ARE ANY OF THE STRINGER SEGMENT HEIGHTS DECISION VARIABLES?   Y
DO YOU WANT TO USE DEFAULT VALUES FOR LOWER AND UPPER BOUNDS
OF STRINGER SEGMENT HEIGHTS?   Y
(LOWER,UPPER) BOUNDS FOR HEIGHTS OF STRINGER SEG. 1=
(6.000E-02 ,   6.000E+02)

IS THE   RING   SPACING A DECISION VARIABLE?   Y
DO YOU WANT TO USE DEFAULT VALUES FOR LOWER AND UPPER BOUNDS
OF RING SPACING?   N
LOWER BOUND FOR   RING   SPACING=   50
UPPER BOUND FOR   RING   SPACING=   300
ARE ANY OF THE   RING   SEGMENT THICKNESSES DECISION VARIABLES?   Y
DO YOU WANT TO USE DEFAULT VALUES FOR LOWER AND UPPER BOUNDS
OF RING SEGMENT THICKNESSES?   Y
(LOWER,UPPER) BOUNDS FOR THICKNESS OF RING SEGMENT   1=
(1.000E-02 ,   1.000E+02)
ARE ANY OF THE RING SEG. HEIGHTS DECISION VARIABLES?   Y
DO YOU WANT TO USE DEFAULT VALUES FOR LOWER AND UPPER BOUNDS
OF RING SEGMENT HEIGHTS?   Y
(LOWER,UPPER) BOUNDS FOR HEIGHTS OF RING SEG. 1=
(1.800E-01 ,   1.800E+03)

196   D. Bushnell

NEXT.   RUN PANCON

$ RUN [BUSHNELL.PANDA]PANCON

    STIFFENED CYLINDRICAL SHELL

PANEL WEIGHT=   6.7499E+05
PANEL WEIGHT=   4.0523E+05
PANEL WEIGHT=   3.4435E+05
PANEL WEIGHT=   3.3100E+05
PANEL WEIGHT=   3.2743E+05
PANEL WEIGHT=   3.3839E+05

DO YOU WISH TO PRINT OUT A SUMMARY OF DESIGN INFORMATION? N
DO YOU WISH TO DO MORE ITERATIONS WITH THE SAME DECISION
VARIABLES? Y

PANEL WEIGHT=   3.3839E+05
PANEL WEIGHT=   3.2169E+05
PANEL WEIGHT=   3.2583E+05
PANEL WEIGHT=   3.1850E+05
PANEL WEIGHT=   3.1849E+05
PANEL WEIGHT=   3.2168E+05
PANEL WEIGHT=   3.1793E+05

DO YOU WISH TO PRINT OUT A SUMMARY OF DESIGN INFORMATION? N
DO YOU WISH TO DO MORE ITERATIONS WITH THE SAME DECISION
VARIABLES? Y

PANEL WEIGHT=   3.1793E+05
PANEL WEIGHT=   3.1822E+05
PANEL WEIGHT=   3.2140E+05
PANEL WEIGHT=   3.1858E+05
PANEL WEIGHT=   3.2177E+05
PANEL WEIGHT=   3.1802E+05
PANEL WEIGHT=   3.2120E+05
PANEL WEIGHT=   3.1853E+05
PANEL WEIGHT=   3.1823E+05

DO YOU WISH TO PRINT OUT A SUMMARY OF DESIGN INFORMATION? Y

    STIFFENED CYLINDRICAL SHELL
****
PANEL WEIGHT =   3.1823E+05

AXIAL LOAD/(LENGTH OF CIRCUMFERENCE), NX = -2.4000E+04
CIRC. LOAD/(LENGTH OF GENERATOR),   NY =   4.0000E+03
IN-PLANE SHEAR STRESS RESULTANT,   NXY =   6.3000E+03

FIXED AXIAL STRESS RESULTANT,   NXFIXED =   0.0000E+00
FIXED CIRC. STRESS RESULTANT,   NYFIXED =   0.0000E+00

DECISION VARIABLES FOLLOW...
THICKNESS OF PANEL SKIN LAYER   1 =   5.0000E-01
STRINGER SPACING, B =   2.5946E+01
THICKNESS OF STRINGER SEGMENT   1 =   2.6833E-01
HEIGHT,BS(I), OF STRINGER SEG. 1 =   4.3072E+00
RING SPACING, A =   7.8736E+01
THICKNESS OF RING SEGMENT   1 =   7.0231E-02
HEIGHT,BR(I), OF RING SEG. 1 =   1.5387E+01

PANDA: Interactive Program for Minimum Weight Design 197

```
LOAD FACTORS FOR VARIOUS TYPES OF BUCKLING FOLLOW...
     (M=AXIAL, N=CIRC.) HALF-WAVES OVER ENTIRE PANEL (AXIAL,CIRC.)
GENERAL INSTABILITY LOAD FACTOR (M,N)=         1.0035E+00(   1,  14)
LOCAL SKIN BUCKLING LOAD FACTOR (M,N)=         1.0028E+00(  42,  87)
BUCKLING BETWEEN RINGS WITH SMEARD STRINGRS=   1.0116E+00(  10,  25)
BUCKLING BETWEEN STRINGRS WITH SMEARD RINGS=   4.8029E+00( 102,  87)
LOAD FACTOR (M) FOR STRINGER SEGMENT NO. 1=    1.0046E+00(  42)
LOCAL ROLLING WITH SKIN BUCKLING BETWN STIF=   1.0183E+00(  42,  87)
ROLLING OF STRINGERS(M,O), NO SKIN BUCKLING=   2.8531E+00( 114,   0)
AXISYMMETRIC ROLLING OF RINGS, NO SKIN BUCK=   9.8531E-01(   0,   0)
BUCKLING(ROLLING) MODE WITH SMEARD STRINGRS=   1.0128E+00(  10,  25)
BUCKLING(ROLLING) MODE WITH SMEARED RINGS  =   4.7815E+00( 102,  87)

LOAD FACTORS REQUIRED TO PRODUCE CRITICAL STRESSES OR STRAINS..
MARGIN FOR EFFECTIVE STRESS IN LAYER 1 =      1.439E+00
MARGIN FOR COMP. STRAIN IN STRINGER SEG. 1=   1.648E+00
***********************************************************


DO YOU WISH TO DO MORE ITERATIONS WITH THE SAME DECISION
VARIABLES? N
DO YOU WANT TO PRINT OUT MORE INFORMATION ABOUT THE
CURRENT DESIGN? Y

   STIFFENED CYLINDRICAL SHELL

CURRENT DESIGN FOLLOWS....
  PANEL RADIUS OF CURVATURE   =   7.200E+02
  PANEL AXIAL, CIRC. LENGTHS  =   8.400E+02   2.262E+03
  PANEL AXIAL PRESTRESS, NXP  =   0.000E+00 (NOT AN EIGENVAL.)
  PANEL CIRC. PRESTRESS, NYP  =   0.000E+00 (NOT AN EIGENVAL.)
  PANEL TOTAL THICKNESS       =   5.000E-01
  PANEL LAYER THICKNESSES     =   5.000E-01
  PANEL LAYER WINDING ANGLES  =   0.000E+00
  PANEL LAYER MATERIAL TYPES  =   1

STRINGER DESIGN DATA FOLLOWS...
  STRINGER SPACING, B            =   2.595E+01
  STRINGER SEGMENT THICKNESSES   =   2.683E-01
  STRINGER SEGMENT HEIGHTS       =   4.307E+00
  STRINGER SEGMENT ANGLES        =   0.000E+00
  STRINGER SEG. MATERIAL TYPES   =   1

RING DESIGN DATA FOLLOWS...
  RING SPACING, A                =   7.874E+01
  RING SEGMENT THICKNESSES       =   7.023E-02
  RING SEGMENT HEIGHTS           =   1.539E+01
  RING SEGMENT ANGLES            =   0.000E+00
  RING SEGMENT MATERIAL TYPES    =   1
  MODULI (E1, E2, G) OF MATL 1 = 2.790E+07   2.790E+07   1.073E+07
  POISSON RAT., DENSITY MATL 1 = 3.000E-01   3.000E-01

  THE TOTAL PANEL WEIGHT       =   3.182E+05
  THE PANEL SKIN WEIGHT        =   2.850E+05
  THE TOTAL STIFFENER WEIGHT   =   3.321E+04

GENERAL INSTABILITY QUANTITIES (STIFFENERS SMEARED OUT)...
  AXIAL STRESS RESULTANT, NX   =  -2.400E+04 (AN EIGENPARAM.)
  CIRC. STRESS RESULTANT, NY   =   4.000E+03 (AN EIGENPARAM.)
  SHEAR STRESS RESULTANT, NXY  =   6.300E+03 (AN EIGENPARAM.)
  AXIAL HALF WAVES OVER PANEL  =   1
```

```
CIRC. HALF WAVES OVER PANEL   =   14
SLOPE OF BUCKLING NODAL LINES=   5.900E-01
GENERAL INSTABILITY MULTIPLIER=  1.004E+00    (EIGENVALUE)

LOCAL INSTABILITY QUANTITIES (BUCKLING BETWEEN STIFFENERS)...
AXIAL RESULTANT IN SKIN      =  -2.195E+04  (AN EIGENPARAM.)
CIRC. RESULTANT IN SKIN      =   3.784E+03  (AN EIGENPARAM.)
SHEAR RESULTANT IN SKIN      =   6.300E+03  (AN EIGENPARAM.)
AXIAL HALF WAVES BETWEEN RINGS=  4
CIRC. HALF WAVES BET STRINGERS=  1
SLOPE OF BUCKLING NODAL LINES =  1.320E-01
LOCAL INSTABILITY MULTIPLIER  =  1.003E+00    (EIGENVALUE)

INSTABILITY WITH SMEARED STRINGERS, BETWEEN RINGS...
AXIAL RESULTANT IN PANEL     =  -2.400E+04  (AN EIGENPARAM.)
CIRC. RESULTANT IN SKIN      =   3.784E+03  (AN EIGENPARAM.)
SHEAR RESULTANT IN SKIN      =   6.300E+03  (AN EIGENPARAM.)
AXIAL HALF WAVES BETWEEN RINGS=  1
CIRC. HALF WAVES OVER PANEL  =   25
SLOPE OF BUCKLING NODAL LINES =  1.091E+01
SMEARED STRINGER EIGENVALUE  =   1.012E+00   (EIGENVALUE)

INSTABILITY WITH SMEARED RINGS, BETWEEN STRINGERS...
AXIAL RESULTANT IN SKIN      =  -2.195E+04  (AN EIGENPARAM.)
CIRC. RESULTANT IN PANEL     =   4.000E+03  (AN EIGENPARAM.)
SHEAR RESULTANT IN SKIN      =   6.300E+03  (AN EIGENPARAM.)
AXIAL HALF WAVES OVER PANEL  =   102
CIRC. HALF WAVES BET STRINGERS=  1
SLOPE OF BUCKLING NODAL LINES =  1.000E-02
SMEARED   RING   EIGENVALUE  =   4.803E+00   (EIGENVALUE)

AVERAGE AXIAL STRAIN         =  -1.653E-03
AVERAGE MIDBAY HOOP STRAIN   =   7.432E-04
AVERAGE SHEAR STRAIN         =   1.174E-03
FIBER STRAINS AT EACH LAYER CENTER=
                                -1.653E-03
STRAINS NORMAL TO FIBERS IN LAYERS=
                                 7.432E-04
SHEAR STRAINS IN MATERIAL COORDS =
                                 1.174E-03
EFFECTIVE STRAINS AT LAYER CENTERS =
                                 1.640E-03
AVERAGE STRAIN IN RINGS      =   7.254E-04
RADIUS(CYL.)/RADIUS(RING C.G.)=  9.894E-01
PRESTRESS(LB/IN) IN STRING.SEGS=  0.000E+00
EIGENSTRESS IN STRINGER SEGS. =  -1.238E+04
PRESTRESS(LB/IN) IN RING SEGS. =  0.000E+00
EIGENSTRESS IN   RING   SEGS. =  1.421E+03

LOCAL ROLLING MODE EIGENVALUE =  1.018E+00
CORRESPONDING TO   4 AXIAL HALF WAVES BETWEEN RINGS AND
CORRESPONDING TO   1 CIRC. HALF WAVES BETWEEN STRINGERS

SMEARED STRINGER ROLLING MODE EIGENVALUE=   1.013E+00
CORRESPONDING TO   1 AXIAL HALF WAVES BETWEEN RINGS AND
CORRESPONDING TO   25 CIRC. HALF WAVES OVER PANEL

SMEARED   RING   ROLLING MODE EIGENVALUE=   4.781E+00
CORRESPONDING TO  102 AXIAL HALF WAVES OVER PANEL AND
CORRESPONDING TO    1 CIRC. HALF WAVES BETWEEN STRINGERS
```

PANDA: Interactive Program for Minimum Weight Design   199

```
STRINGER ROLLING MODE
 (NO BUCKLING OF SKIN)       =   2.853E+00
 NUMBER OF AXIAL HALF WAVES =   114


AXISYMMETRIC(N=0) RING WEB BUCKLING=  9.8531E-01

DO YOU WANT TO STORE OUTPUT DESIGN INFORMATION ON A
PERMANENT FILE?  Y
 NAME OF PERMANENT FILE =  CYLOUT
PERMANENT FILE NAME =CYLOUT    .DAT
DO YOU WISH TO PRINT OUT A SUMMARY OF DESIGN INFORMATION?  Y
DO YOU WANT TO PRINT OUT MORE INFORMATION ABOUT THE
CURRENT DESIGN?  Y
```

****

Next, the following things are done:

1. BEGIN is run again, with the initial design being the final design listed above. However, this time the yield stress of the material is set equal to its real value, 38000 psi. The objective in this investigation is to determine the behaviour under many combinations of axial load Nx and in-plane shear load Nxy in the presence of a constant internal pressure that generates a circumferential tension of Ny = 4000 lb/in. The effects of initial imperfections and plasticity on the (Nx - Nxy) interaction curves governing stress, general instability, local instability, buckling between rings with smeared stringers, and buckling between stringers with smeared rings, are to be determined and are to be plotted. We use the actual yield stress in this investigation rather than twice the yield stress because we are no longer designing a perfect structure here, thereby necessitating the use of artificially high loads (to compensate for unknown imperfection effects) and correspondingly artificially high yield stress (to compensate for the artificially high loads), but we are evaluating an actual structure with expected loads (times a factor of safety of two to compensate for uncertainty in the loads).

2. DECIDE is run after completion of BEGIN. A list of the short interactive session in DECIDE follows this brief discussion.

3. PANCON is run after completion of DECIDE. Some output from the PANCON run appears on the screen (not reproduced here in order to save space), and most of the output is saved on a file. (This file is not reproduced here in order to save space.)

4. PLOTIT is run after completion of PANCON. This program causes to be plotted several interaction curves for (Nx - Nxy) combinations that cause buckling and yielding of both perfect and imperfect shells. The theory for the effect of imperfections is a modified version of that presented in ASME Code Case N-284 by C. D. Miller and J. Tsai.

****

```
LIST FROM EXECUTION OF DECIDE.....

   STIFFENED CYLINDRICAL SHELL

INPUT DATA SUPPLIED INTERACTIVELY BY YOU IN THIS RUN
ARE STORED ON A FILE.  THIS FILE CAN BE USED AS INPUT FOR
FUTURE RUNS OF "DECIDE".  ITS NAME MUST BE DIFFERENT FROM
THE NAME OF THE FILE USED FOR "BEGIN" INPUT DATA.

WHAT IS THE NAME OF THIS FILE?   IMPDEC
PERMANENT FILE NAME =IMPDEC   .DAT
```

```
IS PART OR ALL OF THE INPUT FOR THIS CASE STORED ON
THIS FILE YET? N
DO YOU WISH TO CHANGE THE TITLE OF THIS CASE? N
DO YOU WANT MORE INFORMATION ON THIS PROGRAM? N
DO YOU WISH TO CHANGE THE LOADS OR THE FACTOR PHI? N
DO YOU WANT TO MAKE THE PANEL FLAT? N

YOU CAN DO THREE TYPES OF ANALYSIS WITH PANDA:

    1   AN OPTIMIZATION ANALYSIS;
    2   A SIMPLE BUCKLING ANALYSIS (FIXED DESIGN) WITH A
        SINGLE SET OF IN-PLANE LOADS, Nx, Ny, Nxy;
    3   FIXED DESIGN BUCKLING ANALYSIS TO GET INTERACTION
        CURVE FOR CRITICAL IN-PLANE LOAD COMBINATIONS.

PLEASE RESPOND WITH  1  or  2  or  3.

    3

YOU CAN OBTAIN ANY OF THREE TYPES OF INTERACTION:

    1   (Nx - Ny) INTERACTION     (Nxy IS FIXED)
    2   (Nx - Nxy) INTERACTION    (Ny  IS FIXED)
    3   (Ny - Nxy) INTERACTION    (Nx  IS FIXED)

PLEASE RESPOND WITH  1  or  2  or  3  . . .

    2

FIXED CIRCUMFERENTIAL RESULTANT, Ny = 4000

OUT OF ROUNDNESS...PLEASE PROVIDE THE DIFFERENCE
BETWEEN THE MAXIMUM DIAMETER AND MINIMUM DIAMETER OF THE
IMPERFECT SHELL.  (USE ZERO UNLESS YOU ARE ANALYZING
CYLINDRICAL SHELLS WITH ASME CODE CASE N-284)  15.0

DO YOU WISH TO CHANGE SOME OF THE STRUCTURAL DIMENSIONS
OR WINDING ANGLES?  N

YOU CAN DO BUCKLING OR OPTIMIZATION ANALYSES EITHER

        (1)  NEGLECTING
              or
        (2)  INCLUDING

THE EFFECT OF TRANSVERSE SHEAR DEFORMATION IN THE PLANES OF
THE STIFFENER WEBS FOR PREDICTION OF GENERAL OR PANEL (SEMI-
GENERAL) INSTABILITY MODES (MODES IN WHICH STIFFENERS ARE
SMEARED OUT).  PLEASE NOTE THAT CHOICE (2) INVOLVES MUCH MORE
COMPUTER TIME THAN DOES CHOICE (1).  USE (2) SPARINGLY FOR
OPTIMIZATION RUNS OR RUNS FOR CALCULATION OF INTERACTION
CURVES, AND THEN ONLY AFTER YOU HAVE FIRST PERFORMED THESE

TYPES OF ANALYSIS WITH USE OF CHOICE (1).

NOW PLEASE CHOOSE OPTION (1) OR (2) BY TYPING 1 OR 2:  1

NEXT,   RUN PANCON

    RUN [BUSHNELL.PANDA]PANCON
```

(PANCON generates output on the screen, not shown here, and other list output, also not shown here.)

NEXT,   RUN PLOTIT (to get plots of interaction curves for perfect and
                    imperfect stiffened cylindrical shell.)

$ RUN [BUSHNELL.PANDA]PLOTIT

(PLOTIT produces *.PLV;* files that contain "plots" of the interaction
curves for the perfect and imperfect shells.)

S AND CM: MICROCOMPUTER PROGRAMS FOR STRUCTURAL AND CONTINUUM MECHANICS

C. T. F. Ross

Department of Mechanical Engineering, Portsmouth Polytechnic, U.K.

ABSTRACT

This package contains fourteen computer programs, which have been written in BASIC, for microcomputers. The programs include the static and vibration analysis of two and three-dimensional pin-jointed and rigid-jointed frames and continuous beams.

In addition to this, the package contains programs that can tackle the static analysis of thin-walled cones, grillages, bulkheads, plane stress and plane strain and two-dimensional field problems.

THEORETICAL BACKGROUND

For *static analysis*, the equation relating force and displacement is given by:

$$\{q\} = [K]\{u\} \tag{1}$$

where

$\{q\}$ = a vector of known nodal loads,

$[K]$ = the structural stiffness matrix,

$\{u\}$ = a vector of the displacements that have to be determined.

Solution of (1) can be achieved by inverting the suppressed form of $[K]$ and premultiplying it into (1) to give:

$$\{u\} = [K]^{-1}\{q\}. \tag{2}$$

Equation (2) is not very efficient, particularly for large problems and solution is better achieved by solving the simultaneous equations instead of inverting $[K]$. Furthermore, for many practical problems, $[K]$ is of banded form and considerable savings in space and time can be achieved by considering only those elements of $[K]$ that are contained within one of its half bandwidths.

For *vibration analysis*, the equation of motion is given by:

$$[K]\{u\} - [M]\{\ddot{u}\} = \{q(t)\}, \tag{3}$$

where

$\{\ddot{u}\}$ = a vector of accelerations,

$\{q(t)\}$ = a vector of forcing functions,

For free vibrations, $\{q(t)\} = 0$ and if simple harmonic motion is assumed,

$$([K] = \omega^2[M]) \{u\} = 0,$$

or

$$\left| [K] = \omega^2[M] \right| = 0, \tag{4}$$

where

$[M]$ = the mass matrix of the structure

ω = radian frequency.

Solution of (4) can be achieved through standard eigenvalue techniques, which in this case, was the Power method with Aitken's acceleration.

FIELDS OF APPLICATION

TRUSS. This program calculates the axial forces in the members of plane pin-jointed trusses, subjected to horizontal and vertical loads at the pin-joints. The members can have different sectional and material properties and they can be statically determinate or statically indeterminate. The displacements of the pin-joints are also calculated.

CONTINUOUS BEAMS. This program calculates nodal displacements and bending moments in continuous beams. The applied loads can be of various combinations of point, distributed and hydrostatic loading. Step variation is allowed for in the beam's sectional properties.

PLANE FRAME. This program calculates nodal displacements and bending moments in rigid-jointed plane frames under various combinations of concentrated, distributed and hydrostatic loading. Step variation is allowed for in the sectional properties of the members. The frames can be portal, skew or multi-bay, and there is no difficulty in allowing for sidesway. The program also determines any axial forces that are in the members of the frame.

SPACE TRUSS. This program calculates nodal displacements and member forces in three-dimensional pin-jointed trusses. The applied loads can also be three-dimensional and should be placed at the pin-joints. The members of the truss can have different sectional and material properties, and the trusses can be statically indeterminate.

VIBPTR. This program calculates natural frequencies and eigenmodes of plane pin-jointed trusses, with or without additional masses at the pin-joints. The sectional and material properties of the members of the truss can be different.

VIBCB. This program calculates the natural frequencies and eigenmodes of continuous beams, with or without additional masses at the nodes. Step variation is allowed for in the sectional properties of the beam.

VIBRGPF. This program calculates natural frequencies and eigenmodes of rigid-

jointed plane frames, with or without additional masses at the nodes. The
frames can be portal, skew or multibay, and step variation is allowed for in the
sectional properties of the members of the frame.

VIBSPTR. This program calculates natural frequencies and eigenmodes of three-
dimensional pin-jointed trusses, with or without additional masses at the pin-
joints. The sectional and material properties of the members can be different.

3D-RIGFRAME. This program can determine nodal displacements, axial forces and
bending moments in three-dimensional rigid-jointed frames under concentrated
loads. The frames can be made from members of different sectional and material
properties, and the boundary conditions can be quite complex.

GRILLAGE. This program can determine nodal displacements, bending moments and
torques in orthogonal or skew grids. The geometrical properties of the members
in one direction can be different to those in the other direction. The lateral
loading can be a combination of pressure loading, together with concentrated
loads.

BULKHEAD. This program is a more complex version of "GRILLAGE", in that the
loading can be hydrostatic and step variation is allowed for in the geometrical
properties of the members.

THINCONE. This program can determine axisymmetric nodal displacements and
stresses in thin-walled tapered cones under a uniform axial pressure and a
varying lateral pressure. By joining together a large number of truncated cones,
it is possible to analyse quite complex shapes, such as hour-glass types, domes,
etc.

PLANESTRESS AND PLANESTRAIN. This program can determine displacements and
stresses in plates under in-plane forces. As triangular elements are used, the
boundary shape can be quite complex so that holes and cut-outs can easily be
accommodated. Both plane stress and plane strain conditions can be considered.

LAPLACE. This program solves differential equations of the Laplace and Poisson
type, which appear in problems on the torsion of noncircular sections, steady-
state heat transfer, streamline flow, etc.

A triangular element is used, so that it is possible to analyse two-dimensional
figures with complex boundary shapes and conditions. Worked examples are given
of a torsion problem, a steady-state heat transfer problem and a 52 node sudden
enlargement problem.

PROGRAM DESCRIPTIONS

The method of solution adopted in all the fourteen programs was the finite
element method.

Two end nodes were used to define the skeletal elements and the axisymmetric
conical shell element, and three corner nodes were used to define the in-plane
triangular plate element and the triangular field problems' element. The plane
truss element, the continuous beams' element and the in-plane plate element had
two degrees of freedom per node, whilst the three-dimensional truss element
and the elements representing the grillages and bulkheads, together with the
axisymmetric conical shell element, had three degrees of freedom per node.

The three-dimensional frame element had six degrees of freedom per node, whilst
the field problems' element, had one degree of freedom per node.

All the programs were written in interactive BASIC for microcomputers and the
programs for the Apple II, included a graphical display on the screen.

HARDWARE

The minimum hardware requirements are a 16 k RAM microcomputer, with a TV/monitor display and a home tape recorder.

The programs are available on standard cassette tapes for all machines and on 5¼ inch floppy disks for the CBM PET/64, Apple II +/e, BBC and Tandy TRS 80 I/III.

EXAMPLES OF APPLICATION

TRUSS. This program can be used in trusses, such as shown in Fig. 1. These structures can be statically determinate or statically indeterminate with complex restraints, but the loads must be applied at the joints.

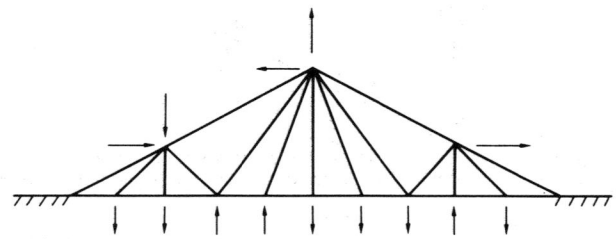

Fig. 1. Plane pin-jointed truss.

CONTINUOUS BEAMS. This program can be used in horizontal beams, such as that shown in Fig. 2. The beams can be statically determinate or statically indeterminate and the loading and supports can be quite complex.

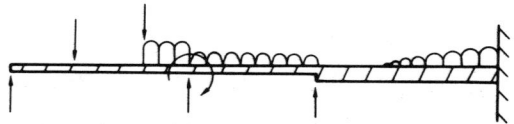

Fig. 2. Continuous beam.

PLANE FRAME. This program can be used in rigid-jointed plane frames, as shown in Fig. 3. The structures can be statically determinate or statically indeterminate and the applied loads, together with the supports, can be quite complex.

SPACETRUSS. This program can be used in three-dimensional pin-jointed trusses, of the type in Fig. 4. The loads must be applied at the joints.

VIBPTR. This program can be used for plane pin-jointed trusses, of the type shown in Fig. 1.

VIBCB. This program can be used in beams, such as shown in Fig. 2. These programs can have concentrated masses added to their nodes, which need not be at the joints.

VIBRGPF. This program can be used in rigid-jointed plane frames, such as shown in Fig. 3. These structures can have concentrated masses added to their nodes,

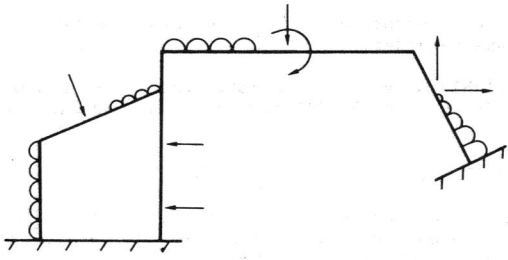

Fig. 3. Rigid-jointed plane frame.

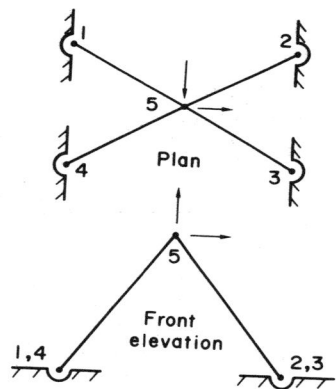

Fig. 4. Pin-jointed space truss.

which need not be at the joints.

VIBSPTR. This program can be used in three-dimensional pin-jointed trusses, such as shown in Fig. 4.

3D-RIGFRAME. This program can be used in three-dimensional rigid-jointed frames, such as shown in Fig. 5.

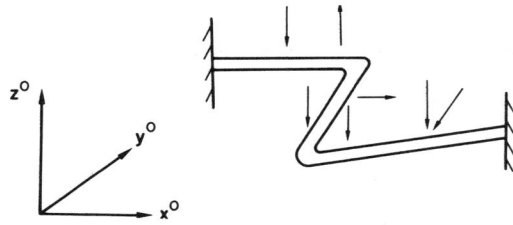

Fig. 5. Three-dimensional rigid-jointed frame.

These structures can be statically determinate or statically indeterminate and the members, which can have different geometrical properties, can be included in three dimensions. Members with unsymmetrical cross-sections, such as angle bars, channels, etc., can be catered for, in addition to those with symmetrical cross-sections.

GRILLAGE. This program can be used in orthogonal or skew grids, such as those shown in Fig. 6.

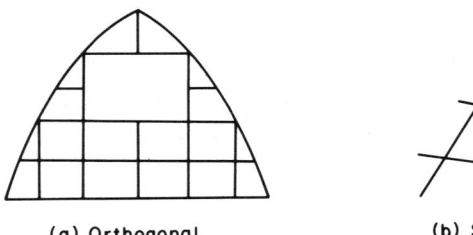

(a) Orthogonal   (b) Skew

Fig. 6. Plan view of grids.

These grids can have complex shapes, with cut-outs and complex supports.

BULKHEAD. This program is a more general form of GRILLAGE, which allows for hydrostatic loading and different member sizes.

THINCONE. This program can be used in thin cones. By joining together, a large number of cones, it is possible to represent cylinder/cone/dome combinations.

PLANESTRESS. This program can be used in plates suffering in-plane deformation. Triangular elements are used, so that it is possible to represent quite complex shapes, such as plates with holes.

LAPLACE. The program is applicable to a number of steady-state field problems including:

(1) Torsion of non-circular sections;

(2) Newtonian flow of fluids;

(3) Heat conduction;

(4) Seepage, etc.

Triangular elements are used, so that it is possible to represent quite complex shapes.

The above programs are particularly suitable for teaching and for smaller firms who cannot afford to purchase large computing systems. They will also be found useful by larger firms for the numerous smaller jobs they encounter, where the use of these programs on micros, will relieve their hard-pressed mainframes.

The programs will be found suitable in most branches of engineering, including mechanical, marine, aeronautical, civil and structural and also, building, shipbuilding, architecture, etc.

The programs can solve quite large problems, but this depends on the bandwidth and the amount of user RAM available. Some program run times are given in Table 1.

TABLE 1 Some Program Run Times.

Program	Degrees of Freedom	Approximate Time
CONTINUOUS BEAMS	398	8 mins
GRILLAGE	75	19 mins
BULKHEAD	66	17 mins–25 s

SESAM '80: A GENERAL PURPOSE STRUCTURAL ANALYSIS SYSTEM

A. Berdal

A. S. Veritec, P.O. Box 300, N-1322 Hovik, Norway

ABSTRACT

SESAM '80 (Super Element Structural Analysis program Modules) is a comprehensive program system employing the finite element method to solve a wide variety of structural analysis problems. Almost any type of structure may be analysed for linear static and dynamic behaviour. Nonlinear analysis is available in FENRIS, a program system coupled to SESAM '80, see separate description of FENRIS. The powerful multilevel superelement technique is a basic feature of SESAM '80. The program is highly user-oriented and incorporates load generation facilities and sophisticated pre- and postprocessing programs. SESAM '80 is actively maintained, supported and updated by A. S. VERITEC, a subsidiary of Det norske Veritas, and from regional offices in Rotterdam, London, Houston and Singapore.

THEORETICAL BACKGROUND

SESAM '80 is based on the linear continuum mechanics formulations in which the basic assumptions of "small" strains and rotations and linear stress-strain relationship (Hooke's law) are made.

FIELD OF APPLICATION

Any type of 3-D structure (beam, bar, shell, membrane, solid) as well as axisymmetric solid structures may be analysed. Linear isotropic, orthotropic and anisotropic material behaviour is offered. A multilevel superlement (substructuring) technique is available for static analysis. This technique is a highly favourable and often necessary approach in the analysis of complex structures, especially in cases of symmetry or recurring parts with identical geometry. Free vibration and forced response analysis may be performed as well. Alternative solution techniques in the time and frequency domain are available. The Master-Slave and Component Mode Synthesis reduction methods may be used to reduce the size of the problem at hand.

Even though the structural analysis is according to linear theory, SESAM '80 offers a nonlinear solution to the highly nonlinear problem of pile-soil interaction as occurring in e.g. the foundation of jacket offshore platforms.

An extensive set of load types is available such as:

- nodal displacements, accelerations and forces
- concentrated forces on beams
- line loads
- pressure loads
- inertia loads (gravity)
- centrifugal loads
- axisymmetric and nonsymmetric loads for axisymmetric elements
- initial stresses and strains (temperature)
- wave forces on jackets, semisubmersibles, gravity platforms, ships, etc.

PROGRAM DESCRIPTION

The finite element method is used in combination with the displacement method. The extensive element library contains all of the commonly used elements such as:

- 2 node bar element
- 2 and 3 node beams
- lower and higher order membrane elements of triangular and quadrilateral shape
- lower and higher order shell elements of triangular and quadrilateral shape
- lower and higher order solid elements of tetrahedron, triangular prism and hexahedron shape
- transition elements for coupling solid elements to shell elements
- axisymmetric solid elements
- spring and dashpot elements, between nodes as well as from node to ground

All higher order elements are based on the so-called Serendipity shape functions, hence there are no nodes in the interior of the elements. Higher order elements are defined here as elements with curved boundaries. The element library contains both conforming and nonconforming elements. All nonconforming elements fulfil the so-called patch test. This test is always fulfilled by conforming elements.

Plane stress as well as plane strain may be specified for relevant elements.

Tables 1-4 present the element library in more detail.

SESAM '80 offers advanced interactive and graphic preprocessors for automatic generation of the finite element models. Likewise, sophisticated interactive and graphic postprocessing facilities are available for evaluating the results. Numerous forms of results presentation including colour visualization are available. The modular program architecture of SESAM '80 facilitates coupling to various CAD systems. At present the EUCLID CAD system is interfaced with SESAM '80.

An automatic optimization of the node numbering with respect to the computer resources required for the equation solving is performed. The labelling of the nodes as specified and referred to by the user will, however, not be altered as a result of such optimization.

SESAM '80 imposes no limitation on the size of the model. Any number of elements, nodes and loading conditions may be employed. In fact, the only practical limitation to the size of the model is dictated by the computer resources available.

TABLE 1

Group	Illustration	Name and description	Number of nodes	d.o.f. pr. node
SPRING		GSPR, Spring to ground	1	2, 3 or 6
SPRING		AXIS, Spring between nodes	2	2, 3 or 6
DAMPER		GDAM, Damper to ground	1	2, 3 or 6
DAMPER		AXDA, Damper between nodes	2	2, 3 or 6
BEAM / TRUSS		TESS, 3-D truss	2	3
BEAM / TRUSS		BEAS, 3-D beam	2	6
BEAM / TRUSS		SECB, 3-D curved beam	3	6

The program design of SESAM '80 is based on up-to-date knowledge in FEM programming:

· Modular program architecture
· Generalized data files interface various program modules
· Thoroughly documented code
· Code (FORTRAN 77) is according to strict standards

The capabilities of SESAM '80 makes the program:

· easy to modify and extend,
· easy to debug,
· suitable for multicomputer usage, e.g. pre- and postprocessing may be performed on dedicated table computers while the time consuming stiffness computation and equation solution are handled by main frame computers or array processors.

SESAM '80 contains approximately 500,000 executable statements.

HARDWARE COMPATIBILITIES

The minimum computer configuration is highly dependent on the problem to be solved. A virtual memory computer with 2 Mb memory and a few hundred Mb disc is sufficient for most practical cases. A magnetic tape unit should be available for permanent storing of model data and results.

SESAM '80 is implemented on the following computer makes, the operating system is indicated within brackets:

TABLE 2

Group	Illustration	Name and description	Number of nodes	d.o.f. pr. node
MEMBRANE		PCST, Constant strain triangle	3	2
		LQUA, Linear quadrilateral	4	2
		ILST, Isoparametric linear strain triangle	6	2
		IQQE, Isoparametric quadratic strain quadrilateral	8	2
AXISYM. SOLID	See illustrations above	Axisym. elements correspond in shape and number of nodes to the membrane elements above: ACST (cor. to PCST) AQUA (cor. to LQUA) ALST (cor. to ILST) AQQE (cor. to IQQE)	3 4 6 8	3 3 3 3

- VAX-11/750/780 (VMS)
- PRIME 550/850 (PRIMOS)
- IBM 43xx (VM/SP)
- IBM 30xx (OS/MVS)
- Cyber 170 (NOS)
- FPS 164

Plotters should preferably be compatible with the Calcomp plotter while graphics screens should preferably be compatible with the Tektronix screens. Most other plotters and screens may, however, be connected to SESAM '80 with little extra work. SESAM '80 is delivered on magnetic tape.

EXAMPLES OF APPLICATION

A detailed and extensive Verification and Examples Manual has been made for SESAM '80. This manual contains numerous test cases in order to cover the most important features of the comprehensive program system. The case presented below is only a selected example, no single case can in any way verify more than a small part of the system.

Cylindrical Barrel Shell Structure

The cylindrical barrel shell problem of Fig. 1 has become a classical problem for

TABLE 3

Group	Illustration	Name and description	Number of nodes	d.o.f. pr. node
SHELL		FTRS, Triangular flat thin shell	3	6
		FQUS, Quadrilateral flat thin shell	4	6
		SCTS, Subparametric curved triangular thick shell	6	6
		SCQS, Subparametric curved quadrilateral thick shell	8	6
TRANSITION	TRS1 TRS2 TRS3	TRSI, Transition element between solid and shell elements. The IHEX/IPRI elements may be coupled with the SCQS/SCTS elements. Three versions of the transition element are needed: TRS1 TRS2 TRS3	18 15 12	3/6 3/6 3/6

comparison of shell elements. The problem has been solved by SESAM '80 using the three different shell type elements:

- FQUS – Quadrilateral flat thin shell element
- FTRS – Triangular flat thin shell element
- SCQS – Subparametric curved quadrilateral thick shell element (parabolic shape)

Analysis data

Geometry. The geometry as it appears in O. C. Zienkiewicz, "The Finite Element Method", is shown in Fig. 1.

Material properties. The shell is made of steel with the following material data:

Specific weight $g = 360$ lb/ft^3

Modulus of elasticity $E = 3 \times 10^6$ psi

Poisson's ratio $\nu = 0$

Boundary conditions. The shell is supported by rigid diaphragms at two ends such that u, w, and θ_y are constrained, see Fig. 1.

TABLE 4

Group	Illustration	Name and description	Number of nodes	d.o.f. pr. node
S O L I D		TETR, Tetrahedron	4	3
		TPRI, Triangular prism	6	3
		LHEX, Linear hexagon	8	3
		ITET, Isoparametric tetrahedron	10	3
		IPRI, Isoparametric triangular prism	15	3
		IHEX, Isoparametric hexahedron	20	3

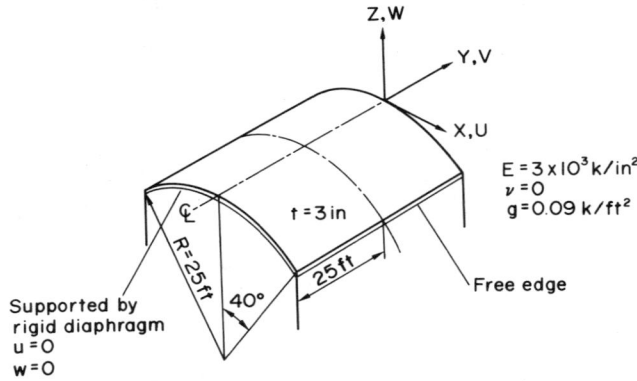

Fig. 1. Cylindrical barrel shell

SESAM '80: A General Purpose Analysis System

Loading conditions. The shell is analysed as loaded by its own weight.

Mathematical model. One quarter of the shell is modelled by two different element grids as shown in Fig. 2. When using triangular thin plate elements, FTRS, two triangular elements replace each quadrilateral element shown in Fig. 2.

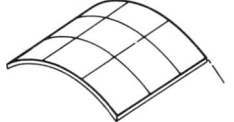 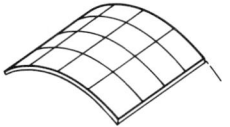

Fig. 2. Element meshes shown for one quarter of the shell.

Linear Static Results. The results obtained were compared with results given in O. C. Zienkiewicz, "The Finite Element Method". The subparametric curved quadrilateral thick shell element, SCQS, performed very well. Both displacements and stresses were within small tolerances equal to the exact solutions.

The thin plate elements, FQUS and FTRS, did not perform as well as the thick shell element. This was expected due to the small radius/thickness ratio of the shell. Also, a model using flat elements cannot represent a curved surface as well as parabolic shape elements. For vertical and longitudinal displacements both FQUS and FTRS performed reasonably well. See also the plots in Appendix A.

Ekofisk Field Water Injection Platform 2/4-K

Brief description. The offshore platform is a very large eight-legged jacket platform for use at the Ekofisk oil field in the North Sea. The jacket shall serve at 71 metre water depth in one of the roughest offshore fields of the world. The design storm wave of 26 metre height causes a horizontal force on the jacket of approx. 100 mega N (1 megaton). The platform shall carry a main support frame which again shall support a living quarter, a flare structure and a helideck in addition to process equipment, a total load of approx. 300 mega N (3 megatons).

Application area. Offshore industry.

Type of problem. Both a static in-place analysis and a deterministic fatigue analysis were performed. Both analyses were performed according to linear theory except for a nonlinear solution of the pile-soil foundation problem performed as part of the static analysis.

Discretization. ·See the enclosed plot, Fig. 3.

Type of elements, number of elements, etc. 2 node beam elements with 6 degrees of freedom in each node were used for the jacket and main support frame (deck). The model consisted of all together 2671 elements and approx. 1000 nodes. The static analysis comprised 10 loading conditions while the deterministic fatigue analysis comprised 320 loading conditions.

Number of degrees of freedom. Approx. 6000 degrees of freedom.

Bandwidth. Approx. 1000

Part of the program, computer and peripherals. A preprocessor (PREFRAME) for generation of frame structures was used to create the model, or rather the 4 superelements (substructures) which when assembled form the complete model. Another preprocessor (PRESEL) was employed for assembling the superelements.

The calculation of wave forces on the jacket was performed by WAJAC while the analysis of the nonlinear pile-soil problem was performed by SPLICE. The linear analysis module SESTRA performed the static as well as the eigenvalue solution, the latter being the basis for the subsequent deterministic fatigue analysis performed by POSTFRAME. All these programs belong to the SESAM '80 suite.

A VAX-11/780 computer with 4 Mb memory was employed together with Tektronix graphics screens and a Calcomp plotter.

Input and output data. Far too excessive to be presented here.

Computation time. Static analysis: approx. 5 hours CPU on VAX-11/780

Deterministic fatigue analysis: approx. 25 hours CPU on VAX-11/780

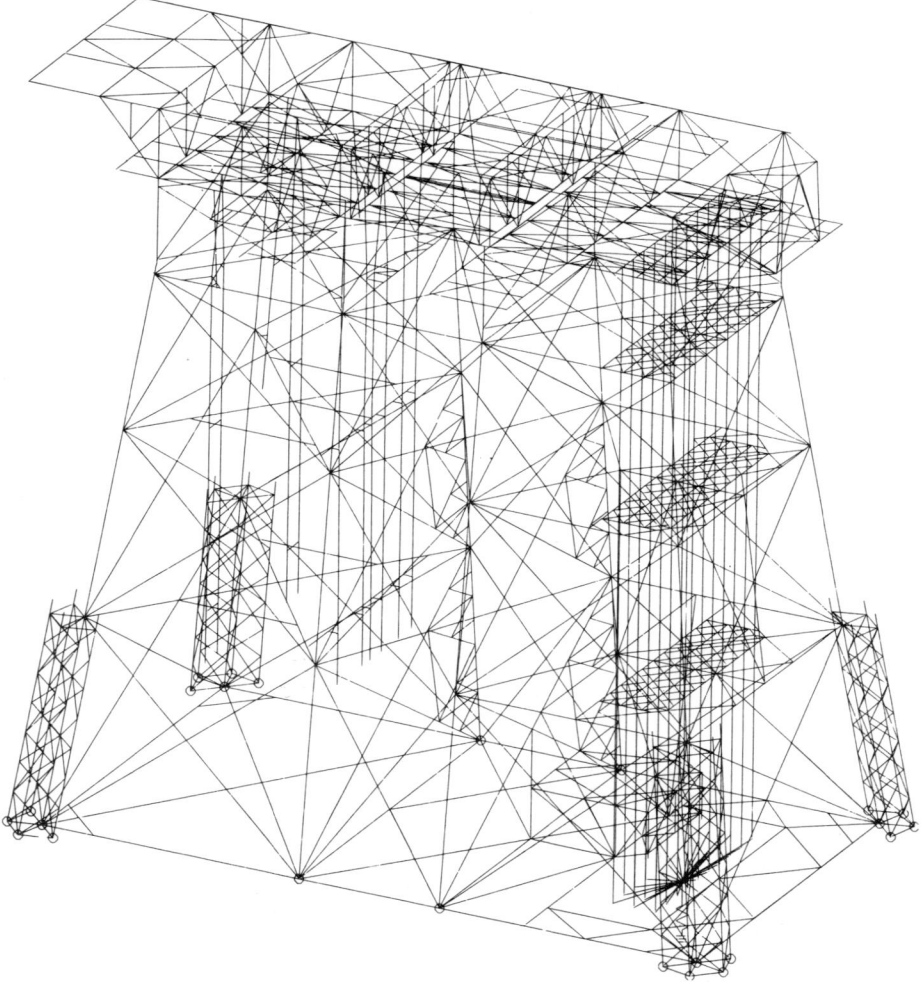

Fig. 3. The water injection platform (2/4-K jacket) of Phillips Petroleum Company Norway, for use at the Ekofisk Field.

SESAM '80: A General Purpose Analysis System

Patented, Alternative Design of Tubular K-joint, Norsk Hydro A/S

Brief description. In offshore structures like jackets and semisubmersibles, welded or casted tubular joints of various designs are extensively used. A new design of a so-called K-joint has been proposed and patented by Norsk Hydro A/S of Norway. The results of the analysis briefly presented here show that the highest appearing stress concentration factor is 1.86 which is very low compared with factors appearing in traditional K-joint designs. The joint is shown in Fig. 4 (drawing by CAD program EUCLID).

Application area. Offshore industry

Type of problem. A linear static analysis employing the superelement technique was performed. 9 different superelements were used.

Discretization. See the enclosed plot, Fig. 5.

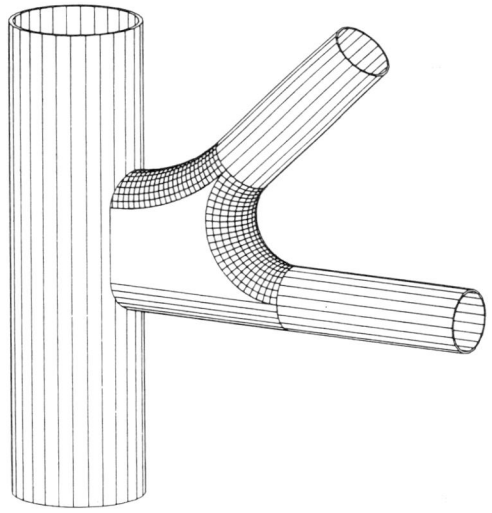

Fig. 4.

Type of elements, number of elements, etc. 8 node quadrilateral thick shell elements with 6 degrees of freedom in each node were used for the whole model which consisted of altogether 1168 elements and 3740 nodes. 8 loading conditions were analysed.

Number of degrees of freedom. Approx. 22000 degrees of freedom

Bandwidth. Ranging from 191 to 1703 for the 9 superelements.

Part of the program, computer and peripherals. Two preprocessors were used for creating the geometry and finite element meshes. One for the tubular parts (PRETUBE) and another for the remaining shell/plate parts (PREFEM). The superelements were assembled by a third preprocessor (PRESEL). The linear analysis module SESTRA performed the static analysis, and the results were presented graphically by POSTFEM. All these programs belong to the SESAM '80 suite.

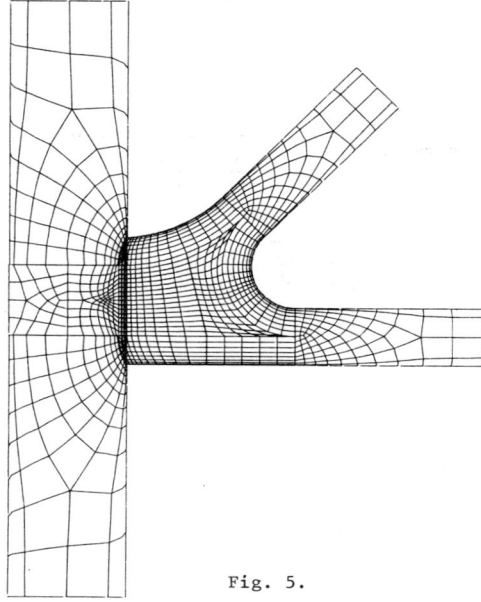

Fig. 5.

A VAX-11/780 computer was employed together with Tektronix screens and a Calcomp plotter.

Input and output data. Far too excessive to be presented here. See, however, Appendix B.

Computation time. Approx. 10 hours CPU on VAX-11/780.

POTENTIAL USERS AND APPLICATIONS

The main area of application is within the offshore and shipping industry. Oil companies, field developers, engineering companies, shipyards and designers and consultants engaged in the corresponding activity are potential users. Other professional areas such as the automotive, civil engineering and machinery industries are also potential users of the general purpose finite element system SESAM '80.

APPENDIX A

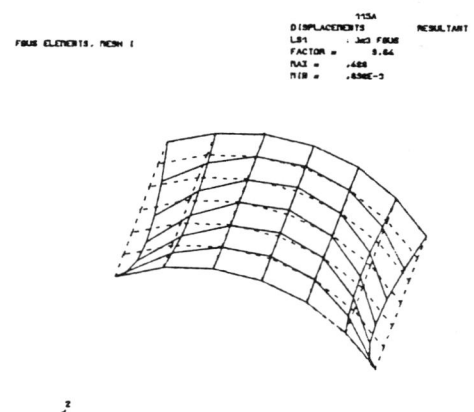

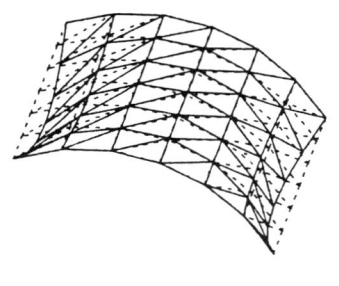

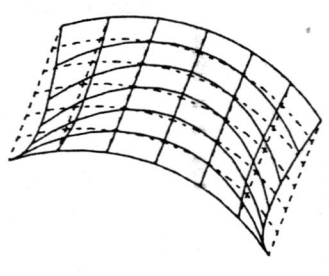

APPENDIX B

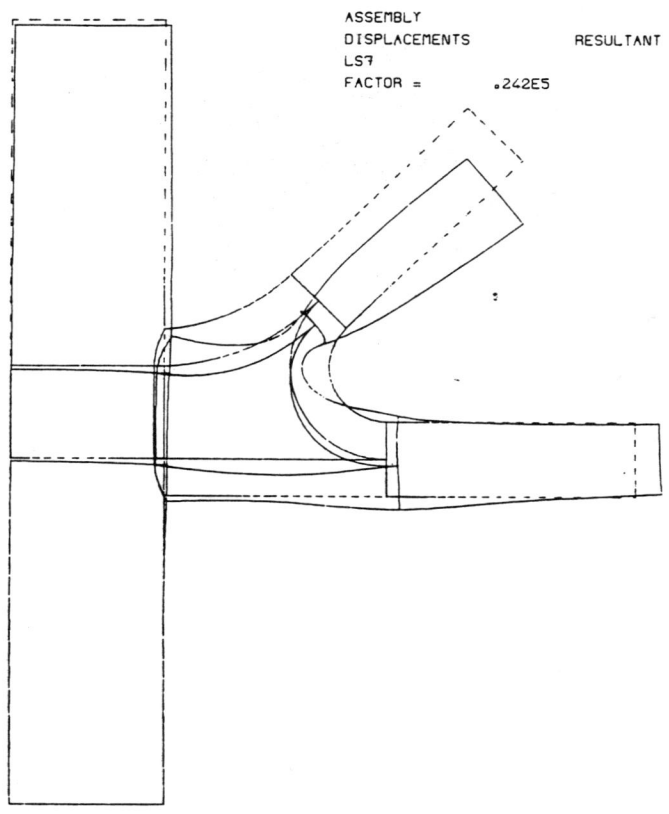

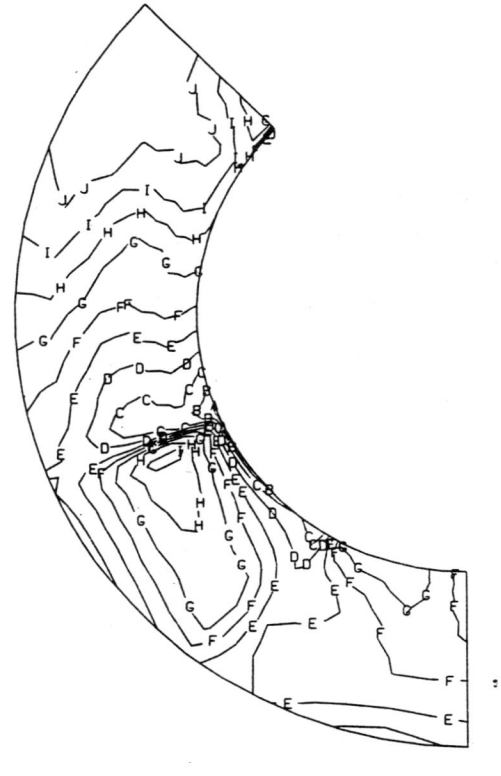

MAX	1.86
J =	1.69
I =	1.53
H =	1.36
G =	1.20
F =	1.03
E =	.869
D =	.704
C =	.540
B =	.375
A =	.210
MIN	.458E-1

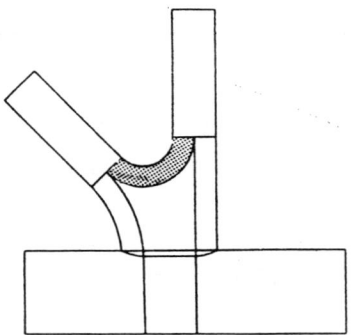

THERMAL: MICROCOMPUTER PROGRAMS FOR THERMAL STRESS ANALYSIS

C. T. F. Ross

Department of Mechanical Engineering, Portsmouth Polytechnic, U.K.

ABSTRACT

This package contains five computer programs, which have been written in BASIC for microcomputers. The programs can solve thermal stress problems i.e. two and three-dimensional pin-jointed trusses and rigid-jointed frames, together with plane strain problems.

THEORETICAL BACKGROUND

For static analysis, including thermal effects, the matrix equations relating forces and displacements is given by:

$$\{q\} = [K] \{u\},$$

where

$\{q\} = \{q_w\} + \{q_T\}$,

$\{q_w\}$ = a vector of applied external loads,

$\{q_T\}$ = a vector of equivalent external loads due to temperature change

$= \Sigma\{P_T\}$,

$\{P_T\}$ = a vector of thermal loads acting on the nodes of an element,

$[K]$ = the structural stiffness matrix,

$\{u\}$ = a vector of nodal displacements.

The vector $\{P_T\}$ depends on the angle of inclination of the element, together with its geometrical and material properties, including its coefficient of linear expansion and temperature change.

FIELDS OF APPLICATION

TRUSST. This program calculates the nodal displacements and member forces in plane pin-jointed trusses, subjected to horizontal and vertical nodal forces, together with temperature change in the elements. The members can have different sectional and material properties and they can be statically determinate or statically indeterminate.

PLANE FRAMET. This program can calculate nodal displacements, together with axial forces and bending moments in rigid-jointed plane frames. Besides thermal loads, due to temperature changes in the elements, the program can cater for a combination of concentrated, distributed and hydrostatic loading. Step variation is allowed for in the geometrical properties of the members. The frameworks can be portal, skew and/or multibay and there is no difficulty in allowing for sidesway.

SPACE TRUSST. This program can calculate nodal displacements and member forces in three-dimensional pin-jointed trusses, subjected to three-dimensional concentrated loads at their pin-joints, together with thermal loads, due to temperature changes in the members of the trusses. The elements of each truss can have different sectional and material properties and the trusses can be statically determinate or indeterminate.

3D-RIGFRAMET. This program can calculate nodal displacements, axial forces, bending moments and torques in three-dimensional rigid-jointed space frames, subjected to three-dimensional concentrated loads, together with thermal loads due to temperature changes in the members of the framework.

PLANESTRESST. This program can calculate nodal displacements and elemental stresses in in-plane plates, subjected to concentrated loads at the nodes, together with thermal loads due to temperature changes in the elements. As triangular elements are used, it is possible to represent quite complex shapes with holes and complex boundary conditions.

PROGRAM DESCRIPTION

The method of solution adopted in all the five programs was the finite element method.

Skeletal elements, with two end nodes, were used for the pin-jointed trusses and the rigid-jointed frameworks, and plane triangular elements with corner nodes, for the in-plane plates.

Two degrees of freedom per node were required for TRUSST and PLANESTRESST and three degrees of freedom per node for PLANE FRAMET. Six degrees of freedom per node were used for the three-dimensional element of 3D-RIGFRAMET.

All the programs were written in interactive BASIC and it was a simple matter to modify the programs by the deletion or introduction of lines, subroutines, etc.

The programs on the Apple II included a graphical display on the screen, before processing, together with the deformed shape of the structure after processing.

HARDWARE

The minimum hardware requirements are a 32 k RAM microcomputer, with a TV/monitor display and a home tape recorder.

Currently, the package can be run on a CBM PET/64 and an Apple II +/e.

The programs are available on standard cassette tapes and also, on $5\frac{1}{4}$ inch floppy disks.

EXAMPLES OF APPLICATION

TRUSST. This program can be used in plane pin-jointed trusses of the type shown in Fig. 1. Each member of the truss can be subjected to a different temperature change, together with concentrated loads applied to the pin-joints.

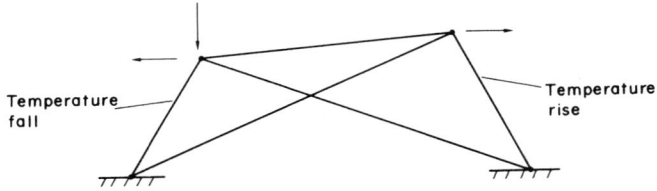

Fig. 1. Plane pin-joined truss.

PLANEFRAMET. This program can be used in rigid-plane frames subjected to a combination of thermal loads due to temperature changes and externally applied loads. The structures can be statically indeterminate (Fig. 2), or statically determinate and they can also be in the form of pipeworks (Fig. 3).

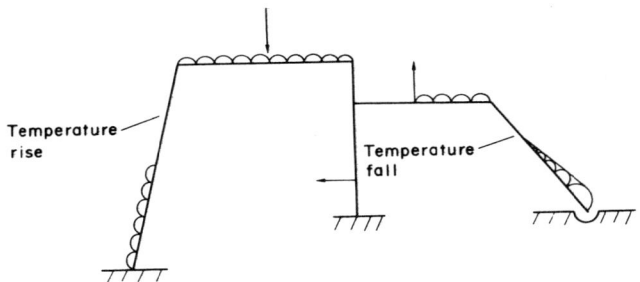

Fig. 2. Rigid-jointed plane frame.

SPACE TRUSST. This program can be used in three-dimensional pin-jointed trusses, as shown in Fig. 4.

3D-RIGFRAMET. This program can be used in rigid-jointed space frames, as shown in Fig. 5. The program can also be applied to three-dimensional pipeworks, undergoing temperature changes, as shown in Fig. 6.

PLANESTRESST. This program can be used in in-plane element stresses, as shown in Fig. 7.

The package is quite suitable as a teaching aid and also for use by smaller firms who cannot afford to purchase large computing systems. They will also be found useful by larger firms for processing smaller jobs, as the programs are, in general, more user-friendly than those found on mainframes and very often, they produce the results more quickly.

The programs will be found suitable in most branches of engineering, including

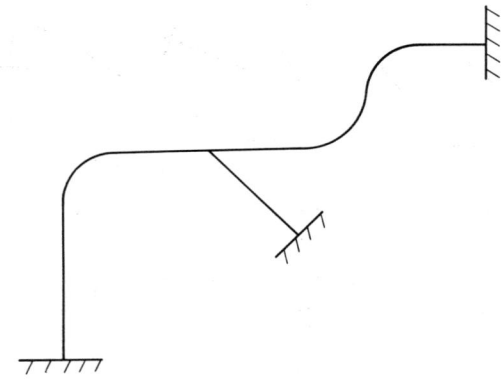

Fig. 3. Two-dimensional pipe system.

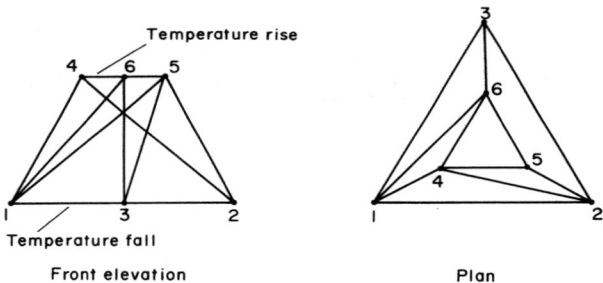

Fig. 4. Pin-jointed space truss (Schwedler dome).

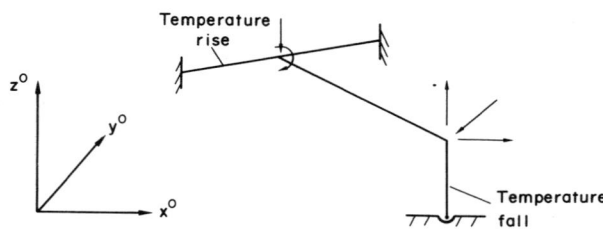

Fig. 5. Rigid-jointed space frame.

mechanical, marine, aeronautical, civil and structural engineering and also, building, shipbuilding, architecture, etc.

Fig. 6. Three-dimensional pipework system.

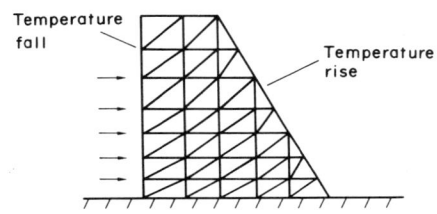

Fig. 7. Earth dam.

The programs can solve some quite large problems, but this depends on the bandwidth and the amount of user RAM available.

TITUS: A GENERAL FINITE ELEMENT SYSTEM

D. Halbronn

Framatome, Centre de Calcul, BP 13, 71380 St Marcel, France

ABSTRACT

TITUS is a general finite element structural analysis system which performs linear/nonlinear, static/dynamic analyses of heat-transfer/thermo-mechanical problems.

One of the major features of TITUS is that it was designed by engineers, to address engineers in an industrial environment. This has resulted in a system easy to use, with a high-level free-formatted problem oriented language, a large selection of pre- and postprocessors and sophisticated graphic capabilities. TITUS has many references in civil, mechanical and nuclear engineering applications.

The TITUS system is available on various types of machines, from large mainframes to minicomputers.

THEORETICAL BACKGROUND

Based upon the displacement formulation, TITUS offers a large choice of algorithms to solve specific problems:

Solution of Static Linear Equations

Several methods - all based upon a gauss triangularization - are currently available, so the user can select the one that fits his problems; they are:

- the sparse matrix technique,
- the variable bandwidth technique.

Solution of Nonlinear Equations

For nonlinear systems of equations, the solution is obtained by iterative methods using different techniques depending on the problem to be solved:

- Newton - Raphson,
- modified Newton - Raphson,

with/without BFGS acceleration.

Solution of Dynamic Equations

Several capabilities are available, depending on the problem to be solved:

- Eigenvalue problem

    - Given's method,
    - inverse power method,
    - subspace iteration method,
    - Lanczos' method.

- Dynamic response

    - modal superposition,
    - direct integration (β - Newmark).

MAIN FEATURES OF THE SYSTEM

Analytical Capabilities

Within TITUS, analytical capabilities - in terms of heat transfer, elasticity, or thermo-mechanical analyses - are classified into different options: plane elasticity, plates, thin/thick shells, beams, 3-dimensional continua, heat transfer, axisymmetric structure with/without axisymmetric loading etc. The choice of the option depends upon the following criteria:

- coordinates used,
- displacements calculated (degrees of freedom),
- loads applied.

The purpose of this choice is to optimize the computations of the specific problem to solve.

Analytical capabilities of TITUS are summarized in Fig. 1. TITUS can also be used to solve:

- fluid/structure coupling problems,
- forced convection,
- gas diffusion.

Nonlinear Analysis

The TITUS system can deal with different types of nonlinearities such as:

- geometrical nonlinearity, e.g. large displacements, large strains, etc. When combined with an eigenvalue algorithm, this type of nonlinearity allows to study instability problems (buckling),
- nonlinearity of material properties: strain dependent Young's modulus, temperature dependent conductivity, etc.
- plasticity: kinematic hardening isotropic hardening, thermo-plasticity,
- creep,
- uncompressible materials,
- special mechanisms: unilateral boundary conditions, gap, friction etc.

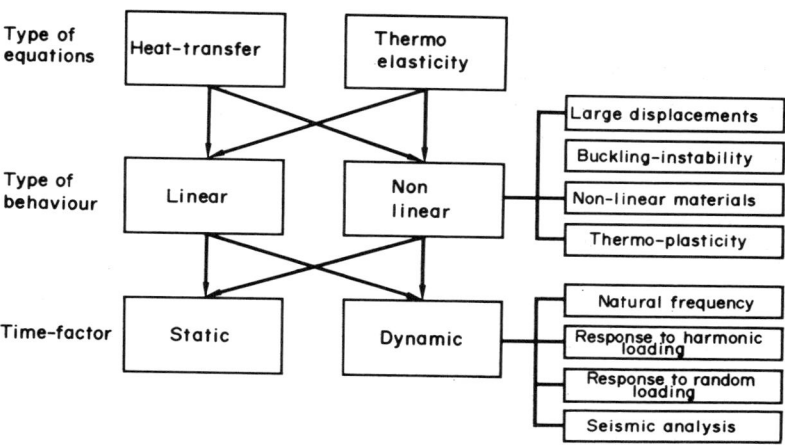

Fig. 1. TITUS: Analytical capabilities (continuum mechanics)

Several types of nonlinearities can be combined together.

Dynamic Analysis

TITUS offers several capabilities in dynamic analyses of structures:

- natural frequency (eigenvalues, natural modes),
- response to harmonic forcing functions (with or without damping)
- response to random forcing functions
  - by modal analysis,
  - by direct integration,
- seismic analysis.

Dynamic analyses can be performed in areas where the structure has a nonlinear behaviour.

Dynamic analyses can be applied to heat transfer problems. Also, input data for a dynamic analysis are fully compatible with a linear or nonlinear static analysis.

A special module performs spectral analysis from natural frequencies.

This procedure is used for seismic analysis; the total response of the structure may take into account:

- either a uniform support acceleration for each direction (mono spectral analysis).

- or a non-uniform excitation inducing pseudo-static stresses due to motion of supports relative to each others (multispectral analysis).

Super-elements

The concept of "Super-elements" (or substructures) allows the TITUS System to perform static analyses of structures with virtually no limitation in the number of degrees of freedom.

From a practical point of view, the analysis of a problem is greatly eased by using the SE method, due to:

- geometry transformation procedures (rotation, symmetry...). Modelling and computing a structure with repetitive elements become an easy task.

- "restart" capability which allows the user not only to analyse a structure step by step but also allows him to modify part of a structure without having to recalculate the lot.

MODELLING CAPABILITIES

Material Properties

Within TITUS, materials can be:

- isotropic,
- orthotropic,
- anisotropic.

And their physical properties can be:

- constant,
- or variable (constitutive nonlinearities).

Boundary Conditions

TITUS accepts the following boundary conditions:

- rigid: $U = Uo$
- elastic: $F = K(U - Uo)$

Boundary conditions can be:

- concentrated at a node,
- or uniform on an element.

Boundary conditions can be defined with reference to any local or general system of axes.

Special Mechanisms

Special mechanisms, or special behaviours can be easily modelized, using the following capabilities of TITUS:

- release of any degree of freedom,
- multipoint constraints,
- contact mechanisms: unilateral support, gap, friction....

DESIGN OF THE TITUS SYSTEM

Element Library

The different types of element geometries available in the TITUS system library are summarized in Table 1. All elements can be mixed one with another, within the same option of analysis; their modelling (shape functions, physical properties, loadings...) depends upon the selected option.

This Table includes "special" elements such as:

- transition elements (e.g.: 5 node-triangle, 7 node-quadrilateral etc.) as used for example in fracture mechanics analysis.
- fluid/structure coupling elements,
- singular elements,
- elements with incompatible modes.

TABLE 1

CURVES (1D - Element)	Beams / Bars	2	3	4	5
SURFACES (2D - Element)	Triangles	3	6	9	
	Quadrilaterals	4	8	12	
VOLUMES (3D - Element)	Tetrahedrons	4	10		
	Prisms	6	15		
	Hexahedrons	8	20		

Users can also program and integrate their own special elements.

General Architecture

TITUS is an integrated set of programs organized into 4 mains functions (see Fig. 2).

Supervisor. A general procedure which manages and organizes the different operations to be performed.

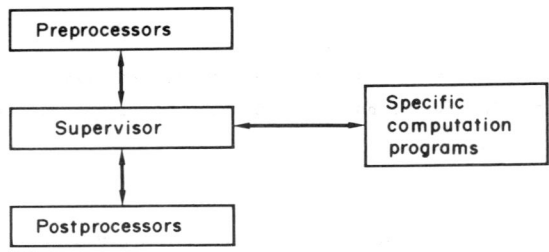

Fig. 2.

Library of specific computation subroutines. Adapted to the type of problem to solve.

Preprocessors. Since it was designed and developed by engineers for engineers, TITUS is definitely user-oriented. The only input it requires from the user is a general definition of the problem to analyse, using logic, words and symbols familiar to every structural engineer. Detailed modelling is performed by preprocessors.

A large number of procedures are available to automatically generate plane and 3-dimensional meshes. Other procedures are also used to copy, alter, modify or complete a predefined mesh.

Models can be displayed on a digital plotter or a VDU with monochrome and shaded colour representations, e.g.:

- plotting whole/part of a structure (with zoom effect, scissoring, windowing, contouring, etc.),
- perspectives,
- plotting with hidden lines removed,
- automatic numbering of nodes/elements...

Some utility programs included in TITUS allow to save, delete, edit... input data files.

Postprocessors. Most of the time, users expect more than the standard results (displacements, supports reactions, internal efforts) given by a finite element program.

With TITUS, the user has access to a wide range of postprocessors which perform selection, combination or special process of results. Of course, all these sophisticated results can be visualized (on a digital plotter or a VDU with monochrome and colour representations) for example:

- deformed structure,
- principal stresses,
- isovalues (isostresca, isodisplacement...)
- curves of variation of results along cross-sections...

Modularity

Modularity is the keyword of TITUS design. Functional programs are organized as a hierarchy, which allows:

- an efficient structure of the whole system,
- easy developments, modifications and add-on of new modules,
- to implement customized versions of TITUS with only a few modules (e.g. on workstations).

Programming Language

TITUS is mostly written in FORTRAN. Only a few subroutines have been written in assembler language, in order to optimize some calculations or in order to perform some operations which are not possible with FORTRAN.

Save and Restart Procedures

The possibility of fractioning a large problem into several small problems has been especially studied.

This is achieved mainly by a flexible automatic file management system.

This capability is not only necessary when using super-elements techniques; it is also very convenient in order to perform sequentially different types of analyses (involving different options), for example:

- Thermo-elasticity after heat-transfer analysis,
- Dynamic after static analysis,
- Nonlinear after linear analysis, etc.

Results Accessibility

Users can access and process results:

- by using standard postprocessors available in TITUS. New postprocessors can be easily developed whenever new requirements arise,
- by using a library of utility subroutines which can access all internal files created by TITUS. This way, engineers can program their own procedures.

GROWTH AND DEVELOPMENT

TITUS is maintained and developed by FRAMATOME who are responsible for continual enhancements based on user feed back.

Current and future developments being made include fluid mechanics, integration with CAD/CAM systems, welding process simulation and residual stresses, expert systems etc.

AVAILABILITY

TITUS is available on various machines, from large mainframes to minicomputers and workstations (CRAY, IBM, VAX, MICROVAX, APOLLO, SUN...).

It can be installed modularly, starting from a basic version (e.g. static linear elasticity) and progressively increased up to the complete system.

TITUS is marketed on a licence fee basis. Quotation on request.

EXAMPLES OF INDUSTRIAL APPLICATIONS

Thermo-mechanical Analysis of a Spray Nozzle in a Pressurizer (Fig. 3.)

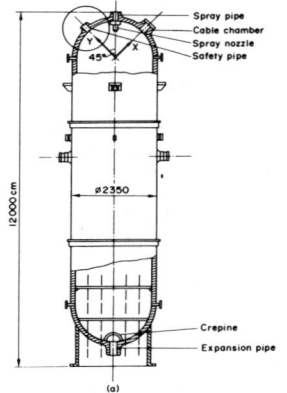

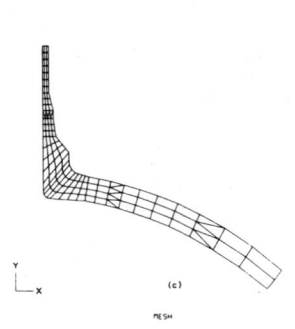

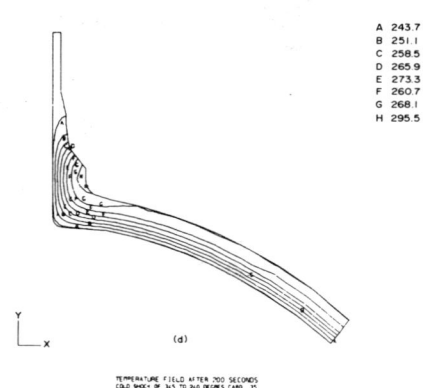

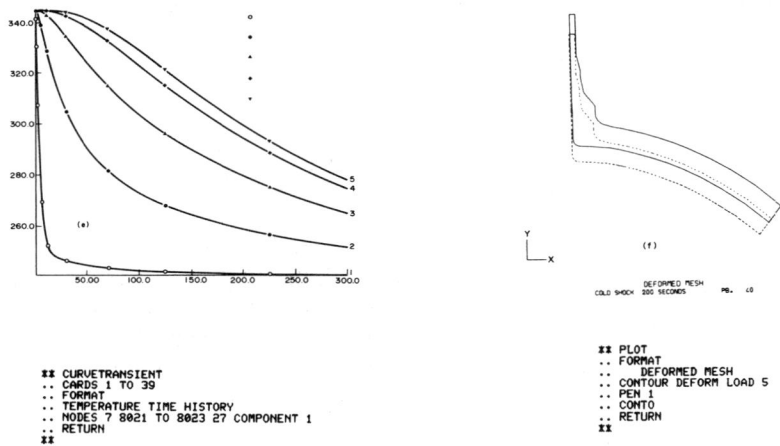

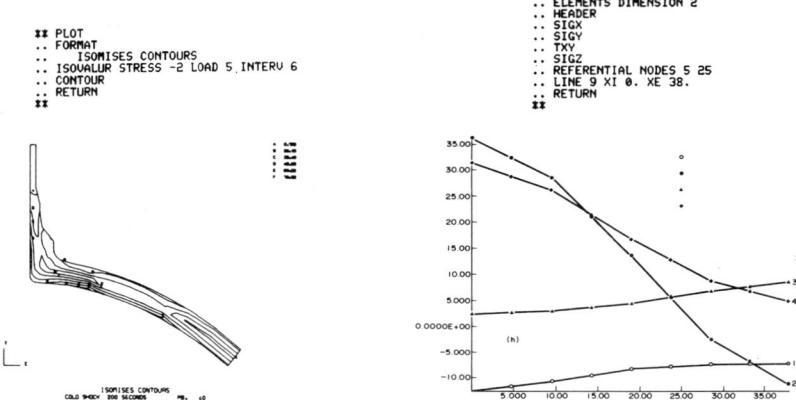

Fig. 3. (a) Scheme of the pressurizer, (b) Example of input data, (c) Mesh, (d) Temperature field 200s after cold shock of 345°C to 240°C (e) Temperature time history, (f) Deformed mesh, (g) Isomises contours (h) Stresses across section

Other Examples

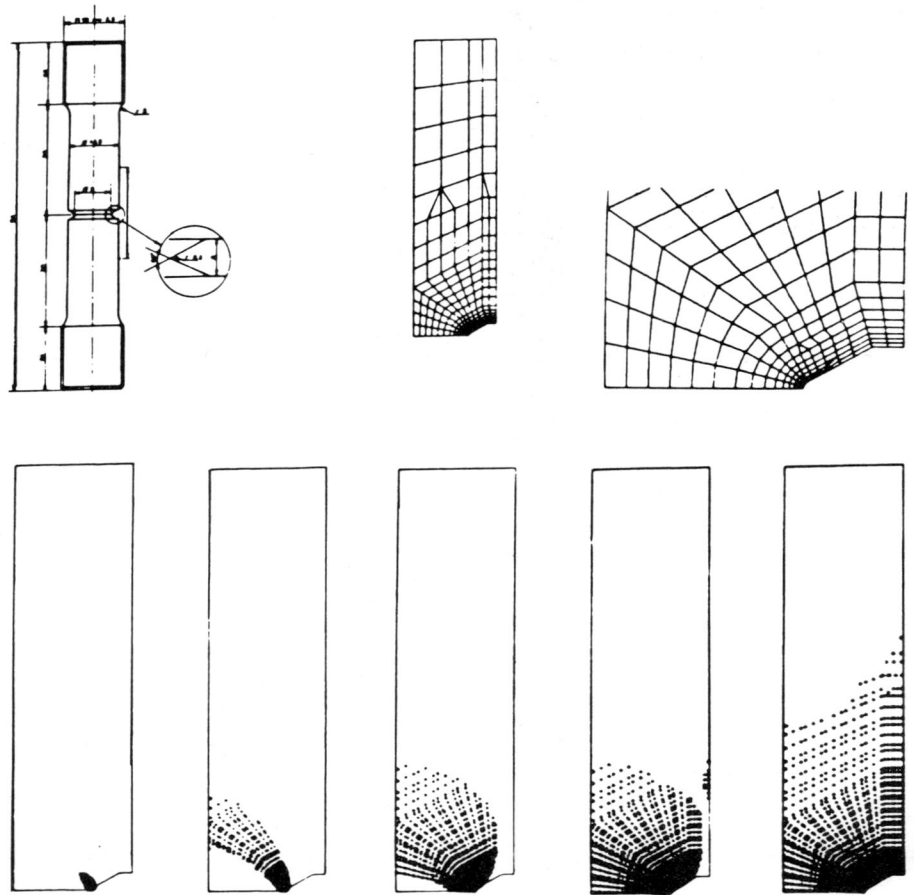

Fig. 4. Propagation of plasticity in a specimen.

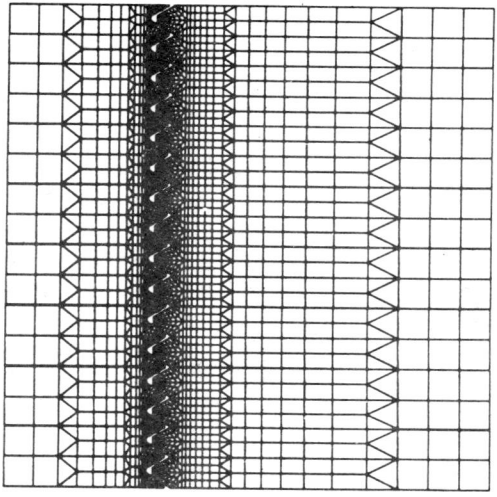

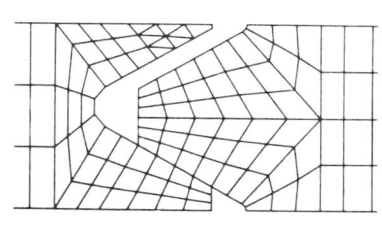
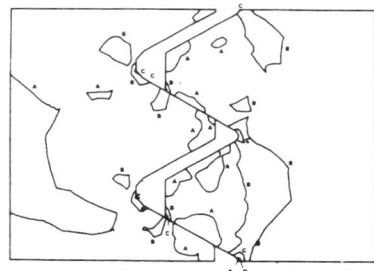

Fig. 5. Stud - bolt analysis.

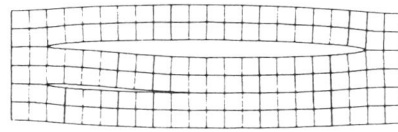

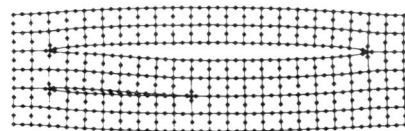

Fig. 6. Defect propagation (fracture mechanics).

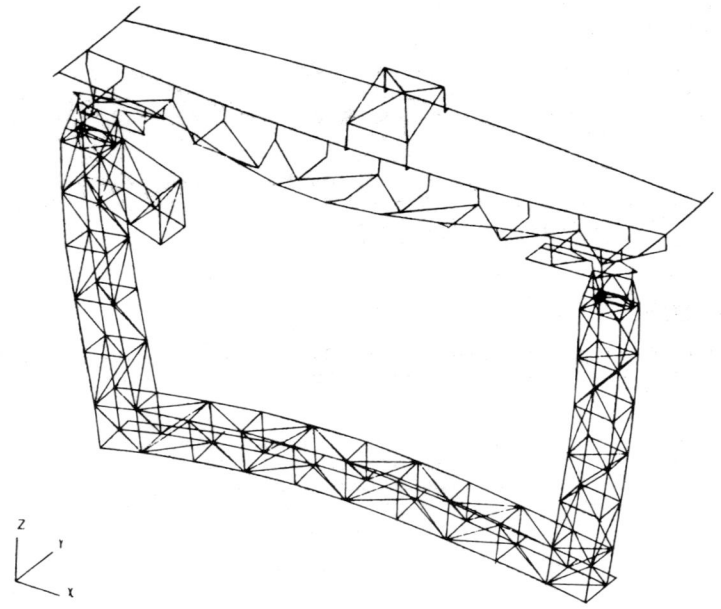

Fig. 7. Seismic analysis (Foot-bridge).

TITUS: A General Finite Element System

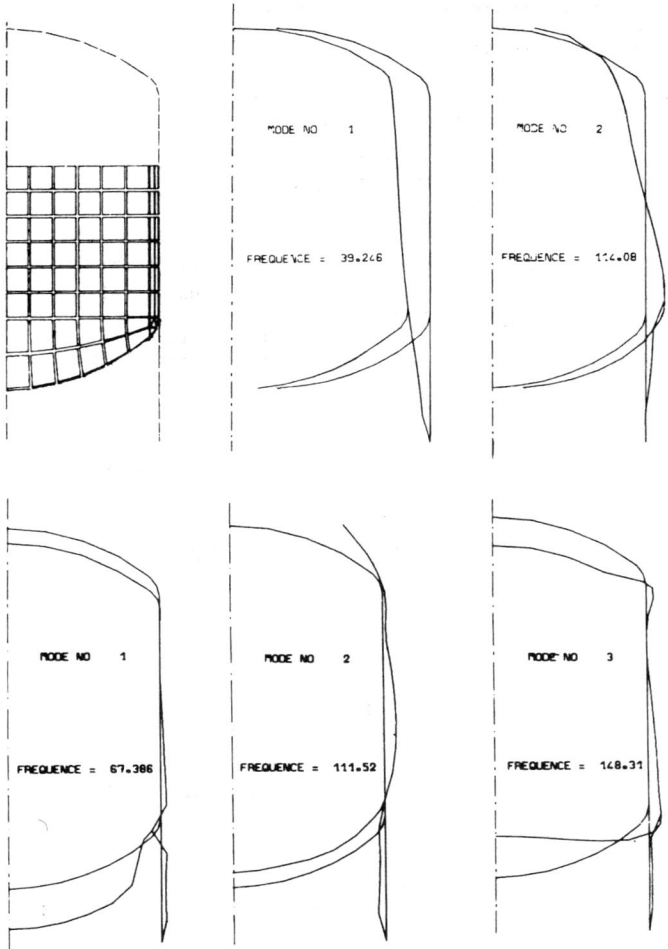

Fig. 8. Fluid/structure coupling (Seismic analysis of a tank partially filled with water).

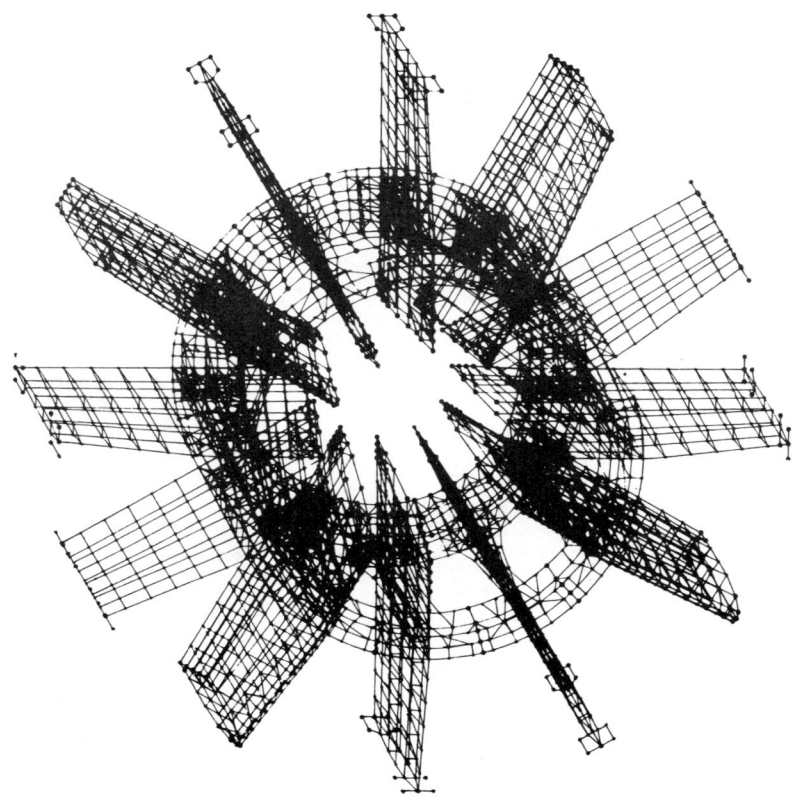

Fig. 9. Turbo-jet inlet analysis (36,000 d.o.f).

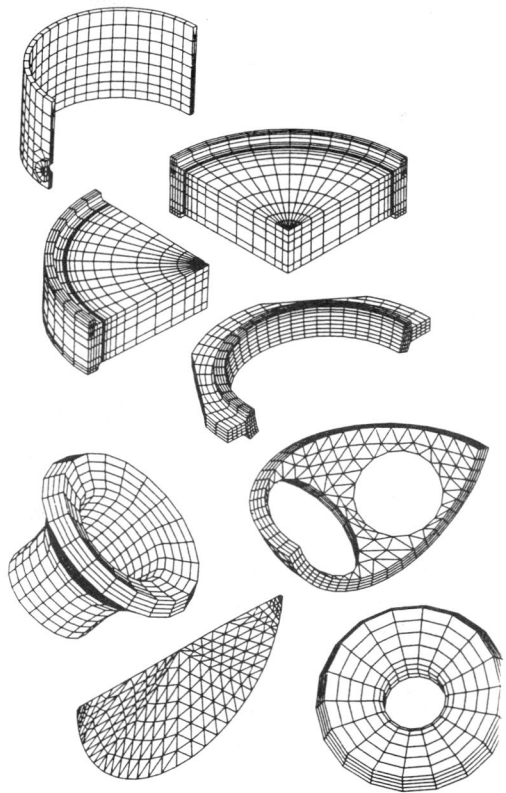

Fig. 10. Bottom head of a steam generator.

Super-element capabilities (Figs. 9 and 10)

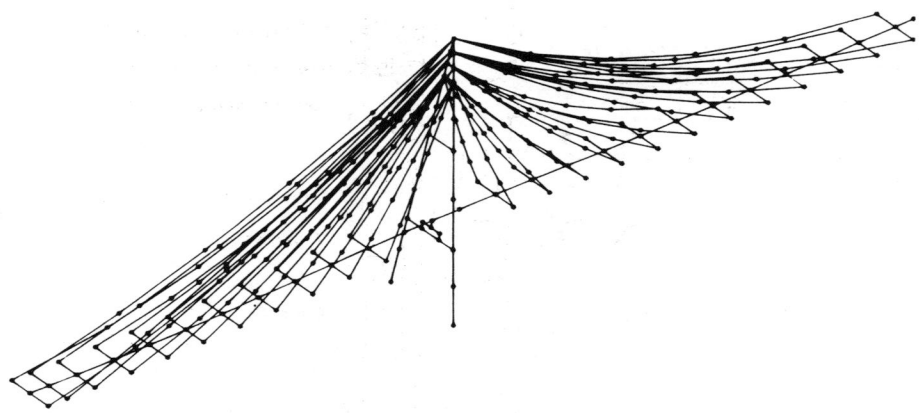

Fig. 11. Large displacement analysis (suspended bridge).

TITUS: A General Finite Element System

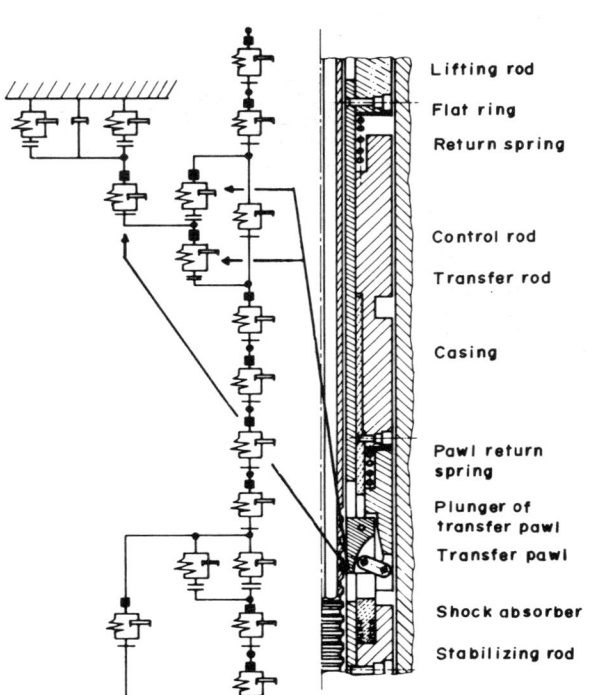

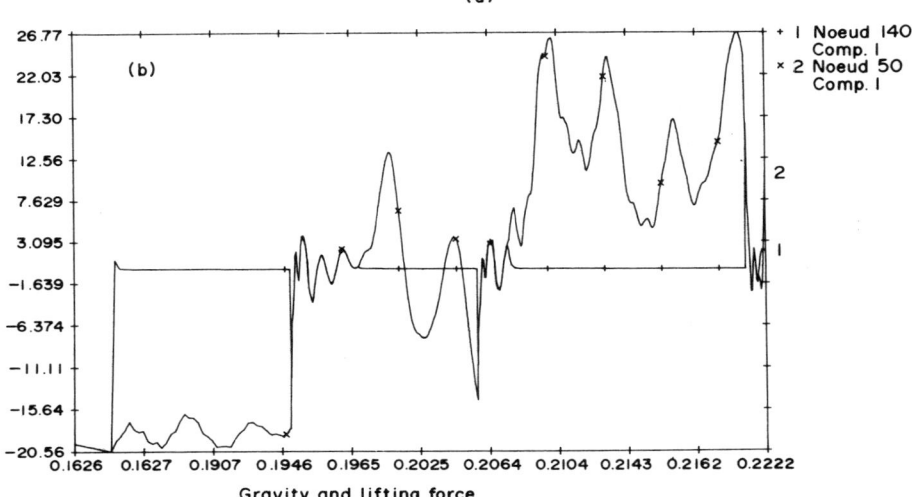

Fig. 12. Time history response of a drive rod (modelling using special mechanisms.

UCIN–GEAR: A FINITE ELEMENT COMPUTER PROGRAM FOR DETERMINING STRESSES IN SPUR GEARS

S-H. Chang and R. L. Huston

Department of Mechanical and Industrial Engineering, University of Cincinnati, Cincinnati, Ohio, USA

ABSTRACT

UCIN-GEAR is a two-dimensional, finite-element computer program developed for determining stresses in standard involute spur gears. It is developed so that the user need supply only a minimum of data defining the gear and tooth geometry, the gear material, and the loading conditions.

THEORETICAL BACKGROUND

UCIN-GEAR employs standard methods of two-dimensional finite element analysis. Quadrilateral isoparametric elements together with automatic mesh generation are used. Stresses and displacements are determined at each node. Centroidal stresses for each element are determined. Mohr's two-dimensional eigenvalue analysis is used to determine the principal stresses, the maximum shear stresses and their locations.

FIELD OF APPLICATION

Geometrical. The program calculates stresses and displacements in spur gear teeth. Number of teeth, diametral pitch, pressure angle, hob cutter radius, and rim thickness to dedendum are user input parameters. Also the user may select from the AGMA Composite System or the AGMA Standard Gear System. The gear tooth rim may be either fixed or free.

Material. The gear material is assumed to be linear elastic. The user need supply values for Young's modulus of elasticity and Poisson's ratio.

Analysis capabilities. Static analysis.

Loading. Forces may be applied at any of 10 equally spaced points along the tooth profile from the tooth tip to the lowest mesh point.

PROGRAM DESCRIPTION

Method. Two-dimensional finite elements.

Type of elements. Two-dimensional quadrilateral.

Program structure. The user supplies the following *input data*:

(1) Title
(2) Number of teeth, diametral pitch, pressure angle, hob cutting radius, ratio of rim thickness to dedendum.
(3) Young's modulus, Poisson's ratio.
(4) Force magnitude, force location, code for driver or follower gear.
(5) Gear tooth system, element mesh density, boundary condition code.
(6) Continuation code for repeated analysis with different loading.

As *output* the program provides both numerical and graphical data. The numerical output data are:

(1) An "echo" or copy of the input data.
(2) A description of the basic tooth proportions.
(3) A listing of nodal coordinates.
(4) A listing of the elements and their corresponding nodes.
(5) A listing of nodes *fixed* by the boundary conditions.
(6) A description of the loading.
(7) A listing of the horizontal and vertical forces on each node.
(8) A listing of the horizontal and vertical displacements of each node.
(9) A listing of the state of stress at the centroid of each element: The horizontal and vertical normal stresses and the shear stresses relative to the horizontal and vertical directions are listed. Also, using Mohr's circle, the maximum and minimum normal stresses and the maximum shear stresses are computed and listed. Finally, the inclination angle of the maximum stresses is listed.
(10) A listing of the state of stress at each node of each element, as in (9). This data is listed by element number.
(11) A listing of the state of stress at each node, as in (9). This data is listed by node number.

By using the input data together with the numerical, finite element results it is possible to develop a graphical representation of the tooth form, the finite element model, the displacements, and the stresses.

Such a representation has been developed for use with the University of Cincinnati Computer Facility. Specifically, the following graphs have been developed:

(1) A representation of the tooth form and the finite element model with element and node numbers provided.
(2) A representation of the undeformed and the deformed gear tooth.
(3) A contour representation of the maximum principal stresses.
(4) A contour representation of the minimum principal stresses.
(5) A contour representation of the maximum shear stresses.

Similar graphical representations can be developed at other computer facilities by using the same algorithms used with the University of Cincinnati facility.

UCIN-GEAR: A Finite Element Computer Program

HARDWARE CAPABILITIES

Configuration. The program is currently active on the University of Cincinnati IBM Computer.

Type of computer. The program should operate successfully on any standard mainframe computer with a Fortran compiler.

Peripherals. A plotter is useful for displaying the output data.

Operating system. Any of the standard operating systems are acceptable.

Media. The program can be transferred between computers by magnetic tape.

EXAMPLE OF APPLICATION

Test case. To illustrate the input data consider a driving gear with 20 teeth, a diametral pitch of 1.0, and cut by a hob with corner radius of 0.2386 in. Let the gear have a ratio of rim thickness to dedendum of 0.1. Let the gear be loaded with a unit load at the pitch point, and let it have a free rim. Finally, let there be 6 vertical finite element divisions.

For this configuration, the input data lines are:

Heading:	20th	1.0DP	0.2386RF	0.10TK	
Gear parameters	20	1.0	20.0	0.2386	0.10
Material properties	3.0D+7	0.3	0.01		
Loading	+1.0	5	1		
Tooth system and Finite element mesh	1	3	0		

Using the above numerical data, together with graphical algorithms prepared for the University of Cincinnati Versatec Plotter, a graphical representation of the results is obtained. For the foregoing input data, the graphical output is shown in Figs. 1-5.

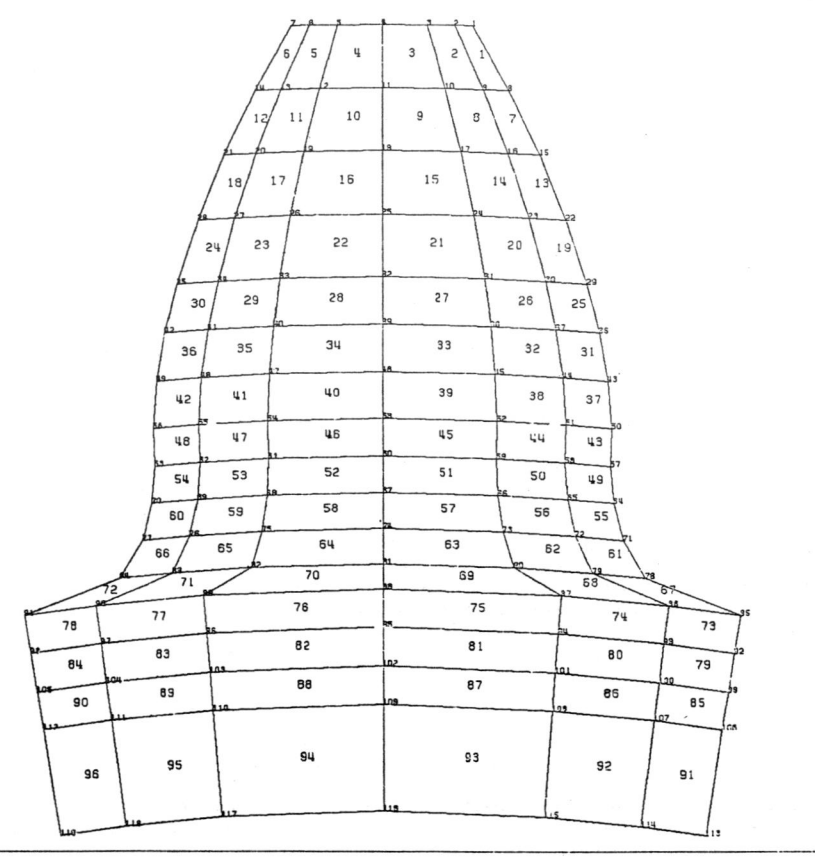

Fig. 1.

DEFORMED STRUCTURE
    EXAMPLE : 20TH 1.0DP 0.2386RF 0.10TK

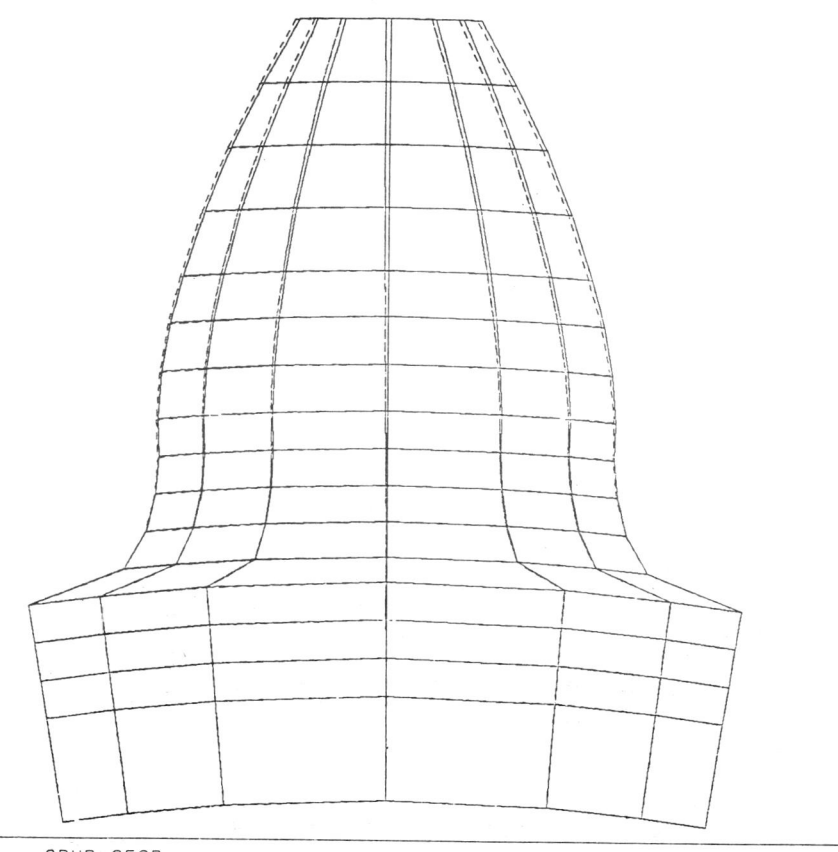

SPUR GEAR

DIAMETRAL PITCH   1.00
NO OF TEETH   20
PRESSURE ANGLE   20.00
HOB CORNER RADIUS   0.2386
RIM THICKNESS (%)   0.1000

Fig. 2.

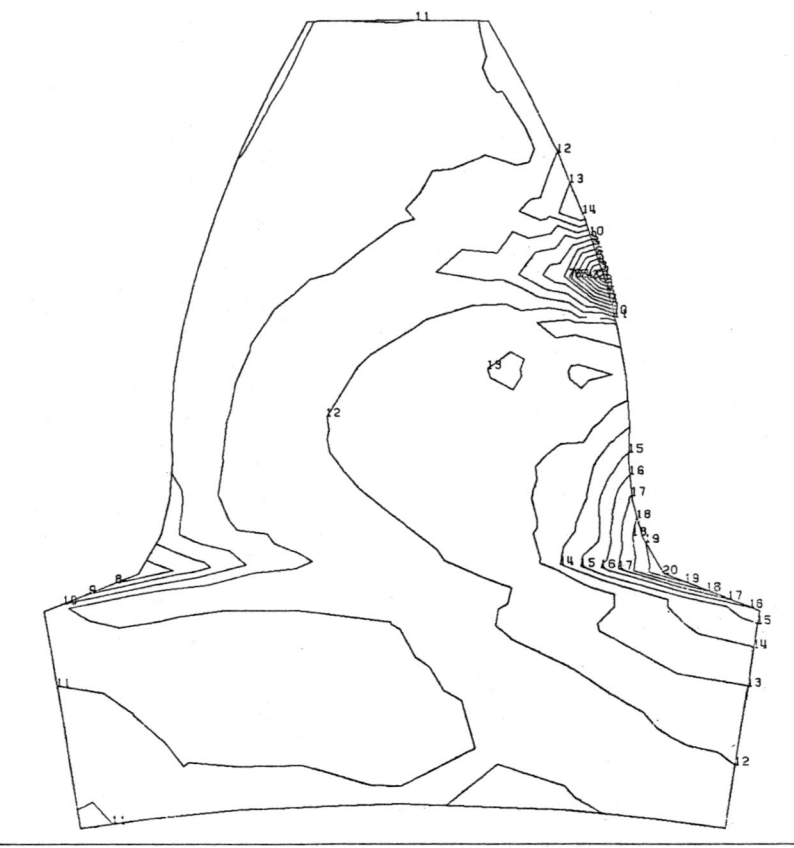

```
CONTOUR PLOT            - SIG MAX -
  EXAMPLE :  20TH  1.0DP  0.2386RF  0.10TK
```

SPUR GEAR			CONTOUR LEVEL		
DIAMETRAL PITCH	1.00	1	−3.017997	11	0.073052
		2	−2.708891	12	0.382157
NO OF TEETH	20	3	−2.399787	13	0.691262
		4	−2.090682	14	1.000368
PRESSURE ANGLE	20.00	5	−1.781577	15	1.309472
HOB CORNER RADIUS	0.2386	6	−1.472473	16	1.618576
		7	−1.163368	17	1.927681
RIM THICKNESS (%)	0.1000	8	−0.854263	18	2.236786
		9	−0.545158	19	2.545891
		10	−0.236053	20	2.854994

Fig. 3.

CONTOUR PLOT — SIG MIN —
EXAMPLE : 20TH 1.00P 0.2386RF 0.10TK

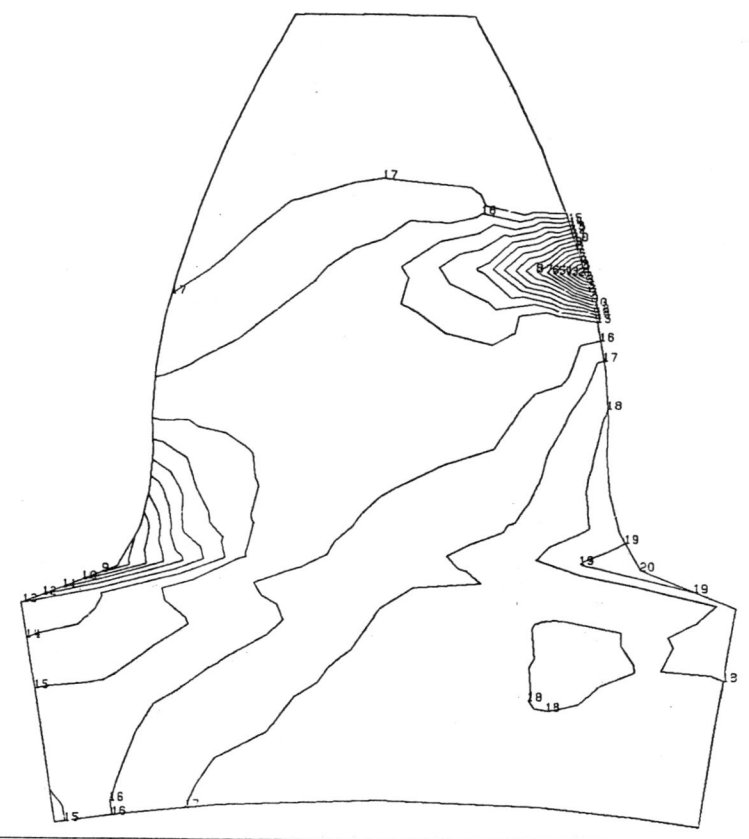

SPUR GEAR		CONTOUR LEVEL			
DIAMETRAL PITCH	1.00	1	-6.093995	11	-2.443473
		2	-5.728945	12	-2.078420
NO OF TEETH	20	3	-5.363890	13	-1.713367
		4	-4.998839	14	-1.348316
PRESSURE ANGLE	20.00	5	-4.633788	15	-0.983263
		6	-4.268734	16	-0.618211
HOB CORNER RADIUS	0.2386	7	-3.903681	17	-0.253159
		8	-3.538630	18	0.111894
RIM THICKNESS (%)	0.1000	9	-3.173576	19	0.476946
		10	-2.808525	20	0.841999

Fig. 4.

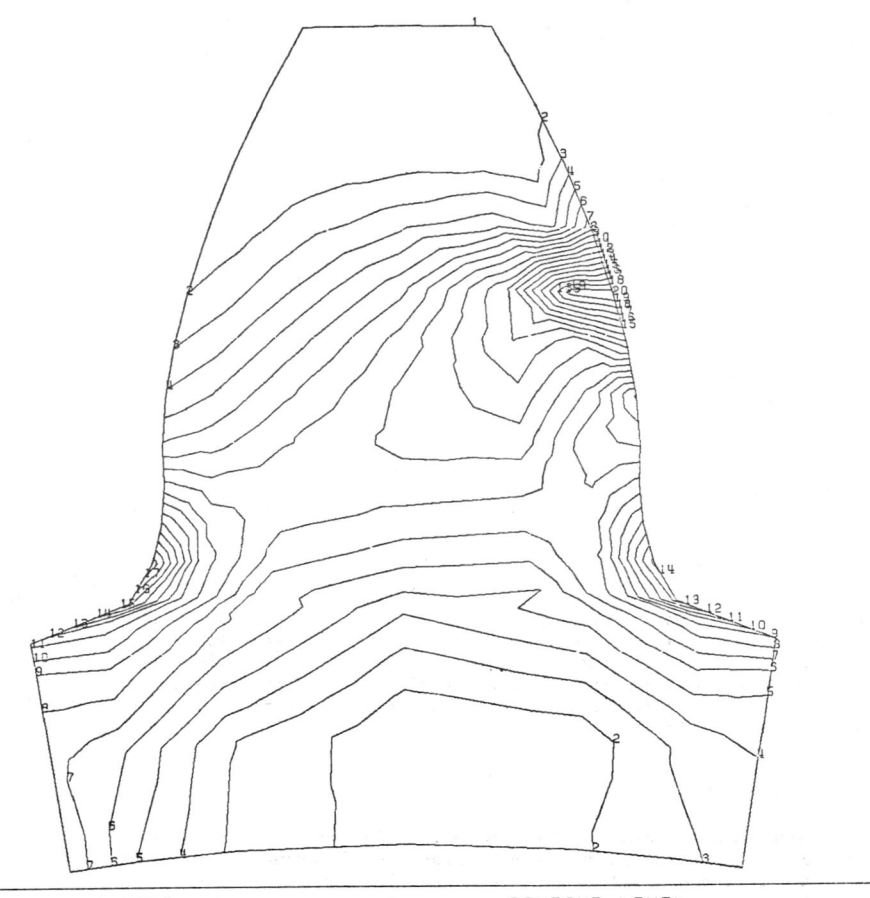

CONTOUR PLOT — TAU MAX —
EXAMPLE : 20TH 1.0DP 0.2386RF 0.10TK

SPUR GEAR		CONTOUR LEVEL			
DIAMETRAL PITCH	1.00	1	0.009740	11	0.814086
		2	0.090175	12	0.894521
NO OF TEETH	20	3	0.170609	13	0.974956
		4	0.251044	14	1.055389
PRESSURE ANGLE	20.00	5	0.331479	15	1.135824
		6	0.411913	16	1.216259
HOB CORNER RADIUS	0.2386	7	0.492348	17	1.296694
		8	0.572782	18	1.377128
RIM THICKNESS (%)	0.1000	9	0.653217	19	1.457562
		10	0.733651	20	1.537997

Fig. 5.

SURVEY OF GENERAL PURPOSE FINITE ELEMENT AND BOUNDARY ELEMENT COMPUTER PROGRAMS FOR STRUCTURAL AND SOLID MECHANICS APPLICATIONS

J. Mackerle

*Linköping Institute of Technology, Department of Mechanical Engineering,
S-581 83 Linköping, Sweden*

ABSTRACT

There is a large number of general purpose finite element computer programs available on the market today and the choice of which program to use is therefore very difficult. In the program evaluation there are too many parameters of objective and subjective characters to take into account. However, each program has a relative strength or weakness when used for analysis of a certain problem. Which program is most suitable and effective for the potential user is very much dependent on his/her working field, type of application, hardware available, problem size, etc

Firstly the potential user is faced with the problems of getting information about programs in existence. This paper is concerned with a review of general purpose finite element and boundary element computer programs that are commonly available for the users community. Programs are ordered alphabetically and their capabilities are presented both in a descriptive and in a tabular form. It is of course not possible to present all the existing programs. For a more complete survey covering several hundreds of programs the reader is referred to the available literature.

INTRODUCTION

After more than a decade of development, a wide variety of finite element computer programs are currently being used in different fields of engineering. Programs developed could roughly be classified into two main categories - general purpose programs and special purpose programs respectively.

General purpose programs are large programs/systems that are oriented towards general use in different application fields. Special purpose programs are relatively small specific programs with enhanced features for certain analysis capabilities. Their use is limited to a specific application area.

The potential user is faced with three main problems. The first is how to find information about existing programs. When the sources of information are known, the second problem is how to sort out the programs of interest. When a number of programs is finally selected the third problem is to sort out the best program for the user's needs.

It is well known that developing a finite element code requires a lot of manpower and professional experience. The price of the commercially available programs seems at a first sight to be rather high. If you, however, compare this to the cost for you to develop your own program, the price in fact is not excessive. You obtain a "product" developed by experts on programming and the finite element technique.

The objective of this paper is to review the commonly available general purpose software for the structural mechanics analysis. Presented programs, all commercially available, are divided into two main groups from the theoretical point of view, namely

(a) general purpose finite element programs
(b) general purpose boundary element programs

The choice of programs for presentation is made from the following criteria: widely used; general purpose; represent recent developments; multinational origin; commercially available; availability of documentation.

The final program selection must, however, be based on a detailed study of program documentation and all available literature about the program. For a partial list of books describing software in more detail, the reader is referred to the Appendix I.

Review of finite element and boundary element programs in this paper consists of two parts: a descriptive part and a tabular one. The tabular presentation is not detailed, it can serve for a more convenient comparison between different programs and for a "quick" dissemination of information.

Programs and their capabilities have been extracted from the structural mechanics data base MAKEBASE being currently developed by the author at the Linköping Institute of Technology, Department of Mechanical Engineering. This data base serves for information storage and retrieval of literature references covering the finite element and boundary element techniques (retrospective back to 1975), information on pre-/postprocessor programs interfacing finite element and boundary element programs, and detailed information on finite element and boundary element software. Selection routine for literature references allows search for records which have a particular author name, journal name, and/or subject defined by one up to ten keywords. Selection routine for the software retrieval allows the user define his/her requirements by defining a search profile. Each item in this profile defines a capability or a characteristic feature that the user needs for the requested program. These requested capabilities are combined into a profile. Output from the data base is a list of programs (including source for program information) which fulfil user's requirements.

GENERAL PURPOSE FINITE ELEMENT COMPUTER PROGRAMS

Descriptive part for each program consists of the sections as follows:

- Program name
- Source for information
- General information
- Program capabilities
- Element library
- Material library
- Boundary conditions and loading
- Notable items
- Solution methods
- Pre/postprocessing

Survey of General Purpose Computer Programs

- Hardware
- Documentation

Pre- and postprocessing capabilities are not presented in detail. For future volumes of this book series we intend to prepare a more detailed review on this subject. The reader is also referred to the list of references for details of other reviews.

The following programs are included in the survey: ABAQUS, ADINA, ANSYS, ARGUS, ASAS, ASKA, BERSAFE, DIAL, EFESYS, FENRIS, FINEL/ICB, LARSTRAN, MAGNA, MARC, MSC/NASTRAN, NEPSAP, NISA, PAFEC, PAM, PLANS, SAMCEF, SAP7, SESAM-80, SPAR, STRUDL, TITUS, and WECAN.

Program Name: ABAQUS

Source: Hibbitt, Karlsson and Sorensen, Inc., 35 South Angell Street, Providence, Rhode Island 02906, USA. Also available through Control Data Cybernet services.

General information: ABAQUS is a relatively new general purpose program available on the market. It is designed specifically to serve for the advanced structural analysis. The program is aimed for applications including nonlinear effects. For linear analysis the simplicity and efficiency is maintained.

Program capabilities: Static and dynamic three-dimensional analysis, material and geometric nonlinear effects can be included. Small/large displacements, rotations and strains are permitted. Fully compatible steady-state and transient heat transfer. Temperature/stress analysis is fully coupled. Coupled seepage flow/stress-displacement analysis for soils. Fracture mechanics evaluations, J-integral calculation on any geometry. Any material model can be used with any element type. No limit on the number of different materials/elements in a model. Modal extraction for frequency determination or eigenvalue buckling load estimation. Transient dynamic analysis. Static preload analysis followed by transient dynamic analysis. Contact problems with friction may be analysed.

Element library: 1D, 2D and 3D elements are available. Stress element library includes: trusses, plane stress/strain membranes, axisymmetric, 3D solids. Beams straight or curved with different cross-section (general, box, I section, hexagonal, pipe, circular, rectangular). General layered shells, elbows, line spring elements to model part-through cracks in shells. Heat transfer element library: 1D, membranes, axisymmetric, 3D solids, and shells. Special elements for coupled temperature/stress problems, fuel rod subassembly interaction modelling, linear/nonlinear springs, interface elements for contact problems. Beam and shell elements are based on penalty methods. Reduced integration elements for efficiency and stress accuracy. Hybrid versions for incompressible cases.

Material library: Elastic- isotropic, orthotropic and anisotropic definitions. Linear elastic, hypoelastic, fully incompressible. Temperature dependence is allowed. Very general elastic-plastic and elastic-viscoplastic models. Isotropic- von Mises yield, anisotropic- Hill's anisotropic hydrostatically independent yield. Isotropic or kinematic hardening. Temperature dependence is allowed. Highly-nonlinear material properties are permitted. Viscoelastic response, creep and volumetric hardening laws, user subroutines for special creep laws. Properties for heat transfer analysis- isotropic, orthotropic or anisotropic conductivity as a function of temperature. Specific heat as a function of temperature.

Boundary conditions and loading: Linear and nonlinear kinematic conditions may be prescribed. Multi-point constraints, built-in mesh refinement, user subroutines, equation definition. Gap and friction conditions, and their combinations. Planar or 3D option. Loading permitted- node forces, distributed loading, thermal loading. Uniform and nonuniform. Pressure, centrifugal, Coriolis forces. Combined loading for multiple linear, elastic load steps. Combined loads during a step.

Notable items: Capability for both symmetric and nonsymmetric matrices. In nonlinear problems- direct user control or automatic control of increment size. Analysis procedures can be arbitrarily mixed. Multiple coordinate systems- Cartesian, cylindrical and spherical with any point of origin. Special translated and rotated systems. Skewed conditions, nodal transformation used. Restart option. Flexibility/user subroutines in modelling and analysis are available.

Solution methods: Displacement method with some hybrid formulations. Wavefront solution algorithm. Elastic reanalysis is based on original stiffness matrix. Geometric nonlinearities- linearized incremental method with Newton-Raphson iteration within an increment. Material nonlinearities- tangent modulus for time-independent response, initial strain for time dependent response. Dynamic analysis- Hilber-Hughes and Newmark Beta, use iteration within an increment, mid-step residual control for automatic step size control, impact equations for contact. Subspace iteration is used for the eigenvalue extraction. Convergence criteria are based on maximum allowable residuals.

Pre-/postprocessing: Free or fixed format, card or external file input. Set definition for easy cross-reference. Extensive input data checking, model geometry plot generation, incremental mesh generation. Numbering need not be sequential or consecutive. Plots for efficient model checking (full or partial), deformed geometry, contour plots, "moment diagram" plots for beams. Time history plots and variable versus variable plots of any variables. Contour plots on surfaces of solid elements.

Hardware: CDC, IBM, VAX, Univac, Prime.

Documentation: User's Manual, Theory Manual, Example Problems Manual, Systems Manual.

Program Name: ADINA

Source: K. J. Bathe, Massachusetts Institute of Technology, Dept. of Mechanical Engineering (3-356), Cambridge, MA 02139, USA. Available through ADINA Engineering, Munkgatan 20D, S-722 12 Västeras, Sweden or ADINA Engineering Inc., 71 Elton Ave., Watertown, MA 02172, USA.

General information: ADINA is a general purpose code developed for structural, heat transfer and fields problems analysis. Development started 1974 on experiences with program SAP IV and NONSAP. Emphasis is placed on the use of reliable and efficient FE techniques, and user conveniences for modelling and performing the anlayses. Suitable especially for solution of nonlinear problems. ADINA system consists of ADINA (displacement and stress analysis), ADINAT (heat transfer and fluid problems), ADINA-IN (input data generation) and ADINA-PLOT (graphical/alpha-numerical display of input/output data). ADINAF (2D fluid dynamics) is under development. Program capabilities are continuously improved. Users group does exist.

Program capabilities: Static and dynamic 3D analysis, material and geometric nonlinearities can be included. Large displacements, rotations and strains are permitted. Buckling and collapse analysis. Linear and nonlinear fracture mechanics. Wave propagation analysis. Linear and nonlinear steady-state and transient heat transfer analysis. Field problems such as seepage, electrostatics, electromagnetics and potential flow may be handled. Static and dynamic 2D and 3D contact problems may be investigated. Multiple, flexible or rigid bodies in contact, Coulomb friction for 2D analysis. Four analysis procedures are available: linear elastic, material nonlinear only, total Lagrangian formulation, and updated Lagrangian formulation. Dynamic analysis capabilities include computation of eigenvalues and eigenmodes, and dynamic response.

Element library: 1D, 2D and 3D elements are available. Stress element library includes: trusses, straight and curved beams, plane stress/strain membrane elements, axisymmetric, 3D solid elements, thick and thin shells. Heat transfer element library: 1D elements, membranes, axisymmetric, 3D solids. Conduction, convection and radiation for nodes, lines and surfaces. Special elements: cable, 2D and 3D fluid elements, pipe elbow. General user-supplied element.

Material library: Isotropic and orthotropic linear elastic. Multilinear isothermal plasticity, von Mises yield condition and flow rule, isotropic and kinematic hardening. Curve description soil and rock model. Soil and rock Drucker-Prager model. Creep for varying and cyclic stress conditions. Thermo-plasticity and creep, temperature-dependence is allowed. Rubber, Mooney-Rivlin model. Concrete model. Heat transfer analysis: constant/temperature-dependent conduction and specific heat, constant/temperature-dependent convection and radiation coefficients, latent heat effects. User-supplied material model is permitted.

Boundary conditions and loading: Multipoint constraints and prescribed displacements. Rigid link. Loading- node forces, distributed surface pressure, gravity, ground acceleration, centrifugal loading, deformation-dependent pressure loading. Concentrated and distributed heat flux, specified temperatures, distributed internal heat generation.

Notable items: Addition/deletion of elements during analysis. Cyclic symmetry. Substructuring in static and dynamic analysis. Skewed conditions permitted. Restart from static or dynamic initial conditions. Restart from any previous load step solution. Response spectrum calculations through ADINA-PLOT. Multiple coordinate systems in ADINA-IN- Cartesian, cylindrical and spherical, any orientation available.

Solution methods: Equation solution by compacted column reduction scheme. Dynamic analysis- mode superposition, implicit time integration (Wilson, Newmark). Wave propagation- explicit central difference method. Extraction of frequencies/mode shapes- Bathe's subspace iteration, Lanczos method for starting subspace, determinant search method. Creep analysis- explicit or implicit time integration. Automatic load incrementation to calculate structural collapse/post-collapse response. Nonlinear static analysis-incremental solution, modified Newton-Raphson method, BFGS method. Heat transfer analysis- explicit Euler forward, implicit trapezoidal or Euler backward method. Equilibrium iteration in steady-state and transient nonlinear analysis, Newton-Raphson method.

Pre/postprocessing: Data is stored for easy and effective retrieval in a specially designed database. ADINA-IN is a preprocessor program, ADINA-PLOT is a postprocessor program. A number of general purpose pre- and postprocessors are used in connection with ADINA, i.e. FEMGEN, GIFTS, INGEN. ADINA-IN-Interactive and batch mode of operation, free-format command language, mesh generation with straight or curved boundaries, generation of constraint equations, all data for substructures can be input as for the main structure, bandwidth minimization, mesh plotting, colour selection for lines and text.

ADINA-PLOT- free-format command language, batch or interactive mode of operation. Plotting- undeformed/deformed meshes, mode shapes, time-history, calculation and plot of user-defined resultants/design variables, vector plots of principal stresses and strains, plotting of variables for user-defined lines and points. Searching for extreme values/exceedances.

Hardware: CDC, IBM, Univac, VAX, Prime, Burroughs, Cray, Cyber.

Documentation: Users Manuals, Theory and Verification Manuals, Book and Video Course, Documentation on Usage.

Program Name: ANSYS

Source: Swanson Analysis Systems, Inc., P.O. Box 65, Houston, PA 15342, USA.

General information: The reader is referred to the separate paper in this book.

Program Name: ARGUS

Source: Merlin Technologies, 977 Town and Country Village, San Jose, CA 95128, USA. Available worldwide through Control Data Co Cybernet service.

General information: ARGUS is a general purpose program developed from the earlier code NEPSAP. Structural linear and nonlinear analysis may be handled. The program is continuously under development, in commercial environment since 1978. ARGUS consists of a series of processors including the input processors, a graphic package, analysis package, modal analysis processor, etc. The program is designed as an overlay structure and uses out-of-core solver with dynamic allocation programming technique.

Program capabilities: 3D static and dynamic analysis can be handled. Extensive capabilities for nonlinear analysis and composite structures are included. Multiaxial failure criteria for composite materials. The program is capable of large displacements, thermo-elastic-plastic and creep analysis of arbitrary structures. Geometric nonlinearities are not limited to small strains. Creep, nonlinear buckling, nonlinear transient thermal stress analysis can be performed. Impact analysis is possible.

Element library: Only stress element library for 1D, 2D and 3D elements is available. It is composed of isoparametric membranes, thick and thin plates and shells, axisymmetric elements, 3D solids, prismatic beam element with eccentricity option. Special elements includes stiffeners, plate/shell anisotropic multilayered elements, isoparametric element for incompressible materials, etc.

Material library: Linear elastic, isotropic and orthotropic, temperature-dependent. Thermoplastic model is based on the von Mises nonisothermal yield function and its associated flow rule, isotropic and kinematic hardening. Creep- Odquist-Norton stationary creep power law with time-hardening and temperature-dependence. Orthotropic elasticity and composite materials. Isotropic incompressible elastic material properties.

Boundary conditions and loading: Imposed zero and non-zero displacements. Multipoint constraint conditions. Boundary conditions can be generated by the ARGUS preprocessor. Permitted loading includes nodal loads, pressure loads, steady-state and transient thermal loads, gravitation. Loads can be defined as an arbitrary function of time. Linear multiload analysis may be handled.

Notable items: Partial and complete restart capability. Arbitrary coordinate

Survey of General Purpose Computer Programs

system can be defined- Cartesian, cylindrical and spherical. Skewed conditions are permitted. Minimum/maximum stress tables can be printed. Node renumbering option.

Solution methods: Displacement formulation is used. Equation solver- variable bond skyline method. Dynamic analysis- Houbolt, Park, Newmark or user-defined integration. Nonlinear analysis- incremental technique, including provisions for equilibrium correction for improved convergence and accuracy; modified tangent stiffness approach with user-specified load and temperature increments. Large displacements- updated Lagrangian formulation.

Pre/postprocessing: Different pre- and postprocessing programs are included. Preprocessor for mesh and boundary conditions generation. Output-volume is user-selectable. Graphics routines for result presentation are included in the graphics module. ARGDYN- postprocessor for dynamic modal analysis.

Hardware: CDC, Univac.

Documentation: User's Manual (2 Volumes).

Program Name: ASAS

Source: Atkins Research and Development, Woodcote Grove, Ashley Road, Epsom, Surrey KT18 5BW, England.

General information: Development started in 1969, the first commercial version was released at the end of 1971. In present, several versions are available. The most used one includes capabilities for linear elastic static and dynamic analysis, where heat transfer part is also included. Nonlinear version, ASAS-NL, was released in 1980. COMPAS (Component ASAS) is a multilevel substructure analysis system which embodies all the features of the ASAS program. In addition, various pre- and postprocessing programs can be included into the system. ASAS consists of several modules, each performing one logical process and communicating with the steering program through a single interface. User's group does exist. User's conferences are arranged.

Program capabilities: The linear version of ASAS consists of ASAS-G (statics, dynamics, steady-state heat), ASAS HEAT (transient heat transfer), ASDIS (graphics), and various pre-/postprocessors. ASAS-NL is a quite separate program for nonlinear analysis. ASAS can solve linear elasto-static isotropic and anisotropic problems. Eigenvalue, linear response and spectral analysis of dynamic systems may be handled. Steady-state and transient heat transfer problems can be analysed. Time steps for transient analysis can be calculated automatically by the program or may be defined by the user. Field problems such as electrostatics, seepage, etc. can be solved. Nonlinear capabilities include material or geometric nonlinear static analysis. Material nonlinearity models- plasticity, creep and swelling. Large deflection and buckling analysis is possible.

Element library: Displacement, stress and force equilibrium formulations. Stress element library includes ties, beams, membranes, plates, shells, axisymmetric and 3D solid elements. Semiloof element is available. Most of these element types can also be used for the heat transfer analysis. Special elements- crack elements for stress intensity factor calculations, warped elements with stiffeners, shear panel, general element for model with generalized stiffness and mass. For dynamic analysis, both lumped and consistent mass formulations are available.

Material library: Linear elastic, isotropic and anisotropic. Temperature-dependent material properties. Thermal analysis- isotropic or anisotropic material properties for all elements. For transient analysis, prescribed temperatures, surface fluxes, internal heat sources, film coefficients and surface conductivities may randomly varied with time, continuous/ discontinuous variations are allowed.

Boundary conditions and loading: Suppressed and prescribed displacements in any direction for any DOF. Kinematic supports for dynamic response. Constrained relation possibilities. Nodal, body force or gravitational, distributed, constant and varying pressure, temperature and centrifugal loading are allowed. Initial stresses. For dynamic analysis- prescribed time dependent forces, prescribed accelerations. Any number of loading cases is permitted.

Notable items: Multilevel substructuring. Comprehensive range of pre- and postprocessor programs. Extensive data-checking procedures. Automatic restart option, currently with 24 restart stages. Skewed conditions allowed.

Solution methods: Equation solver- partitioned Cholesky, out-of-core, frontal solution technique for smaller problems. The program uses the frontal solution, and switches, if necessary, to the hypermatrix scheme. The user can force "only" the hypermatrix solution by means of a control option. Condensation- Guyan's reduction. Eigenvalues- subspace iteration, Jacobi, Householder, QL, Householder Sturm sequence. Dynamic response- Newmark Beta method. Nonlinear problems- any required combination of iterative/incremental procedures.

Pre/postprocessing: Pre- and postprocessors available can perform i.e. graphical representation of geometry, topology and results; mesh and input data generation, automatic load calculation; checking against design rules/codes. Some of the programs available: ASBAND- bandwidth optimization, ASDIS- mesh display, ASPECT- stress contour plotting, MESHGEN, MESHMOD- 2D mesh and data generator, TUBEGEN- mesh generation for multibranch intersecting tubes of different sizes, COSET- factoring and combination of results from different load cases, ASAS-WAVE- evaluates statically equivalent loads on marine structures, BEAMST- check the strength of offshore structures against AISC, API. Interfaces are available to the general purpose pre/postprocessors PDA/PATRAN, SUPERTAB, FEMGEN and FEMALE.

Hardware: CDC, IBM, Univac, Honeywell, DEC, Prime, Xerox, Sigma, ICL.

Documentation: ASAS User Handbook, ASAS User Manual, Programmers Manual, Theoretical Documentation, Element Evaluation Reports.

Program Name: ASKA

Source: ASKA Group, Institute for Statics and Dynamics, University of Stuttgart, D-7000 Stuttgart 80, West Germany. Available through the IKOSS, IKO-Software Service GmbH, Albstadtweg 10, D-7000 Stuttgart 80, West Germany.

General information: Development started at the University of Stuttgart in cooperation with the Imperial College in London, approximately in 1965. Later on the system was commercially implemented with the aid of IKOSS. Developing is still continuing. The system consists of main programs and special-purpose programs. Each year the user's conference is being held. ASKA can be purchased, but rental and licensing is possible as well. ASKA 80 version is under development.

Program capabilities: A wide range of static and dynamic, linear and nonlinear problems may be solved. The following programs are included in the ASKA system- ASKA I (linear elasto-static analysis), ASKA II (linear dynamic analysis), ASKA III-1 (elasto-plastic problems and creep), and ASKA III-2 (linear buckling). Special programs include- ASKA-T (steady-state and transient temperature field analysis), ASKA-HS (harmonic analysis), ASKA-CS (cyclosymmetric structures and loadings), ASKA-FM (linear elastic and elasto-plastic fracture mechanics), ASKA-PIPE (static and dynamic analysis of piping systems), ASKA-CA (linear contact problems), etc. Material nonlinearity models include elastic-plastic, visco-plastic and creep. Thermal effects may be incorporated. Dynamic analysis options- eigenvalue, eigenfrequencies and modes, linear response, impact, and spectral analysis. Modal, non-modal and proportional damping. Linear fracture problems- energy release rate, stress intensity factor. Elasto-plastic fracture mechanics- crack-opening displacement, J-integral. Rock and soil problems may be handled.

Element library: Stress element library includes 1D, 2D and 3D elements such as flange, extensive range of beams, plates, thin and thick shells, axisymmetric and 3D solid elements. Heat transfer element library- heat conduction isotropic and anisotropic elements, membranes, axisymmetric and 3D solid elements. Heat transfer elements are compatible with heat conduction elements. Special elements: shear panel, elements for St. Venant torsion, etc. Users-defined elements can be easily incorporated into the program.

Material library: Linear elastic, isotropic and anisotropic. For elastic-plastic material the von Mises yield criterion with associated Prandtl-Reuss flow rule is adopted. Isotropic, kinematic and mixed hardening. A Drucker-Prager generalization of the linear and parabolic Mohr-Coulomb laws with non-associated flow rule may be adopted. Arbitrary temperature-dependent material properties. Creep- incompressible flow, isotropic time or strain hardening, arbitrary uniaxial material laws. Concrete material model.

Boundary conditions and loading: Local, external, prescribed, and suppressed freedoms. Constrained relations are possible. Special boundary conditions- sliding, lift-off. Skew boundary conditions are permitted. Loading- nodal loads, distributed loads, dead weight, acceleration forces, gravity, thermal loads. Load generators for i.e. water pressure, wind load, soil pressure are available. Analysis of axisymmetric structures with nonaxisymmetric loads can be handled. Loading cases can be combined. Initial stresses and strains are allowed. Up to 720 load cases in one run can be performed.

Notable items: Free-format data input. A multilevel substructuring technique for all types of analysis. Global and local coordinate systems. ASKA has several "rigid formats" for a specific range of applications. For problems that cannot be solved by means of the rigid format, the user can write own steering program to perform the appropriate analysis steps. Comprehensive data checking. Restart option is available. General hypermatrix operations. Own defined subroutines can be incorporated into the program. User access to internal data structure makes it possible to apply ASKA as an effective research tool. Complete compatibility between different programs for static, dynamic and heat transfer analysis.

Solution methods: Displacement formulation. Equation solver- partitioned Gauss elimination. Condensation- Guyan's reduction. Eigenvalues- simultaneous vector iteration, Householder, Jacobi. Dynamic response- modal superposition with the option of analytical integration of harmonic linear and delta function, numerical integration of user specified functions, direct integration. Nonlinear problems- initial strain/stress method. Harmonic analysis is based on Fourier series.

Pre/postprocessing: FEPS (Finite Element Plotting System) is the only program that goes as the standard graphics package with ASKA. Plotting of undeformed/deformed structure, isocurves, stress intensity curve over previously defined lines, nodal point stresses in a diagram form, displacements and nodal point stresses in a vector form. Interfaces are available to the general purpose pre-/postprocessors FEMGEN and FEMVIEW.

Hardware: CDC, IBM, Univac, Cyber, ICL, Burroughs, Honeywell, Prime, Amdahl, Robotron, Cray, VAX.

Documentation: Extensive range of documentation is available. Users Manuals, Theoretical Manuals, Processor Descriptions, Application Experiences.

Program Names: BERSAFE

Source: BERSAFE Advisory Group, CEGB, Berkeley Nuclear Laboratories, Berkeley, Gloucestershire, GL13 9PB, England.

General information: Development started 1967 at the Berkeley Nuclear Laboratories. The program was originally developed for the nuclear industry but its versatility has permitted its application to a large number of other problems in different fields of engineering. The program system consists of many modules which can be classified into group of data generation, analysis, and data presentation. BERSAFE system is built up of FLHE (heat transfer), BERSAFE (static analysis), BERDYNE (dynamic analysis), and various pre- and postprocessor programs. Different modules are in a continuous state of development. BERSAFE system is available also to users outside of the CEGB. Licensing agreement for "in-house" usage, or using the hardware of CEGB Computing Bureau. A user's group does exist. FELSET is a computer aided design package built up around the BERSAFE- developed by J. Lucas Ltd.

Program capabilities: Linear and nonlinear static, linear dynamic, as well as steady-state or transient heat transfer analyses may be performed. Limited geometric and material nonlinearities may be included in the static analysis. Geometric nonlinearities- large deformations only. Material nonlinearities- elastic-plastic properties and creep. Plasticity and creep may also be performed with thermal transient loading. BERDYNE can handle linear dynamic problems of arbitrary 2D and 3D structures subjected to free, forced undamped or forced damped vibrations. Extensive capabilities are available for the solution of fracture mechanics problems. Evaluation of the stress intensity factors, several techniques are available (energy methods, substitutions methods). ELOPPER module calculates J and J* type integrals. Field problems, i.e. electromagnetic potential calculations may be handled.

Element library: Some 60 different elements are available in the stress and heat transfer element library. These are membranes (plane stress/strain), axisymmetric elements, beams (Euler, Timoshenko), thin and thick shells, plates, and 3D solid elements. Semiloof elements are available. Heat conduction elements. Special reinforcing elements, 2D and 3D crack-tip elements, torsion elements, pipe elements. User-specified stiffness matrix.

Material library: Linear elastic isotropic and orthotropic. Elasto-plastic material properties- Tresca, von Mises yield criterion. Isotropic and kinematic hardening. Time or strain hardening for creep problems. Temperature-dependence is permitted. Special laws are included to deal with concrete creep and graphite. Routines for special rock behaviour.

Boundary conditions and loading: Zero or non-zero DOF can be prescribed. Generalized constraints facility. Arbitrary form of loading. Point loads, line loads with facial pressures and applied strains defined within an element, body loads, cetrifugal loading, nodal temperatures. Heat transfer

Survey of General Purpose Computer Programs 267

analysis- time-dependent loadings as internal heat generation, surface flux, forced convection, free convection and radiation. Up to 50 loading cases/run. Loading for dynamic analysis- harmonic forces, forces expressed as analytic functions or a Fourier series, impulsive forces, forces expressed as sections of cubic polynomials, support accelerations.

Notable items: Either fixed or free-format input. Multilevel substructuring is included in the program. Substructures may be mixed with any standard elements. Automatic bandwidth optimization. Special techniques for fracture mechanics applications are available. Cartesian, spherical or cylindrical coordinate system. A decoupling feature enables any number of DOF at any node to be independent between different elements meeting at that node. A useful option for i.e. shear slip or crack problems. A roundoff criterion to check on the deterioration of solution accuracy. Restart is possible.

Solution methods: Displacement formulation. Equation solver- frontal solution method. Eigenvalues- Sturm sequence. Dynamic response- direct time integration. Nonlinear problems- initial stress method, initial strain method, tangential stiffness method. Transient heat transfer problems- Crank-Nicholson procedure.

Pre/postprocessing: Many pre- and postprocessor programs are available. To name some of them: BERPLOT- general purpose plotting, BERMAGIC- interactive/batch mesh generator and editor, ELOPPER- selection and manipulation of analysis results, PLOPPER- processing of results obtained during a plasticity or creep computations. POINTA is an interactive pre- and postprocessor for mesh generation, editing, mesh display and 3D result interpretation. Loads and boundary conditions can also be graphically checked. Colour graphics is included in the BERMAGIC.

Hardware: CDC, IBM, Univac, Burroughs, VAX, Prime, GEC, Amdahl, ICL.

Documentation: Users Manuals, guides to all modules, theoretical reports for some of the modules, application experiences.

Program Name: DIAL

Source: Lockheed Missiles and Space Co., P.O. Box 504, Sunnyvale, CA 94086, USA. Developed by N. A. Cyr and G. H. Ferguson. For information contact Manager of Structural Department, 81-12, Building 154.

General information: DIAL was released in 1975. The system consists of several independent processors. These communicate via a generalized data base and file management system. Each processor is driven by a free-field command language. All processors can be run in an interactive or batch mode. Because of the modular database architecture is used, the user control is versatile and economical. Program is undergoing continual development. It is also available for users outside the Lockheed Missiles and Space Co.

Program capabilities: Static and transient linear/nonlinear analysis of arbitrary structures may be performed. Geometric (large displacements, large rotations, large strains) and material nonlinearities (plasticity, creep) are allowed. Also vibration and bifurcation buckling analysis is possible. Both these analyses can be performed from either a linear or nonlinear prestress state. Lumped or consistent mass matrix generation. Contact problems can be handled. Analysis of layered composite materials is possible.

Element library: Element library contains 2D and 3D isoparametric elements (linear, parabolic, cubic, mixed order) such as membranes (plane stress/plane strain), plates, thick and thin shells, straight and curved beams, axisymmetric elements (with and without Fourier modes), and 3D solid elements. Shells and curved beams are degenerated from the 3D solid isoparametric elements. Special elements- gap and contact element, constraint elements, lumped nonlinear springs, dashpot elements, stiffeners. User-defined stiffness.

Material library: The following models are available- isotropic and general anisotropic linear elastic, isotropic elastic-plastic with kinematic or isotropic strain hardening, small strain isotropic incompressible, large strain incompressible hyperelastic, isotropic high temperature creep, nonlinear elastic orthotropic-different tension and compression properties, layered orthotropic shells and plates, general nonlinear material and damping properties for use with the nonlinear spring and dashpot elements, including sliding friction. All materials can be temperature-dependent.

Boundary conditions and loading: Prescribed and suppressed boundary conditions are allowed. Skewed boundary conditions. Symmetry planes are defined easily. Load generators are available. Consistent load vectors for pressures, tractions, body forces, inertia loads, and temperatures. Conservative or nonconservative normal pressure loads can be specified. Concentrated nodal loads, displacements and initial conditions for transient response are permitted.

Notable items: Input may be any mixture of free-field commands and FORTRAN subroutine calls. Substructuring technique is included- static condensation or a general Rayleigh-Ritz approach for dynamic substructuring. Axisymmetric structures under nonsymmetric loadings may be analysed by using Fourier decomposition. Global and local coordinate systems can be defined. Bandwidth optimization routine. Restart capability at almost any analysis step. Intersection boundaries between shells can be generated. A standard Cartesian coordinate system is used but if desired a mesh can also be defined in any of several shell curvilinear coordinate systems. Multiple meshes can be defined.

Solution methods: Displacement formulation. Eigenvalue extraction- inverse iteration method. Geometric nonlinear analysis- total Lagrangian formulation. Nonlinear static analysis- many variations of Newton-type methods with automatic load step selection.. Nonlinear transient analysis- several implicit schemes using Newton methods, and an explicit central difference scheme. Automatic time-step selection.

Pre/postprocessing: Several modules are available. MESH- automatic mesh generation. Two methods are used, namely i-j-k and isoparametric mesh generation. LOAD is a module for load generation. SPSYST are two processors for substructure generation. BAND- bandwidth optimization. PLOT/SCOPE is a processor for the output printing/plotting. The following types of plots can be handled- undeformed/deformed meshes, contour plots, load/time history plots, stress/strain/displacement section plots, x-y plots, node quantity display plots.

Hardware: Univac, CDC Cyber, VAX.

Documentation. DIAL Users Manual.

Program Name: EFESYS

Source: R. Dungar, Motor-Columbus Consulting Engineers Inc., Baden, Switzerland.

General information: Program development started in early seventies. Presently version, EFESYS II, is a general purpose package where special capabilities

for the analysis of dams and offshore structures have been incorporated.
The program uses an efficient and comprehensive system of data storage.
All input data, results, etc, are placed into one data bank. The storage
is structured so that the use is made of one internal disc stream/one large
in-core virtual memory array block. Thus the complete data bank may be
easily stored and re-used in other stages of analysis. This data
organization enables the EFESYS to be run on a variety of both a virtual
memory, or on nonvirtual memory machines with limited core size.

Program capabilities: Analysis of arbitrary 2D and 3D structures and their
foundations, subjected to a variety of static and dynamic loads can be
handled. Material nonlinearity may be included. Elastic-plastic and no-
tension analysis. Dynamic analysis- eigenvalue and eigenvector analysis,
linear response elastic, nonlinear analysis, dynamic analysis with radiation
damping of foundation. Fluid/structure interaction problems can be solved.
Seepage analysis, seepage-structure coupling, fluid add-mass coupling for
the linear and nonlinear dynamic analysis.

Element library: Structural element library contains isoparametric membrane and
3D solid elements. Fluid elements. New elements can easily be included
into the program.

Material library: Linear elastic, nonlinear elastic, elastic-plastic.

Boundary conditions and loading: Zero and nonzero boundary conditions can be
prescribed. Radiating boundary conditions in a dynamic calculation can be
introduced. Mechanical loading is accepted. Earthquake loading.

Notable items: Restart facility. New solution techniques for different types of
problems may easily be added.

Solution methods: Displacement formulation. Equation solver- sky-line method.
Elastic-plastic and no-tension analysis- initial stress or modified force
method. Eigenvalue analysis- inverse iteration. Linear elastic dynamic
analysis uses an eigenvalue-vector solution.

Pre/postprocessing: 2D interactive mesh generator and special-purpose 3D mesh
generators (i.e. for arch dams) are included. Routines for 2D and 3D stress
and displacement plots are available. Obtained results may be interactively
displayed and then plotted. Plotting either vectors or a series of contour
lines is possible.

Hardware: EFESYS is a machine independent program. It can be implemented on
machines ranging from a CDC 7600 to a small minicomputer. Presently is
the program running on a CDC and on a virtual memory Prime P400.

Documentation: Users Manual.

Program Name: FENRIS

Source: The Norwegian VERITAS, Hovik, Norway. The Norwegian Institute of
Technology, Division of Structural Mechanics, Trondheim, Norway.

General information: FENRIS is a new nonlinear finite element program system
developed as a joint project between the Norwegian Institute of Technology,
SINTEF and the Norwegian VERITAS. The program is a highly efficient also
for the linear analysis problems. The development started in 1980. The
program system is organized in four different levels- the system level,
substructure level, element level and material point level. Program
architecture is highly modular, which allows for easy modifications and

extension of the system. The various computational steps in the program are isolated as much as possible from each other.

Program capabilities: FENRIS is a general package for large scale static and dynamic nonlinear analyses. The formulation of motion used in the program accommodates for translations and rotations of unlimited size. The program also allows for introduction of finite size, rigid bodies at the nodes- these are denoted "potatoes". This concept is useful and more efficient than the conventional approach of introducing linear constraints between nodal DOF. Large deflections, large rotations. Buckling, postbuckling, plastic collapse studies. Reinforced concrete structures may be analysed. Solution of contact problems is possible. Fluid-structure interaction problems may be solved.

Element library: Nonlinear finite element subroutines are organized into the element library. This is continuously being extended. It contains bars, beams, membranes, thin plate/shell element, thick plate/shell element, axisymmetric and 3D solid elements. Special elements- springs, cable element, element for reinforced concrete, contact. The element nodes are standardized to have up to 3 translational and 3 rotational DOF. Spatial elements are compatible. New elements can be incorporated into the program.

Material library: A menu of material laws is included. User may add any particular law that he might want to use. Material models available- linear and nonlinear elastic, elasto-plastic with various types of hardening, an "overlay" model and a special model for concrete.

Boundary conditions and loading: Prescribed displacements or rotations are allowed. General load history concept which allows for arbitrary combinations of prescribed displacements, concentrated loads, conservative and nonconservative element loads, inertia load associated with acceleration histories, buoyancy forces, etc. Hydrostatic and hydrodynamic forces, initial stress/strain, temperature fields. Constant or deformation-dependent reference load vectors. Generation of load histories, combination by adding, by multiplying, by shifting. Any number of load cases or prescribed displacements may be specified.

Notable items: Global and co-rotated local coordinate systems. A structural system analysed may include rigid parts of finite dimensions denoted "potatoes". Interaction with other FE programs is possible, i.e. connection to the SESAM '80 is available. Results from new research can easily be incorporated into the system. Restarting is possible.

Solution methods: Equation solver- skyline block method. Static problems- the various incremental-iterative solutions are available. The load incrementation process can also be controlled automatically. Another nonlinear algorithm available is based on principle of minimizing unbalanced forces through adjustment of the loading level. Both total Lagrangian and updated Lagrangian description. Dynamic problems- The dynamic algorithm is a constant average acceleration method which may be combined with various types of equilibrium iterations. An algorithm for automatic computation of time steps is available.

Pre/postprocessing: Various types of pre- and postprocessor programs with graphical display are under preparation. These interact on the level 2 with other satellite programs.

Hardware: VAX, IBM, Prime, Cyber.

Documentation: Users Manual. Theoretical Reports.

Program Name: FINEL/ICB

Source: Principia Systems Ltd., Genesis Centre, Birchwood Science Park South, Warrington WA3 7BH, England. Developed at the Imperial College of Science and Technology, London in conjunction with Babcock Power Ltd.

General information: FINEL is a general purpose program capable of solving the stress analysis, thermal analysis and fluid flow problems. A fully integrated modular program design was implemented. The program consists of a series of libraries- module library, element library, load library, analysis library, region library, results library, and plotting library. The analysis is provided by selecting and executing a series of entries in these libraries. The analysis steps are controlled by the FINEL Executive which remains resident in a core. Modules do not communicate directly with each other. A single central Data Base contains all data, a random access. A fully automatic block storage scheme is used to utilize as much machine core storage as is available to the current job.

Program capabilities: 2D and 3D problems can be solved. Static, dynamic and heat transfer analysis. Both linear and nonlinear problems under either steady-state or time varying conditions may be handled. Material nonlinearities include plasticity models. Limit load analysis. Dynamic analysis- eigenvalue extraction, response spectrum analysis, impact. Fatigue and fracture analysis problems can be solved.

Element library: Element library contains more than 60 element types. Most elements can be used for both the stress and the conduction analysis. Beams, bars, membranes, plates, axisymmetric thick and thin shells, general shells, and 3D solid elements are available. Special elements- heat conduction element, element for dynamic damping, layered elements for composites.

Material library: Linear elastic, elastic-plastic.

Boundary conditions and loading: Zero and nonzero prescribed nodal values. Multipoint constraints conditions. Loading types available- nodal forces, pressures, temperatures, arbitrary acceleration, initial strain, initial stress. Heat transfer analysis- surface heat transfer, internal heat generation, defined heat flux. The load routines apply to all of the relevant elements in the program. There is a limit on max. 2000 load cases.

Notable items: Bandwidth optimization. Comprehensive error-checks. Various output and data dumping modules. New procedures can be added easily with all facilities of the system being immediately available to this new procedure. Substructuring technique. Restart at any step within the FINEL Executive is possible. The Cholesky factorization module can be stopped/restarted at any point within its execution. Results of the analysis may be selectively printed/plotted.

Solution methods: Not known.

Pre/postprocessing: Free-format data input, data checking and editing, interactive mode of operation. Semiautomatic 2D and 3D mesh generators are based on the division of the structure into simple regions which are then automatically divided into elements. Plotting options include undeformed/deformed mesh plots, contour plots, rapid coarse curve plotting or slower smooth curve plotting, plotting of variables along a user-defined line. A result surface can be plotted over a face. The plot module can be called at any point in the execution sequence. Combination of various load cases can be done. Stress recovery as a postprocessor operation.

Hardware: The program can be run on most computers.

Documentation: FINEL Programming Manual. Theoretical Project Reports.

Program Name: LARSTRAN 80

Source: Institute for Statics and Dynamics, University of Stuttgart,
    Pfaffenwaldring 27, 7000 Stuttgart 80, West Germany.

General information: Development started since 1977, first version released in
    1981. LARSTRAN is a general purpose program oriented to the solution of
    nonlinear problems in structural mechanics. The system consists of four
    parts- a library of standard computational methods to handle nonlinearity
    problems, a data base manager, a library of structural hypermatrix
    operations, and library of finite elements. The program is continuously
    under development. Maintenance and user support is provided on a long-term
    basis.

Program capabilities: Static elastic and inelastic problems, as well as linear
    and nonlinear stability, and dynamic analysis problems can be solved.
    Nonlinear elastostatics can include large displacements and/or large strains.
    Large strains plasticity/viscoplasticity. Special range of applications
    includes fracture mechanics, heat transfer, thermomechanically coupled
    problems, contact problems, analysis of wide-span lightweight structures, etc.

Element library: Flange elements, beam elements, axisymmetric, plane stress/
    strain membranes, 3D solids, plates and shells are available. All elements
    include large displacements, some of them large strains and material
    nonlinearities respectively. Special elements- curved cables. New elements
    can easily be added.

Material library: Linear elastic, nonlinear elastic, elastic-perfectly plastic,
    elastic-plastic strain hardening, high temperature creep. Temperature
    dependence is allowed. Isotropic and anisotropic material properties.

Boundary conditions and loading: Zero and nonzero boundary conditions can be
    prescribed. Arbitrary mechanical and thermal loads are permitted.
    Deformation-dependent loads, cyclic loading, contact loading, nonproportional
    loading.

Notable items: The user can access into all levels of the program hierarchy. An
    open concept design. Other FE programs can easily be attached to the system
    through a unified concept. Restart facility is included. Extensive
    diagnostic system (over 1000 diagnostics). No limits on a problem size.

Solution methods: Equation solver- Cholesky decomposition, QR-decomposition for
    nonsymmetric matrices. Conditioning information. Eigenvalue: several
    versions of the subspace vector iteration. Linear dynamics- simultaneous
    vector iteration. Static nonlinearity- modified Newton-Raphson iteration,
    implicit and explicit time integration, total and updated Lagrangian method.
    Dynamic nonlinearity- third order Hermitian nonlinear dynamic algorithm,
    explicit nonlinear dynamic algorithm, Newmark. Pre- and post buckling by
    iterative solutions.

Pre/postprocessing: LINDA is a free-field format input utility. Simple data
    generators are available. Selected results can be plotted. Interface is
    available to the general purpose interactive graphics package INGA.

Hardware: The system is machine-independent, presently running on CDC, IBM and
    Univac.

Documentation: LARSTRAN 80 User's Manual, LARSTRAN 80 Element Library, MM User's

Manual, LINDA User's Manual, LARSTRAN 80 User's System Manual, LARSTRAN 80 Error Manual.

Program Name: MAGNA

Source: University of Dayton Research Institute, Analytical Mechanics Group, Aerospace Mechanics Division, Dayton, OH 45469, USA.

General information: The program was released in 1979. It has been developed mainly for the nonlinear analysis of 3D structures. Modular program concept design is used Development was sponsored by the Air Force. MAGNA is a proprietary product of the University of Dayton, distribution and use of the code are restricted.

Program capabilities: MAGNA is a general purpose program for the materially and geometrically nonlinear analysis of 3D structures subjected to static and transient loading. Large displacements, large rotations, large strains, and plastic deformations may be handled. Elastic-plastic analysis is performed directly in terms of Piola stress and Green's strains. Static and dynamic instability problems can be solved. Layered structures may be analysed. Dynamic capabilities- eigenvalue analysis, damped forced vibration analysis, impact studies. Steady-state harmonic analysis. Natural frequency analysis with prestress effects. Contact problems without friction can be handled. Fracture mechanics analysis is possible. Analysis of composite materials.

Element library: Element library involves bars, membranes, plates, shells, axisymmetric solids and 3D isoparametric solids. 3D solid elements have either a constant or variable-number-of-nodes scheme. A variable scheme permits the use from 8 to 27 nodes/element. Individual 3D solid elements can be treated selectively as "tangent stiffness" elements, "pseudo-force" elements or "averaged-tangent-stiffness" elements. Shell isoparametric elements are based upon the penalty function formulation. All nonlinear elements are fully compatible. Any two elements of different types having similar nodal configurations may be joined directly. Special elements- 1D continuum elements, layered shells, surface contact element, shear panel, crack-tip element.

Material library: Isotropic linear elastic, initially isotropic elastic plastic, orthotropic linear elastic. Isotropic hardening, kinematic hardening and combined hardening Temperature-dependence is allowed.

Boundary conditions and loading: Zero and nonzero prescribed boundary conditions are permitted. Linear multivariable constraints. Skewed boundaries are accepted. Nonlinear boundary conditions due to surface contacts, including sliding, can also be considered. Applied loadings may consist of concentrated nodal forces, distributed surface pressures, body forces and line loads, "live" pressures such as fluid loading, etc. Deformation-dependent forces may be defined in user-written subroutine.

Notable items: Particular attention is given to the problems of modelling flexibility (virtually any combination of elements) and efficiency of element-level computations. Restart is possible for all nonlinear analysis options. User written subroutines can be added into the program. Cartesian, cylindrical and spherical coordinate systems. User-defined coordinate systems.

Solution scheme: Out-of-core skyline matrix storage and factorization scheme. Eigenvalues- simultaneous iteration method. Dynamic response- generalized Newmark's method, an implicit method of time integration. Nonlinear analysis- subincremental method, full Newton-Raphson iteration. Geometric nonlinearity incorporates a total Lagrangian scheme.

Pre/postprocessing: A number of data generation and model editing facilities are available. Mesh and boundary condition generators. Plotting utilities for deformed/undeformed geometry plotting, contour and relief plotting, variable versus variable.

Hardware: Cyber, Cray, VAX.

Documentation: Users Manual.

Program Name: MARC

Source: MARC Analysis Research Corporation, 260 Sheridan Ave., Suite 200, Palo Alto, CA 94306 USA. Available also via Control Data Corp., University Computing Co., Babcock and Wilcox data center, Boeing data center.

General information: Program development started at the end of the sixties. Since then, several versions were released. MARC is a very general program for the linear and nonlinear analyses. New features are added to the program continuously. The MARC Analysis Corp has representatives in many countries, the main offices are in Palo Alto, Tokyo and The Hague. MARC is a proprietary code, monthly or paid-up license available. The program-usage courses are offered.

Program capabilities: Linear and nonlinear static and dynamic problems can be handled. Material and geometric nonlinearities may be included in both the static and dynamic analysis. All nonlinear material models may be combined with geometric nonlinearities. Steady-state and transient heat transfer analysis is included in the package. MARC is built-up of three different libraries- element, material and structural. Structural library contains procedures for linear static analysis, nonlinear statics, linear dynamics large displacements and finite strains, buckling, elastic-plastic buckling, creep buckling, etc. Impact problems, response spectrum analysis. Automatic time-step control. Modal and proportional damping and lumped masses for dynamic analysis. Heat transfer part can treat nonlinearities such as nonlinear flux and radiation boundary conditions, changes of phase, etc. Fluid-solid interaction problems can be analysed. Analysis of contact problems with friction. Concrete cracking may be studied. Fracture mechanics applications- J-integral evaluation.

Element library: All element types are represented, more than 60 different elements are incorporated. Stress element library contains trusses, beams, isoparametric membranes, axisymmetric elements (symmetric and non-symmetric loading), plates, shells and 3D solid elements. Special elements- pipe bend element, truss/gap element in space, gap and friction link element in space, generalized plane strain element, axisymmetric torsional element, linear and nonlinear springs, concentrated masses, incompressible Hermann element, linear shear panel, reduced integration element, etc. Joining of shells and 3D solid elements is possible, fine and coarse mesh. Heat transfer library uses element types of which a structural analog exists.

Material library: Extensive material library contains more than 30 different models. Linear elastic, generalized Maxwell and Kelvin viscoelasticity, incompressible and nearly incompressible formulations, Mooney large-strain elasticity, hypoelasticity, von Mises and Mohr, Coulomb plasticity, isotropic, kinematic or combined hardening, associated or nonassociated flow rules, temperature and strain rate dependent yield stress, von Mises type of creep, pure volumetric and pure deviatoric creep, anisotropic behaviour for elasticity-plasticity-creep, viscoplasticity, low tension, cracking, etc.

Boundary conditions and loading: Zero and non-zero boundary conditions can be prescribed. Multipoint constraints description. Tying degrees of freedom. Thermal and mechanical loading is permitted. Mechanical loads may be described as nodal loads, uniformly and nonuniformly distributed loads such as variable pressures or centrifugal forces. Loads may be constant or time-dependent.

Notable items: The program checks the mesh quality from an energy point of view. "User subroutines" in which the user describes a very general problem by means of a short subroutine. A restart facility permits restart at any increment or time-step.

Solution methods: Equation solver- Gaussian elimination. Eigenvalues- inverse power sweep. Dynamic response- modal superposition, Duhamel exact integration, implicit Newmark's Beta method, explicit central difference method, Houbolt operator, step-by-step using tangent stiffness. Nonlinear problems- tangent stiffness, initial strain. Heat transfer- backward differences (Crank-Nicholson).

Pre/postprocessing: Many pre- and postprocessing programs are available. MENTAT is an interactive pre- and postprocessor developed for the MARC system. This program may also be used in conjunction with any finite element program on most computers. MENTAT utilizes an in-memory data base for faster access. The program is menu drived. Hierarchical organization, free-format input, multiple commands per line are permitted. Input data generation and display, on-line editing, plotting. The following plots are available: undeformed/ deformed geometry, vector plots, contour plots. Time and load history plots of selected variables at given locations may be generated. Other programs: MARCMESH3D- mesh generation, MARCOPT- bandwidth optimization, MARCPIPE- mesh generation for piping systems, MARCPLOT- time series plotting.

Hardware: IBM, CDC, Univac, Prime, Cray, Cyber, VAX, Data General, HP, Harris.

Documentation: User's Information Manual, Program Input Manual, Demonstration Problems Manual, Theoretical Reports, etc. Quarterly Newsletter.

Program Name: MSC/NASTRAN

Source: MacNeal-Schwendler Co., 7442 North Figueroa Str., Los Angeles, CA 90041, USA. Also available on most large data centers.

General information: NASTRAN is probably the most known finite element program in existence today. Development started in 1966 and was sponsored by NASA. There exist several versions of NASTRAN. The official version is distributed through COSMIC. MSC/NASTRAN is a proprietary version of the MacNeal-Schwendler Corp. and is maintained exclusively by MSC. This version is an advanced version of NASTRAN. There is a continuous development directed towards the inclusion of new capabilities. NASTRAN is a large-scale general purpose program system that can solve a wide variety of engineering analysis problems. The package consists of the analysis program MSC/NASTRAN, model generation program MSGMESH and the graphics package MSGVIEW. Modular design. Functional modules communicate with each other only via the Executive System. This system establishes and controls the sequence of calculations, allocates files and supports a restart. The main offices of MSC are in Los Angeles, Munich and Tokyo. "Hot line" service is available. Users' conferences are being held each year in USA and Europe. The educational courses are offered.

Program capabilities: MSC/NASTRAN can solve linear and nonlinear static and dynamic problems, heat transfer and field problems (such as acoustic, electromagnetic, aeroelastic, etc). Material and geometric nonlinearities are permitted. Stability analysis can be handled. NASTRAN contains a number of "rigid formats" which are prepacked programs for static, dynamic and

buckling analysis. For problems that cannot be solved by these formats the
user may write own program performing the appropriate analysis steps. In
this case the user writes so called DMAP program. Pre-formatted programs
available- linear static analysis, static analysis with differential stiffness,
static analysis with large displacement geometric nonlinearity and material
nonlinearity, buckling analysis, vibrational analysis, direct and modal
complex eigenvalue analysis, direct and modal frequency analysis and random
response, direct and modal transient analysis including response spectral
analysis, linear steady-state heat transfer, nonlinear steady-state heat
transfer, transient heat transfer, aeroelastic response, and aeroelastic
flutter. All these solution sequences are also available for problems
requiring substructuring.

Element library: All element types are available. Stress element library contains
rods, beams, membranes, plates, thick and thin shells, axisymmetric, and 3D
solid elements. Not all these elements can be used for the nonlinear
analysis. Heat transfer element library. Special purpose elements- shear
panel, out-of-plane membrane, singular plastic element, rigid elements,
thermo-fluid element, scalar elements, general elements described by
influence coefficients, mass element, damping elements, curved pipe element,
etc.

Material library: Linear and nonlinear material analysis can be performed.
Isotropic and anisotropic material properties are allowed. Temperature-
dependent elastic properties. Material models available include linear
elastic, nonlinear elastic, elastic-plastic and viscoelastic material.

Boundary conditions and loading: Zero and nonzero boundary conditions may be
prescribed. Multipoint constraints. Skewed boundary conditions are permitted.
Automatic singularity suppression. Arbitrary mechanical and thermal loading
may be applied, i.e. nodal loads, line loads, surface loads, gravity loads,
initial stresses/strains, centrifugal loading, deformation-dependent loading,
cyclic loading, random, gyroscopic, contact, and nonproportional loading.

Notable items: Multilevel substructuring technique. DMAP processor. Extensive
checking and error diagnostics. Cartesian, cylindrical or spherical
coordinates can be used. Very large problems can be solved. Sparse matrix
routines. Generalized dynamic reduction. Cyclic symmetry- currently
available for static, vibrational and buckling analysis. Automatic
resequencing. Restart facility.

Solution procedures: Equation solver- partitioned LDL decomposition for symmetric
matrices, LDU decomposition for nonsymmetric matrices. Condensation- Guyan's
reduction. Eigenvalues- determinant method, inverse power with shifts,
Givens and QR-method. Dynamic response- modal superposition, numerical
integration of uncoupled systems, direct integration by Newmark Beta-method.
Nonlinear problems- user-selected combinations of incremental, initial stress,
Newton-Raphson and modified Newton-Raphson methods.

Pre/postprocessing: Many pre- and postprocessor programs are available. MSGMESH
is a preprocessor for automatic mesh generation. MSGVIEW is a postprocessor
program for the graphical display of undeformed meshes in both batch and
interactive modes of operation. MSGSTRESS is a postprocessor for stress
interpolation. MSC/GRASP is an interactive pre- and postprocessor system
currently under development. Postprocessing functions will include- deformed
and modal plots, element stress contours, output scanning, x-y plots, key-
frame animation. Other plotter programs as NASPLOT, TECK are available.
Interface is available to the pre- and postprocessor program MENTAT.

Hardware: IBM, Amdahl, Itel, Fujitsu, CDC, Cyber, Univac, VAX, Cray, Siemens.

Survey of General Purpose Computer Programs   277

Documentation: Very extensive. User's Manual, Theoretical Manual, Programmer's Manual, Application Manual, Demonstration Problem Manual, Handbook for Linear Static Analysis, MSC/NASTRAN Primer, etc. NASTRAN Conference Proceedings.

Program Name: NEPSAP

Source: S. Nagarajan, Lockheed Missiles and Space Co., Inc., Dept. 81-12/Bldg. 154, 1111 Lockheed Way, Sunnyvale, CA 94086, USA.

General Information: NEPSAP is a proprietary code of the Lockheed Missiles and Space Co. The system is available for potential users under licensing agreements for either binary version or both binary version and source code.

Program capabilities: The program is similar to the ARGUS code. The reader is referred to the description of ARGUS.

Program Name: NISA

Source: Engineering Mechanics Research Corp., P.O. Box 696, Troy, Michigan 48099 USA. Honeywell Corp. is marketing NISA program internationally.

General information: Released in 1976. NISA is a general purpose, proprietary program for analysing of various problems encountered in different fields of engineering. The nonlinear program NON-NISA is available for the analysis of nonlinear problems. Programs are being continuously updated and developed to reflect the state-of-the-art of the FE technology. NISA programs are available through the computer network on royalty basis, or a binar and source versions can be leased or purchased directly from EMRC.

Program capabilities: NISA program can analyse static, dynamic, steady-state and transient heat transfer, and various field problems. The nonlinear version can handle both the material and geometrical nonlinearities. Geometric nonlinearities include large deformations, material nonlinearity incorporates plasticity. Stability problems may be investigated. Dynamic analysis- eigenvalue and eigenvector extraction, linear and nonlinear transient dynamic analysis, shock spectrum analysis, frequency response, random vibration analysis. Contact problems with friction can be handled. Analysis of composite materials is possible.

Element library: An extensive element library is available. Stress linear element library contains isoparametric (linear, parabolic, cubic, etc.) elements for plane stress/strain problems, axisymmetric (symmetric and nonsymmetric loading), general shells, thick shells, beams, trusses and 3D solid elements. Elements for nonlinear analysis- isoparametric elements for plane stress/ strain problems, axisymmetric elements, thick shell and 3D solid elements. Special elements- laminated composite or sandwich shells, springs, mass element, rigid element. Heat transfer element library.

Material library: Isotropic, orthotropic and anisotropic properties can be specified for all elements. Orthotropic material may be specified on the global or element level. The direction of orthotropy may be different for each element or each node. Temperature-dependent material properties are allowed for static, dynamic and heat transfer analysis. Material models available include linear elastic, nonlinear elastic-plastic and nonlinear elastic incompressible (Mooney-Rivlin). In elastic-plastic model the von Mises yield criterion is used with several hardening rules such as isotropic, kinematic, combined, mechanical sublayers.

Boundary conditions and loading: Zero and nonzero nodal displacements. Coupled
    displacements. Multipoint constraints equations. Skewed boundary conditions
    are allowed. Type of loading permitted includes concentrated nodal forces,
    uniform and nonuniform distributed loads, body forces, nodal temperatures and
    gradients. Dynamic loads- time dependent concentrated nodal forces, pressure
    loads, base excitation. Thermal analysis- concentrated nodal heat flow,
    distributed heat flux, convective heat flow at surfaces, radiation boundary
    conditions, internal heat generation. Multiple load cases can be solved in a
    single run.

Notable items: Coordinates can be described in the Cartesian, cylindrical or
    spherical coordinate system. Global and local coordinate systems may be
    defined. Flat or curved shells of constant or varying thickness can be
    converted into equivalent solid elements. Substructuring technique is
    included in the program. Gaps and interfaces including friction can be
    simulated. Cyclic symmetry. Extensive restart capabilities. Capabilities
    to simulate sandwich and layered composite structures.

Solution methods: Equation solver- wavefront method. Condensation- Guyan's
    reduction. Eigenvalue- subspace iteration, inverse iteration with Sturm
    sequence, QR-Householder. Dynamic analysis- Wilson's Theta and Newmark's
    time integration scheme. Large displacements- total Lagrangian formulation.
    Transient heat transfer- unconditionally stable, time integration technique.

Pre/postprocessing: Extensive range of pre- and postprocessor programs is
    available. DISPLAY/DIGIT is a general purpose interactive preprocessor to
    generate, verify and modify finite element models. It is built on menu
    structures, editors and various mesh generators and can be used for any
    finite element program. RAPIDRAW is an interactive postprocessor, TEKPIC is
    a time-sharing postprocessor program. NISAWFR- automatically minimize and/or
    checks the wavefront. ISOCROS- calculation of cross-sectional properties and
    stress analysis of beams with arbitrary cross-sectional properties and stress
    analysis of beams with arbitrary cross-section. JIP- automatic mesh
    generation for intersecting pipes. NISA/AISC/ASCE- checking program for the
    design of transmission towers building frames, etc. Combination of results
    from various loading cases can be handled.

Hardware: Cray, Amdahl, IBM, CDC, Univac, Honeywell, VAX, Prime, Harris, HP.

Documentation: User's Manual, Demonstration Problem Manual, Verification Manual,
    Pre- and Post-processing Manual, NISA Manager Manual, etc.

Program Name: PAFEC

Source: PAFEC Ltd., Strelley Hall, Strelley, Nottingham NG8 6PE, England.

General information: For details see the paper in this book.

Program Name: PAM

Source: Engineering Systems International, S A, 20 rue Saarinen, Silic 270,
    94578 Rungis Cedex, France.

General information: PAM system is a modular system consisting of several
    programs- PAM-GSD for linear static, dynamic and thermal analysis, PAM-AX3D
    for static and dynamic analysis of axisymmetric structures, PAM-TH for
    nonlinear thermal analysis, PAM-NL for nonlinear static and dynamic analysis,
    PAM-GEOM for nonlinear static and dynamic analysis of geological media and
    soil-structure interaction, PAM-MASL for nonlinear static and dynamic analysis

of lightweight and flexible structures, and PAM-HYDCAB for nonlinear static and dynamic analysis of fluid-surrounded systems of cables and beams. Various mixed finite difference/finite element codes are also available for special applications. The program system is maintained and continuously extended by ESI. ESI is a firm of consulting engineers in engineering mechanics and applied sciences. Branch offices are located in Frankfurt, West Germany and Boulder, Colorado. Implementation, training and maintenance at clients' facilities are offered.

Program capabilities: Linear and nonlinear, static and dynamic, and thermo-mechanical analyses can be handled. Geometrical nonlinearities include large displacements. Material nonlinearities- isotropic thermo-elasticity and plasticity, thermo-elastoviscoplasticity and creep. Stationary and transient heat conduction. Buckling and incremental collapse analysis. Dynamic capabilities- eigenvectors and eigenfrequencies, spectral, harmonic and random response analysis. Calculation of the dynamic behaviour of anchoring systems under arbitrary motion in space. Contact problems, impact/penetration analysis can be handled. Soil-structure interaction. Fluid-structure interaction problems may be solved.

Element library: Stress element library includes truss, beam, thin and thick shells, plane stress/strain membranes, axisymmetric, and 3D solid elements. Thermal element library has the same elements as in linear stress element library. Special elements: straight and curved pipes, nonlinear spring, contact element, fluid element, cable, sandwich and thick-walled composite shell element, etc. Direct input of mass and stiffness matrices.

Material library: Linear elastic, elasto-plastic, thermo-viscoelastic, thermo-viscoplastic, thermoplasticity. Von Mises plasticity, isotropic and kinematic hardening. Thermo-creep analysis, Norton-Odqvist law. Complex material models for geological media and concrete, including plasticity (with strain hardening and softening), viscoplasticity, creep, incompressibility. Soil and rock constitutive models- Mohr-Coulomb, Drucker-Prager, orthotropic plasticity, visco-elasto-plasticity, post peak behaviour.

Boundary conditions and loading: Zero and nonzero displacements are allowed. Kinematic conditions and node tying options. Fluid support. Loading permitted- nodal forces and moments, pressure, gravity, centrifugal, thermal. Base acceleration for earthquake loading. Initial loads. Conservative and nonconservative general loading. Combination of load cases is possible.

Notable items: Bandwidth reduction. Restart option.

Solution methods: Equation solver- skyline technique. Dynamic analysis- mode superposition, direct integration according to Wilson, Newmark, Houbolt. Large displacements- total Lagrange formulation. Nonlinearity- different solution algorithms incremental/iterative with automatic load correction, complete and modified Newton/Raphson iteration.

Pre/postprocessing: Plotting routines are available. Plotting of deformed/undeformed structure, stress contours, time histories, etc.

Hardware: CDC, IBM.

Documentation: Users Manuals, Theoretical Manuals.

Program Name: PLANS

Source: Grumman Aerospace Corp., Research Dept., Bethpage, NY 11714, USA. Several modules of PLANS are available through the COSMIC, University of Georgia,

Athens, GA 30601, USA.

General information: PLANS is a system of nonlinear analysis programs. The initial version was funded by NASA, the program developed by the Grumman Aerospace Corp., first release in 1972. PLANS is a highly modularized general purpose package especially suitable for nonlinear structural analysis. Each module is an independent finite element program with its associated element library, each module is directed for a distinct physical problem class. The following modules are available- BEND (plastic analysis for structures where bending and membrane effects are significant), FAST (fracture mechanics analysis), HEX (3D elastic, plastic analysis), REVBY (plastic analysis of axisymmetric structures), O-PLANE (plastic analysis of structures, where membrane effects predominate) and DYCAST (nonlinear dynamic analysis). Each module can be individually loaded or be integrated into a single system by making each module callable from an executive program. Programs are continuously updated.

Program capabilities: Different modules of PLANS can handle the static and dynamic, linear and nonlinear analyses of arbitrary structures. Material nonlinearities include plasticity and combined elasticity-plasticity-creep. Combined material and geometric nonlinearities are allowed. Geometrical nonlinearities are limited to large deformations, small strains. Vehicle crashworthiness can be studied. Fracture analysis problems may be analysed.

Element library: Different modules contain different element types- beams with various cross section (solid rectangular, circular, Z section, I section, L section, hollow circular, rectangular, etc.), trusses, membranes, bending plate elements, axisymmetric ring and shell elements, isoparametric solid elements. Special elements- nonlinear spring, stiffeners, shear panel. New elements can also be added.

Material library: Linear elastic, nonlinear elastic, elasto-plastic. Isotropic or orthotropic material properties. Perfectly plastic, linear strain hardening or nonlinear strain hardening laws. The cyclic plastic behaviour are treated by Prager-Ziegler kinematic hardening theory, the cyclic creep behaviour is treated according to the ORNL auxiliary rules for stress reversal.

Boundary conditions and loading: Zero and nonzero boundary conditions may be prescribed. Multipoint constraint capability is incorporated. Arbitrary mechanical and thermal loads may be applied. Centrifugal loading, cyclic loading, nonproportional loading.

Notable items: Substructuring technique is incorporated into several modules. Global and local coordinate systems. Use of multistress points within an element has been incorporated into the modules (Gauss or Lobatto points), each element can be specified differently. Restart procedure is available.

Solution methods: All the programs in PLANS employ the initial strain concept within an incremental procedure to account for the effect of plasticity. Geometric nonlinearities use the updated Lagrangian approach or convected coordinate approach. Equation solver- in-core or out-of-core compacted solver. Eigenvalues- automatic reduction of matrix to tridiagonal form. Lanczos method. Nonlinear static analysis- incremental technique, modified Newton iteration, incremental solution with equilibrium correction, incremental predictor-corrector iterative or noniterative methods. Nonlinear dynamic analysis- implicit Newmark or Wilson, explicit central difference or variable time step modified Adams.

Pre/postprocessing: There are limited capabilities in pre- and postprocessing. A number of special programs exist for different modules, i.e. for particular mesh generation, plotting, etc. One common program is available for data checking, bandwidth optimization and plotting of the undeformed meshes.

Hardware: IBM, CDC, Univac, Cyber, CDC Star, VAX.

Documentation: Theoretical and User's Manuals.

Program Name: SAMCEF

Source: University of Liege, LTAS, Rue E Solvay, 21, 4000 Liege, Belgium.

General information: Program development started in 1965, first release in 1970. Since then the program has been updated/completed several times. SAMCEF was first developed for research activities. Now it is being used also in an industrial environment. The program has a highly modular organization. The independent modules can communicate information and results to each other through standardized communication files. These files can be easily accessed by the user for non-standard applications. SAMCEF is an university code, commercially available conditions on request. The price includes the source code installation, documentation and training. Maintenance and updates available at LTAS. User's group does exist.

Program capabilities: Static and dynamic, transient linear and nonlinear mechanical analysis, steady-state and transient linear and nonlinear heat transfer, and shape and dimensions optimization problems can be handled. The nonlinearities may be due to large displacements, large strains and nonlinear material behaviours. Dynamic analysis- eigenvalues and eigenmodes, dynamic response, spectral analysis. Stability analysis including buckling and postbuckling. Macromechanic and micromechanic analysis of composite materials. Weight optimization studies are possible. Fluid-structure interaction problems can be analysed.

Element library: Very comprehensive element library, about 100 different element types are available. Displacement, equilibrium, hybrid and mixed formulation. In the stress element library all element types are represented. Heat transfer library- bar, membrane, axisymmetric, isoparametric volume, heat flux model. Special elements- local stiffness element, shell stiffener, special convection elements, tying element, fluid elements, multilayered axisymmetric and shell elements. New element types can be incorporated by the user.

Material library: The material can be isotropic, anisotropic, temperature-dependent, linear elastic, nonlinear elastic or elastic-plastic.

Boundary conditions and loading: Zero and nonzero boundary conditions may be specified. Skewed boundary conditions are allowed. Multipoint constraints equations. Loading permitted- concentrated, distributed, line loads, surface loads, volume loads, gravity, centrifugal, thermal, deformation-dependent, contact loading. Nonproportional loading. Initial stress/strain.

Notable items: Local coordinate systems may be introduced. Multilevel substructuring technique is included. Failure criterions for the analysis of composite materials are incorporated into the program. Restart option. Program can be used for optimization studies.

Solution methods: Equation solver- frontal solution method. Condensation-Iron's and Guyan's reduction. Eigenvalues- simultaneous iteration. Dynamic analysis- modal superposition, Newmark Beta method. Nonlinear problems-incremental solution, Newton-Raphson. Optimization- fully stressed design method.

Pre-postprocessing: Interactive programs are available for mesh generation and input data generation. Postprocessors as well as graphics display for the data and results. Plotting options include deformed/undeformed structure plots, modal shapes, isocurves, principal stresses, transient responses, contour plots, etc. Interface is available to a general purpose preprocessor CAE SUPERTAB.

Hardware: IBM, CDC, VAX, Univac, Siemens, Cray.

Documentation: User's Manual, Theoretical Manual, Comparative Sample Problem Solution. Theoretical reports.

Program Name: SAP7

Source: Structural Mechanics Computer Laboratory DRC 394, University of Southern California, Los Angeles, CA 90007, USA.

General information: SAP7 is one of many versions of the SAP program available on the market today. SAP is one of the early general purpose linear finite element programs originally developed at the University of California at Berkeley. SAP7 is a general purpose code based on the SAP4 and NONSAP programs. It means that linear and nonlinear analyses may be performed. SAP7 consists of three independent programs interfaced together MODEL is a preprocessor, SAP7 is a main analysis code and POST is the postprocessor. Modular architecture allows quick modifications and simple additions, particularly in the nonlinear analysis range. The program is continuously being developed. The package can be licensed for in-house use for a one-time initial fee and an annual maintenance fee for each program is used.

Program capabilities: Linear and nonlinear static and dynamic analysis is possible. The nonlinearities may be due both to geometry and material, or their combination. Dynamic analysis- natural mode and frequencies, linear modal and response spectrum analysis, nonlinear transient response of structures include the effects of plasticity, large displacements and gap. Contact problems can be solved. Analysis of layered structures is possible.

Element library: Stress element library contains truss, plate stress/strain membranes, axisymmetric shell and solid elements, thick shell, general isoparametric shell, 3D solid and beam elements. Special elements- gap element, layered plate/shell element, pipe element, constraint elements, general stiffness element. Different element types can be combined.

Material library: Linear elastic, nonlinear elastic, elastic-plastic (von Mises or Drucker-Prager yield conditions), Mooney-Rivlin model, curve description model, variable tangent module type. Isotropic and orthotropic linear elastic. Layered sandwich material may be analysed.

Boundary conditions and loading: Zero and nonzero boundary conditions may be prescribed. Elastic foundation. Mechanical and thermal loads may be applied. Concentrated and distributed loads.

Notable items: Substructuring is implemented for static problems. User-written subroutines can easily be incorporated into the program. Local coordinate systems may be defined. Free-format input. Restart capability. An out-of-core active column solver for static and dynamic analysis. Both Tsai-Hill and Tsai-Wu failure criteria are implemented for the analysis of composite materials.

Solution methods: Equation solver- skyline storage method, Gaussian elimination and profile reduction technqiue. Eigenvalue- subspace iteration, determinant search technique. Dynamic analysis- Wilson Theta method, Newmark step-by-step integration. Nonlinearity- incremental solution procedure. Updated and total Lagrangian formulation.

Pre/postprocessing: Free-field input. MODEL is an interactive preprocessor used to generate the required data, i.e. mesh, boundary conditions, element properties. The program can also be used to calculate the centre of gravity and total mass of the discretized system. The MODEL has a machine independent nonrestrictive data input. POST is an interactive graphics postprocessor program. The program is used to graphical check meshes generated by the MODEL and to produce graphic displays of the results. POST provides a direct communication between user and computer and allows on-line decisions/editing. Analysis program can also be interfaced with a general purpose pre- and postprocessor PATRAN.

Hardware: IBM, CDC, VAX, Prime, Data General.

Documentation: Users Manual for all programs in the SAP7 package. Theoretical Manual, Problem Verification Manual.

Program Name: SESAM '80

Source: Computas A/S, P.O. Box 300, N 1322 Hovik, Oslo, Norway.

General information: For more detailed information see the paper included in this book.

Program Name: SPAR

Source: Engineering Information Systems, Inc. (EISI), 5120 Campbell Ave., Suite 240, San Jose, CA 95130, USA. Available through COSMIC, 112 Barrow Hall, University of Georgia, Athens, GA 30602, USA.

General information: SPAR development was sponsored by NASA. The program was released in 1976 and is being continuously developed. SPAR contains over 33 processors that store and retrieve information, and communicate with each other through a unified data base. As each module is executed, its output is stored into the direct access library. The large linear elastic problems can be solved.

Program capabilities: SPAR can analyse stress, buckling, vibration and thermal problems of large elastic structures. Dynamic capabilities include modal analysis and transient dynamic linear response. Laminated composites can be handled. Fluid-structure interaction problems can be solved. Hydroelastic vibration modes may be computed.

Element library: Hybrid formulation. Stress element library contains beams, membranes, plates, axisymmetric elements and 3D solids. 3D solids use the assumed stress-complementary energy formulation. Several elements may be warped. Special elements- shear panel, fluid elements. Fluid elements are compatible with structural elements.

Material library: Linear elastic, isotropic and anisotropic materials.

Boundary conditions and loading: Zero and nonzero initial boundary conditions may be prescribed. Mechanical and thermal loads are permitted. Concentrated and distributed, centrifugal, etc. Dynamic loading- transient applied forces, transient base acceleration. Initial stress/strain. Symmetric structures may be nonsymmetrically loaded.

Notable items: Substructuring technique is available. Multiple coordinate system. User control over the output. User control over the analysis steps. Restart after any processor function makes it possible that if a mistake is encountered, usually only the last processor need be re-executed. The large problem size capability is achieved by extensive use of secondary storage, sparse hypermatrix storage, and dynamic core allocation. The arithmetic utility system module provides a possibility to perform matrix algebra operations. Ability to allow multiple analysis to access the data base parallel.

Solution methods: Not known to the author.

Pre/postprocessing: Free-field input. Interactive graphics utilities are available. Mesh generators are included. Graphical capabilities- undeformed/deformed mesh plots, stress resultants plots, mode shape plots, etc.

Hardware: CDC, Univac, Prime, VAX, Cyber.

Documentation: SPAR Reference Manual- Program Execution, Theory, Demonstration Problems.

Program Name: STRUDL

Source: Messerschmitt-Bolkow-Blohm GmbH, Zentralabteilung, Datenverarbeitung Postfach 801109, D8012 Ottobrunn, West Germany.

General information: STRUDL is a subsystem of the ICES program system. The basic concept of ICES was formulated at MIT, Cambridge, MA, USA (1962-1970). There are several versions of the STRUDL used in the different parts of the world (i.e. McAuto STRUDL, GTSTRUDL, PSDI STRUDL). The version described in this paper is that developed and continuously being updated att MBB Corp in West Germany. Here the development started in 1968. The STRUDL is a modular system. ICES-BASIC is a main system. A large number of ICES subsystems there are in existence, i.e. TABLE, TOPOLOGY, STATS, GRAPHIC, INFO, UGH, DGM and others. Pre- and postprocessor programs are an integrated part of the package. STRUDL is a general purpose program for the linear analysis of general structures. A problem-oriented language is incorporated which can be used for a variety of data manipulations. MAN/ICES-STRUDL is avaialble also for the potential users outside the MBB corporation. Source code/object code and documentation can be leased. ICES User Group does exist. Seminars are being held.

Program capabilities: Linear static and dynamic analysis of general structures subjected to arbitrary mechanical and thermal loads may be handled. Linear and nonlinear steady-state and transient analysis, as well as geometric nonlinear static analysis (large displacements, small strains) have been incorporated into the program. Dynamic capabilities- modal, harmonic, shock spectrum, transient response. Diagonal and nondiagonal mass matrices. Linear buckling analysis. Analysis of piping systems. Torsion problems may be handled. Optimization procedures are available.

Element library: Isoparametric and hybrid formulation. Stress element library contains trusses, beams, plane stress/strain membranes, plates, shells, axisymmetric shell and solid elements (symmetric and nonsymmetric loads), and 3D solid elements. An extensive library of thermal elements for conductivity, heat transfer, heat flux, heat source elements. Special elements: straight and curved pipes.

Material library: Linear elastic, isotropic and anisotropic. Temperature dependence is allowed.

Boundary conditions and loading: Zero and nonzero boundary conditions may be prescribed. Multipoint constraint equations. Coupling of selected DOF.

Elastic foundation. Mechanical and thermal loads are allowed. Uniform and nonuniform. Nodal, line, surface, volume. Initial stress/strain. Generators for dead loads, accelerations, wind loads. Loading combination is possible.

Notable items: Substructuring technique for static and dynamic analysis. Cartesian, cylindrical, conical and spherical coordinate definition. Global and local coordinate systems. Material data can be stored in the material properties table. Special treatment for large problems- pre-allocation of input arrays, decomposition, bandwidth reduction. Static and dynamic results may be combined for sizing and checking of steel and reinforced concrete members. Cyclic symmetry. Restart option is included in the program.

Solution methods: Eigenvalues- Householder tridiagonalization, inverse iteration, Sturm sequence, Jacobi, subspace iteration. Dynamic analysis modal superposition, Newmark implicit integration. Nonlinear analysis-incremental technique, Newton-Raphson iteration.

Pre/postprocessing: Pre- and postprocessor programs can be used in batch or interactive mode of operation. The problem is defined in the STRUDL-Command-Language, free-field input. MBB-MAN/ICES-TOPOLOGY is a subsystem for 2D and 3D mesh generation. MBB-MAN/ICES-GRAPHIC is a problem oriented language for the graphical presentation of data from the different ICES-subsystems. Error checking and on-line correction of input data with an editor. Different plots can be obtained- undeformed/deformed shape, contour plots, eigenvectors, generated time-history plots, 3D plots.

Hardware: MAN version- IBM, VAX, Siemens, Philips. Other versions- CDC, IBM, Univac.

Documentation: User's Guide (2 Volumes), Verification Manuals, ICES-STRUDL Introduction Book, Theoretical Manual, etc.

Program Name: TITUS

Source: FRAMATOME, Computational Mechanics Center, B.P. 13, 71380 Saint Marcel, France.

General information: Fore more detailed information see the paper included in this book.

Program Name: WECAN

Source: Westinghouse Computer Service, Advanced System Technology, 777 Penn Center Boulevard, Pittsburgh, PA 15235, USA.

General information: The program was developed by Westinghouse R&D and other Westinghouse user divisions. It is a proprietary code, first released in 1972. WAPPP- WECAN Auxiliary Pre- and Postprocessors are FIGURES II- Finite Element Interactive Graphics User Routines are two programs incorporated into the WECAN system. All programs are updated/modified regularly, new developments are funded by user organizations. WECAN is a general purpose code for the linear and nonlinear static and dynamic analysis. Expert consulting services are provided. One-week training courses are offered. An annual User's Colloquium is being held each fall. Maintenance is funded by a surcharge. WECAN is available for general use at the Westinghouse Power Systems Computer Center in Pittsburgh, PA.

Program capabilities: 2D and 3D linear and nonlinear static and dynamic problems

can be solved. Steady-state and transient heat transfer problems may be
handled. Flow analysis. Material nonlinearities include plasticity, creep
and irradiation-induced deformation. Geometric nonlinearity- large
deformations, moderate rotations, small strains. Linear elastic buckling
analysis may be performed. Dynamic capabilities- mode and frequency analysis
(standard or reduced modal analysis), harmonic, linear and nonlinear transient
response, seismic response spectrum analysis. Lumped or consistent mass
matrices. Either Rayleigh or viscous damping may be imposed. Composite
materials may be analysed. Steady-state hydraulic analysis. Fracture
mechanics analysis.

Element library: Element library contains over 70 different elements.
   Displacement and mixed formulation. Stress element library includes spars,
   beams, thin shells, membranes (plane stress/plane strain), axisymmetric
   elements, plates, 3D solids. Heat conduction elements, convection and
   radiation elements, hydraulic conducting elements. Special elements-
   straight and curved pipes, elbows, gap elements, friction interface elements,
   springs, masses, dampers, cable, fluid coupling elements, torsion bars,
   rotating disc, wedge element, etc. General matrix input. New elements
   can easily be added.

Material library: Linear elastic, elastic-plastic, creep. Kinematic and
   isotropic hardening. For multiaxial stresses, the plastic and creep response
   is based upon the von Mises equivalent stress and strain definitions and the
   Prandtl-Reuss flow equations. Isotropic, orthotropic and anisotropic
   material properties. Temperature dependence is allowed.

Boundary conditions and loading: Zero and nonzero prescribed boundary conditions.
   Generalized linear constraints. Boundary conditions for thermal/flow
   analysis- specified temperatures, specified heat flow rates. Loading
   accepted- forces, pressures, temperatures.

Notable items: Substructuring technique is available, two different methods are
   included. An automatic convergence check for the inelastic analysis. An
   automatic time step selector for the creep analysis. Coupled thermal/stress
   analysis. Global and local coordinate systems. A dummy element capability.
   User-defined FORTRAN Subroutine for optional use to define user's creep and
   swelling constitutive laws and in heat convection analysis the surface film
   coefficients. General restart option.

Solution methods: Equation solver- wave front method. Condensation- Guyan's
   reduction. Thermal and flow analysis- Crank-Nicholson-Galerkin integration
   scheme, quadratic integration scheme. Eigenvalue- Householder-QL, Jacobi
   method, inverse power with shifts method. Linear dynamic analysis- modal
   superposition, Houbolt or cubic integration method. Nonlinear dynamic
   response- implicit time integration (Newmark or Houbolt) or modal
   superposition. Nonlinear static analysis- method of successive elastic
   approximations with extrapolation.

Pre/postprocessing: WAPPP is a collection of batch pre- and postprocessor
   programs for i.e. mesh generation, calculation of equivalent material
   properties, wave front optimization, plotting, J-integral calculation,
   calculation of Fourier coefficients, superposition of results, etc. FIGURES
   II is a collection of interactive preprocessors for the input data
   preparation.

Hardware: WECAN and WAPPP on CDC, FIGURES on Data General Eclipse.

Documentation: User's Manual (4 Volumes), WAPPP User's Manual (2 Volumes), WECAN
   Verification Manual, FIGURES II User's Manuals, Theoretical Reports,
   Conference Proceedings, etc.

Survey of General Purpose Computer Programs   287

GENERAL PURPOSE BOUNDARY ELEMENT PROGRAMS

The main attraction of the boundary element method is that only the boundary/ surface of the domain analysed needs to be discretized. The process of modelling is not so tedious as by the finite element method. The dimensionality of the problem analysed is reduced by one. The application of this method offers a versatility and economy when compared with finite elements.

The developments of the boundary element technique are still in early stages. There are not so many commercially available programs on the market today.

Presented boundary element programs are described in the same way as were the finite element programs in the previous section.

The following programs are presented: BEASY, BEFE, BETSY, BISON, CASTOR, and KYOKAI. Some of them are combined finite element-boundary element codes.

Program Name: BEASY

Source: Computational Mechanics Centre, Ashurst Lodge, Ashurst, Southampton, SO4 2AA, England.

General information: For the detailed description see the paper included in this book.

Program Name: BEFE

Source: Dept. of Civil Engineering, University of Queensland, St. Lucia 4067, Australia.

General information: BEFE is an extension of the finite element program MASS, to which subroutines for the boundary element method have been added. The finite and boundary element methods are fully integrated. Modular program design. Four modules are available: MESH for the mesh topology input, BEFE for the linear and nonlinear analysis, PLOT (GPLOT) for plotting/displaying mesh and the results, and PRINT for the printout of the results. All modules are independent of each other.

Program capabilities: The program can analyse static 2D and 3D problems using boundary element and/or finite element discretizations. Material nonlinearities with finite elements. Geomechanical problems can be handled.

Element library: The library contains 2D and 3D finite and boundary elements. These can be used separately or in combination depending on the type of problem to be analysed. Available elements: linear and parabolic boundary elements, linear and parabolic finite elements, shell finite elements. Special elements- truss, contact elements, infinite elements. Isoparametric elements are exclusively used.

Material library: Linear elastic, elasto-plastic or viscoplastic. Yield conditions include von Mises, Drucker-Prager, Mohr-Coulomb, critical state and multilaminate model with either associative or nonassociative flow law.

Boundary conditions and loading: Zero and nonzero boundary conditions may be prescribed. There are several loading types which allow to define nodal, distributed, gravity, thermal loads, etc. For problems in geomechanics an initial stress field (constant or varying with depth) can be inputted. Excavation forces.

Notable items: Nodes can be defined in Cartesian, cylindrical or spherical
coordinates. Global and local system of axes. Boundary, material data and
loading may be generated. Output is controlled by the user. An extensive
error check. Special facilities for the geomechanical problems are available.

Solution methods: Equation solver- partitioned frontal solution method, Gaussian
elimination. Nonlinear analysis- increment-iteration method. Solution in
iteration steps with residual forces being redistributed at each step.

Pre/postprocessing: Input from cards, files or interactive mode of operation.
Several pre- and postprocessors are available. Mesh generators. The results
can be plotted or displayed on the colour graphics terminal. Plots available-
deformed mesh in plane or axonometry, deformation vectors, principal stress
vectors, contour plots, plots of plastic zones.

Hardware: VAX.

Documentation: User's Manual.

Program Name: BETSY

Source: Lehrstuhl für Technische Mechanik, Universität Erlangen, Pestallozziring
20, 8520 Erlangen, West Germany. Available through T.Programm GmbH,
Oskar-Kalbfellplatz 8, D-7410 Reutlingen, W. Germany.

General information: BETSY consists of five programs- BETSY-2D for plane
problems, BETSY-AXT for axisymmetric torsion, BETSY-AXO for axisymmetric
problems with symmetric loading, BETSY-AX1 for geometrically axisymmetric
problems with loads constituting the first term of a Fourier expansion of a
load function, and BETSY-3D for general spatial problems. Thermoelastic
problems can be analysed.

Program capabilities: 2D and 3D thermoelastic static problems can be handled.

Element library: For 2D problems- linear, quadratic or cubic functions, linear,
quadratic or circle boundary elements. For 3D problems- linear or quadratic
functions and boundary elements, description of the surface by Coon's method.

Material library: Linear elastic, isotropic. Material changes are possible by
substructuring.

Boundary conditions and loading: Single point constraints at the boundary
(displacements and/or tractions) with stationary temperature fields and
nonvanishing volume forces.

Notable items: Substructuring is available.

Pre/postprocessing: Mesh generator is available. Plot routines are being
developed in connection with the finite element program TPS-10 plot routines.

Hardware: IBM, CDC, Siemens, Prime.

Documentation: Two dissertations, internal reports, published papers.

Program Name: BISON

Source: PAFEC Ltd., Strelley Hall, Strelley, Nottingham NG8 6PE, England.

General information: BISON is a program for the thermoelastic static stress analysis. The program is fully operational, new options are continuously being added. The program can also be used in connection with the finite element program PAFEC. Boundary elements and finite elements can be mixed. BISON is marketed as a separate package.

Program capabilities: The program can handle 2D and 3D thermal and linear elastic problems.

Element library: One-dimensional line segments with three nodes, two-dimensional triangular and quadrilateral elements with six and eight nodes respectively. Quadratic interpolation for all elements.

Material library: Linear elastic, isotropic.

Boundary conditions and loading: Temperature, heat flux, force, pressure, prescribed displacements. These are approximated by the use of quadratic interpolation functions. Generalized constraints.

Notable items: The pertinent field variables can also be evaluated at any number of user-specified internal points. Substructuring technique is available. Global and local coordinate systems. Different generators developed for PAFEC are also available for BISON.

Pre/postprocessing: Free-field input. Plot routines are being developed.

Hardware: Most 32-bit virtual machines.

Documentation: User Manual. System Manual.

Program Name: CASTOR

Source: CETIM, 52, Avenue Felix, Louat, 60300 Senlis, France.

General information: For more details see the paper in this book.

Program Name: KYOKAI

Source: Applied Mathematics Department, Fukuoka University, Fukuoka 814-01, Japan.

General information: For more details see the paper in this book.

TABULAR PROGRAM PRESENTATION

The finite element and boundary element programs earlier described are now presented in tabular forms for ease of reference and rapid retrieval of the information provided. Both types of programs are included in the same table. All programs are ordered alphabetically. A dot in the tables means that the appropriate option/function is available in the program.

There are six tables describing the packages:

    Table 1 - Range of applications. Formulation.

    Table 2 - Phenomena. Data input/output.

    Table 3 - Element library.

    Table 4 - Material library. Notable items.

    Table 5 - Boundary conditions. Loading types.

    Table 6 - Hardware.

TABLE 1

① RANGE OF APPLICATIONS / FORMULATION

	Linear statics	Nonlin statics	Linear dynamics	Nonlin dynamics	Heat transfer	Fluid-struct int	Soil-struct int	Other field prob	Fracture mech	Contact problems	Composite mater	Optimization	Displacement	Force	Hybrid	Mixed	
ABAQUS	•	•	•	•	•		•		•	•	•		•		•		
ADINA	•	•	•	•	•	•	•		•	•	•		•				
ANSYS	•	•	•	•	•			•	•	•	•		•				
ARGUS	•	•	•	•					•	•			•			•	
ASAS	•	•	•		•			•	•	•	•		•	•			
ASKA	•	•	•		•		•		•	•	•		•				
BEASY	•	•			•			•	•								BEM
BEFE	•	•					•			•							BEM
BERSAFE	•	•	•		•			•	•	•		•		•			
BETSY	•								•								BEM
BISON	•				•												BEM
CASTOR	•	•	•		•				•	•			•				BEM
DIAL	•	•	•	•					•	•	•		•		•	•	
EFESYS	•	•	•	•		•	•	•		•			•				
FENRIS	•	•	•	•		•	•			•			•				
FINEL/ICB	•	•	•		•	•			•		•		•				
KYOKAI	•		•		•	•		•									BEM
LARSTRAN	•	•	•	•	•				•	•	•		•				
MAGNA	•	•	•	•					•	•	•		•				
MARC	•	•	•	•	•	•	•	•	•	•	•		•		•	•	
MSC/NASTRAN	•	•	•	•	•	•		•	•	•	•		•	•			
NEPSAP	•	•	•	•					•	•			?			•	
NISA	•	•	•	•	•			•		•	•		•				
PAFEC	•	•	•	•	•	•			•	•	•		•		•		
PAM	•	•	•	•	•	•	•			•	•		•				
PLANS	•	•	•	•					•	•	•		•				
SAMCEF	•	•	•	•	•	•			•	•	•	•	•	•	•	•	
SAP7	•	•	•	•			•			•	•		•				
SESAM 80	•	•	•	•	•	•	•		•	•			•				
SPAR	•		•		•	•					•				•		
STRUDL	•	•	•		•								•		•		
TITUS	•	•	•	•	•	•		•	•	•			•				
WECAN	•	•	•	•	•	•			•	•	•		•				

Survey of General Purpose Computer Programs

TABLE 2

PHENOMENA IN/OUTPUT	Small displacem.	Small strains	Large displacem.	Large strains	Thermal effect	Modal analysis	Dynamic response linear	Dynamic response nonlinear	Dynamic spectral analysis	buckling	Post buckling/collapse	Free format input	Mesh generation	Load generation	Plot routines	Interactive graphics	User defined output
ABAQUS	•	•	•	•	•	•	•	•		•			•	•	•	•	
ADINA	•	•	•	•	•	•	•	•	•	•	•	•	•		•	•	•
ANSYS	•	•	•		•	•	•	•	•	•	•		•	•	•	•	•
ARGUS	•	•	•	•	•	•	•	•		•	•		•	•	•		•
ASAS	•	•	•		•	•	•		•	•	•	•	•	•	•	•	•
ASKA	•	•			•	•	•		•	•	•	•	•	•	•		•
BEASY	•	•			•								•	•	•		
BEFE	•	•			•								•	•	•		
BERSAFE	•	•	•		•	•	•		•			•	•	•	•	•	•
BETSY	•	•			•								•		•		
BISON	•	•			•							•	•	•	•		•
CASTOR	•	•	•		•	•	•		•	•			•	•	•		•
DIAL	•	•	•	•	•	•	•	•		•	•	•	•	•	•	•	•
EFESYS	•	•				•	•	•					•		•	•	•
FENRIS	•	•			•	•	•	•		•	•			•			
FINEL/ICB	•	•			•	•	•		•	•		•	•	•	•	•	•
KYOKAI	•	•			•								•	•		•	•
LARSTRAN	•	•	•	•	•	•	•	•		•	•	•	•	•	•		•
MAGNA	•	•	•	•	•	•	•	•		•	•		•	•	•		•
MARC	•	•	•	•	•	•	•	•	•	•	•	•	•	•	•	•	•
MSC/NASTRAN	•	•			•	•	•	•	•	•	•	•	•	•	•	•	•
NEPSAP	•	•	•	•	•	•	•	•		•	•	•	•		•	•	•
NISA	•	•	•	•	•	•	•	•	•	•		•	•	•	•	•	•
PAFEC	•	•	•		•	•	•	•		•	•		•	•	•	•	•
PAM	•	•	•	•	•	•	•	•	•	•	•		•	•	•		•
PLANS	•	•	•		•	•	•	•		•			•		•		•
SAMCEF	•	•	•	•	•	•	•	•	•	•	•	•	•	•	•	•	•
SAP7	•	•	•		•	•	•	•	•	•			•	•	•		•
SESAM 80	•	•	•		•	•	•	•	•	•	•	•	•	•	•	•	•
SPAR	•	•			•	•	•			•			•	•	•	•	•
STRUDL	•	•	•		•	•	•		•			•	•	•	•		•
TITUS	•	•	•	•	•	•	•	•	•	•	•	•	•	•	•	•	•
WECAN	•	•	•		•	•	•	•	•	•			•	•	•	•	•

TABLE 3

③ ELEMENT LIBRARY

	Rod/bar	Beam	Membrane	Plate	Thin shell	Thick shell	Axisymmetric	3D solid	Thermal pr	Fluid elem	Shear panel	Crack-tip	Gap, contact	Pipe, elbow	Layered	Stiffener	User-defined	
ABAQUS	•	•	•		•		•	•	•				•	•	•			
ADINA	•	•	•	•	•	•	•	•	•	•	•		•	•		•	•	
ANSYS	•	•	•	•	•	•	•	•	•	•	•	•	•	•	•		•	
ARGUS	•	•	•	•	•	•	•	•			•		•		•	•		
ASAS	•	•	•	•	•	•	•	•	•		•	•	•	•		•	•	
ASKA	•	•	•	•	•	•	•	•	•		•				•	•	•	
BEASY																		BEM
BEFE	•		•		•			•					•					BEM
BERSAFE	•	•	•	•	•	•	•	•	•			•				•	•	
BETSY																		BEM
BISON																		BEM
CASTOR	•	•	•	•	•	•	•	•	•					•				BEM
DIAL	•	•	•	•	•	•	•					•		•	•	•		
EFESYS			•			•			•								•	
FENRIS	•	•	•	•	•	•	•	•				•			•	•		
FINEL/ICB	•	•	•	•	•	•	•	•	•					•				
KYOKAI																		BEM
LARSTRAN	•	•	•	•	•	•	•	•						•		•		
MAGNA	•	•	•	•	•	•	•			•	•	•		•				
MARC	•	•	•	•	•	•	•	•	•		•		•	•	•	•		
MSC/NASTRAN	•	•	•	•	•	•	•	•	•	•	•	•	•	•	•	•	•	
NEPSAP	•	•	•	•	•	•	•	•			•		•	•	•	•		
NISA	•	•	•	•	•	•	•	•	•				•		•			
PAFEC	•	•	•	•	•	•	•	•				•	•		•		•	
PAM	•	•	•		•	•	•	•	•	•			•	•			•	
PLANS	•	•	•	•	•	•	•	•			•				•	•	•	
SAMCEF	•	•	•	•	•	•	•	•	•	•	•		•		•	•	•	
SAP7	•	•	•	•	•	•	•	•					•	•	•		•	
SESAM80	•	•	•	•	•	•	•	•										
SPAR	•	•	•	•		•	•		•	•			•					
STRUDL	•	•	•	•	•	•	•	•					•					
TITUS	•	•	•	•	•	•	•	•	•	•		•					•	
WECAN	•	•	•	•	•	•	•	•	•			•	•	•	•	•	•	

Survey of General Purpose Computer Programs

TABLE 4

④ MATERIAL LIBRARY NOTABLE ITEMS

	Isotropic	Anisotropic	Temperature dependent	Linear elastic	Nonlinear elast.	Elastic-plastic	Viscoelast., viscoplastic	Creep	Incompress	Geomech	Substructuring	Restart	Cyclic symmetry	User's subrout.	Multiple coordin. syst.	Bandwidth optim.
ABAQUS	•	•	•	•	•	•	•	•			•			•	•	•
ADINA	•	•	•	•	•	•	•	•	•	•	•	•	•		•	
ANSYS	•	•	•	•	•	•	•	•		•	•	•			•	•
ARGUS	•	•	•	•		•	•	•	•		•	•			•	•
ASAS	•	•	•	•	•	•	•	•			•	•			•	•
ASKA	•	•	•	•	•	•		•		•	•	•	•	•		
BEASY	•			•		•			•		•	•				
BEFE	•			•		•				•						
BERSAFE	•	•	•	•	•	•		•		•	•	•			•	•
BETSY	•			•							•					
BISON	•			•							•					
CASTOR	•	•	•	•		•	•				•	•	•		•	•
DIAL	•	•	•	•	•	•		•	•		•	•			•	•
EFESYS	•			•	•	•					•	•		•		
FENRIS	•			•	•	•				•	•	•		•		
FINEL/ICB	•	•		•		•		•			•	•		•		•
KYOKAI	•			•							•					
LARSTRAN	•	•	•	•	•	•	•	•			•	•			•	
MAGNA	•	•	•	•	•	•					•			•	•	
MARC	•	•	•	•	•	•	•	•	•	•	•			•	•	
MSC/NASTRAN	•	•	•	•	•	•					•	•	•	•	•	•
NEPSAP	•	•	•	•		•	•	•	•		•	•			•	•
NISA	•	•	•	•	•	•			•		•	•	•		•	•
PAFEC	•	•		•	•	•		•			•	•	•	•	•	
PAM	•	•	•	•	•	•	•	•	•	•	•	•				•
PLANS	•	•		•	•	•		•			•	•				•
SAMCEF	•	•	•	•	•	•		•			•	•				
SAP7	•	•		•	•	•		•	•		•	•		•		
SESAM 80	•	•		•		•					•	•				•
SPAR	•	•		•							•	•			•	•
STRUDL	•	•	•	•							•	•	•		•	•
TITUS	•	•	•	•	•	•		•	•		•	•		•		•
WECAN	•	•	•	•		•		•			•	•		•		

293

J. Mackerle

TABLE 5

⑤ BOUNDARY CONDITIONS LOADING	Presc. displacem.	Elastic foundat.	Multipoint constr	Contact probl.	Skewed	Concentrated	Line loads	Surface loads	Volume loads	Gravity	Initial stress/strain	Thermal	Centrifugal	Deformat. dependent	Time dependent	Load generation
ABAQUS	•	•	•	•	•	•	•	•	•	•	•	•	•	•	•	•
ADINA	•	•	•	•		•	•	•	•			•		•	•	
ANSYS	•	•	•	•		•	•	•	•	•	•	•	•		•	•
ARGUS	•	•	•	•	•	•	•	•	•	•		•		•	•	
ASAS	•	•	•	•	•	•	•	•	•	•	•	•	•	•	•	
ASKA	•	•	•	•	•	•	•	•	•	•	•	•		•	•	
BEASY	•	•				•	•	•	•			•	•		•	•
BEFE	•			•		•	•	•				•				•
BERSAFE	•	•	•			•	•	•	•		•	•	•	•	•	
BETSY	•					•						•				
BISON	•					•	•	•				•				•
CASTOR	•	•		•		•	•	•		•		•			•	•
DIAL	•	•	•	•	•	•	•	•	•	•	•	•	•		•	•
EFESYS	•	•		•		•		•	•			•			•	
FENRIS	•	•		•		•	•	•	•		•	•		•	•	•
FINEL/ICB	•		•			•	•	•			•	•			•	•
KYOKAI	•					•	•	•							•	
LARSTRAN	•			•		•	•	•	•	•	•	•		•	•	•
MAGNA	•	•	•	•	•	•	•	•	•				•	•	•	•
MARC	•	•	•	•		•	•	•	•	•	•	•	•	•	•	•
MSC/NASTRAN	•	•	•	•	•	•	•	•	•	•	•	•	•	•	•	•
NEPSAP	•	•	•	•	•	•	•	•	•	•	•	•	•	•	•	•
NISA	•	•	•	•	•	•	•	•	•			•	•	•	•	•
PAFEC	•	•	•	•	•	•	•	•	•	•	•	•	•	•	•	•
PAM	•	•		•		•	•	•	•	•	•	•	•	•	•	•
PLANS	•	•	•	•		•	•	•			•	•	•	•	•	
SAMCEF	•	•	•	•	•	•	•	•	•	•	•	•	•	•	•	•
SAP7	•	•	•	•		•	•	•	•			•			•	•
SESAM 80	•	•	•	•		•	•	•	•	•	•	•	•		•	•
SPAR	•					•		•			•	•	•		•	•
STRUDL	•	•	•			•	•	•		•	•	•			•	•
TITUS	•	•		•		•	•	•			•	•		•	•	•
WECAN	•	•	•	•		•	•	•	•	•	•	•			•	•

Survey of General Purpose Computer Programs

TABLE 6

(6)

HARDWARE	CDC Star	Cray	CDC	IBM	Univac	Amdahl	Burroughs	Cyber	Fujitsu	Data General	HP	Harris	Honeywell	ICL	Siemens	Prime	VAX
ABAQUS			•	•	•			•								•	•
ADINA		•	•	•	•		•	•								•	•
ANSYS		•	•	•	•	•	•	•	•	•		•	•			•	•
ARGUS			•		•												
ASAS			•	•	•							•	•			•	
ASKA		•	•	•	•	•	•	•					•	•		•	•
BEASY		•	•	•	•											•	•
BEFE																	•
BERSAFE			•	•	•	•	•							•		•	•
BETSY			•	•											•	•	
BISON		•	•	•	•		•				•		•	•		•	•
CASTOR			•														•
DIAL					•			•									•
EFESYS		•														•	
FENRIS				•				•								•	•
FINEL/ICB																	
KYOKAI				•												•	•
LARSTRAN			•	•	•												
MAGNA		•						•									•
MARC		•	•	•	•			•	•	•						•	•
MSC/NASTRAN	•	•	•	•	•	•		•	•				•		•		•
NEPSAP			•	•	•	•											
NISA		•	•	•	•	•					•	•	•			•	•
PAFEC		•	•	•	•		•			•			•	•		•	•
PAM			•	•													
PLANS	•		•	•	•		•										•
SAMCEF		•	•	•	•										•		•
SAP7			•	•							•					•	•
SESAM 80					•		•									•	•
SPAR		•			•		•									•	•
STRUDL		•	•	•											•		•
TITUS		•			•		•										•
WECAN		•									•						

ACKNOWLEDGEMENT

The information presented is based on different questionnaires sent to the developers of each program. Description of some programs is based only on the literature available to the author. The author thanks all companies which responded to the questionnaire. He would appreciate to receive any comments/suggestions on further improvements on "organization" for the program descriptions and their tabular form presentation.

REFERENCES

1. Mackerle, J. (1983), Review of pre- and postprocessor programs included in the major commercial general purpose finite element packages. *Adv. Eng. Software*, 5 (1), 43-53.

2. Mackerle, J. (1983), Review of general purpose pre- and postprocessor programs for the finite element applications, *Adv. Eng. Software*, 5 (3), 148-159.

3. Noor, A. K. (1981), Survey of computer programs for solution of nonlinear structural and solid mechanics problems. *Computers & Structures*, 13, 425-465.

4. Mackerle, J. and T. Andersson (1984), Boundary element software in engineering, *Adv. Eng. Software*, 6 (2), 66-102.

APPENDIX I

Some of the sources of reviews on available structural mechanics computer software are presented.

* <u>Structural Mechanics Computer Programs. Surveys, Assessments and Availability.</u>

Edited by W. D. Pilkey, K. Saczalski and H. G. Schaeffer.

The book contains reviews of structural mechanics computer programs in civil engineering, mechanical engineering, nuclear engineering, applied mechanics, aerospace engineering and marine engineering. Programs are presented by summary of their capabilities, details of availability and source for more detailed information is given. Included chapters also deal with subjects as programs for building analysis and design, bridge design, structural members, stability, nonlinear analysis, structural optimization, fracture mechanics, piping systems, plastic analysis, shock wave propagation, transient analysis, etc.

The second part of the book contains papers discussing pre- and postprocessors, computer graphics and future trends in software and developments.

Availability: University Press of Virginia, Box 3608, University Station, Charlottesville, Virginia 22903, USA.

* <u>Structural Mechanics Software Series.</u>

Edited by N. Perronne and W. Pilkey.

The objective of this series of books is to provide access for the technical community to structural analysis and design computer programs. The first five volumes are available. Each volume contains sufficient documentation of several of the programs available on nationwide commercial computer networks that can be accessed by remote terminal devices connected via phone lines. Another role of this series is to inform the reader in form of reviews what is

Survey of General Purpose Computer Programs

new in software of different branches of engineering.

Vol 1- Presented programs: BOSOR4, GIFTS, preprocessor for SAP, TOTAL, BEAM BEAMSTRESS, SHAFT. Review articles: Computer-aided building design, bridge design, floor analysis, crash simulation.

Vol 2- Presented programs: SAP V, UCIN, WHAMS, DISK, TWIST, GRILL, TABS77. Review articles: FE-programs for pressure vessels, European finite element programs, cantilever retaining wall design, bridge rating systems.

Vol 3- SAP6, BOSOR5, GIFTS/SAP interface. Review articles: Nonlinear analysis programs, programs for multipoint boundary value problems, rotor dynamics programs, fracture mechanics programs and mixed methods analysis.

Vol 4- Book is divided into three sections: Sources of information and programs, Review and summaries of available programs, and Reviews of computational mechanics technology.

Availability: University Press of Virginia, Box 3608, University Station, Charlottesville, Virginia 22903, USA.

* Structural Mechanics Program Catalogues.

   Compiled by B. Fredriksson and J. Mackerle.

   The following publications are available: Structural Mechanics Finite Element Computer Programs (last is the 4th Edition, 1983)- Compact summary about 800 programs developed in USA, Japan and Europe, presented in table and descriptive form. About 3800 references is included. The book is arranged in three sections. The first part gives in a graphical form information about program type, range of application, area of application and type of computers on which programs operate. The second part gives detailed information about respective programs, including short abstract and address of program developer. The third part contains literature references which describe program documentation and application experiences.

   Other publications: Structural Mechanics Pre- and Postprocessor Programs, Finite Element Review, Stress Analysis Programs for Fracture Mechanics, Finite Element Review (A Bibliography).

   Availability: AEC, Advanced Engineering Co., Box 3044, S-580 03 Linköping, Sweden.

* Finite Element Systems, A Handbook.

   Editor: C. A. Brebbia.

   Description of different programs is given, including programs based on techniques other than finite elements. Description of some pre- and postprocessor systems is included. The book consists of a series of papers and some tables to present the capabilities of each program included in the book.

   Availability: Computational Mechanics Centre, Ashurst Lodge, Ashurst, Southampton, SO4 2AA, England.

* Engineering Software.

   Editor: R. A. Adey.

   These books contain an edited version of the papers presented at the

International Conferences and Exhibitions on Engineering Software, which are held at Southampton University.

Each book is divided into sections covering major application areas and some sections on general techniques. To give some examples: Finite element systems, Structures and stress analysis, Fluid mechanics and water resources, Computer aided design, Civil engineering, Mechanical engineering, etc.

Availability: Computational Mechanics Centre, Ashurst Lodge, Ashurst, Southampton, SO4 2AA, England.

* Structural Analysis Systems.

Editor: A. Niku-Lari.

Software, hardware, capability, compatibility, applications. The aim of the book series is to further understanding between research and industry. Each chapter gives detailed information about one package, its capability, its limitations and several practical examples from industry with computer and user cost. Published by Pergamon Press Ltd., Headington Hill Hall, Oxford OX3 OBW, UK.

CASE STUDY INDEX

The following case study index only includes the industrial examples described by the authors in their papers. Most of the programs have been used worldwide to solve a broader range of industrial problems. However, the results of such investigations are not always readily available for publication as they remain the property of the users. The present non-exhaustive study index should therefore not be considered by the reader as an indication of a program's capability, but only as a subject index.

Actuator arm, FIESTA
Aculator carriage, ANSYS
Air systems, CASTOR
Aircraft wheel leg, SAMKE
Arch, ELASTODYNAMICS (2D)
  bridge arch, AFAG, RECAFAG
  concrete, ASE
  dam, FIESTA, MODULEF, ZERO-4
Automobile body structure, ALSA

Bar, SURFOPT
Beam, S AND CM PACKAGE
Birfurcated duct, FIESTA
Bolt, ANSYS, SURFOPT, TITUS
Bracket, FIESTA
Branched structure, BOSOR 4
Bridge arch, AFAG, RCAFAG
Building, FLASH

Casks, FEMFAM
Chip/chip carrier, FEMPAC
Chopper, FEMFAM
Church, FIESTA
Coke oven, INFESA
Compressor casing, CASTOR
Concrete
  arch dam, ASE
  plate, KYOKAI
  slab, ADINA, ASE
  wall, INFESA

Connecting
  flange, CASTOR
  joint, MEF/MOSAIC
  rod, BEASY
Containment vessel, PANDA
Continuous beam, S AND CM
Cooling
  hole, FIESTA
  tower, LASSAQ
  water, AXISYMMETRIC PACKAGE
Crane, FLASH
Crank arm, CASTOR
Crankshaft, BEASY
Cryogenic cooler, BOSOR 4
Cylinders, CASTEM
  ring-stiffened, BOSOR 5
Cylindrical
  panel, PANDA
  shell, BOSOR 5, LASSAQ

Dam, FIESTA
  arch, FIESTA, MODULEF, ZERO-4
    concrete, ASE
  foundation, FIESTA
Domes, AXISYMMETRIC PACKAGE, STDYNL, THERMAL PACKAGE

Earthquake, PAID, TITUS
Electric engine, MODULEF
Ellipsoidal tank, BOSOR 4
Excavator, FEMPAC

Case Study Index

Fibre reinforced plastics, LASSAQ
Floating frame, FENRIS
Floor panel, ALSA
Fluid structure, ADINA,
    AXISYMMETRIC PACKAGE, MODULEF,
    TITUS
Flywheel, ROBOT
Food processor, PDA/PATRAN
Foundation, OSTIN
Fracture mechanics, CASTEM, TITUS
Frame, AFAG, DEFOR, FENRIS, RCAFAG,
    S AND CM, STDYNL, THERMAL PACKAGE

Gear, CASTOR, UNIC GEAR
  case, FIESTA

Heat
  exchanger, ROBOT
  generator, CASTEM
Hexagonal bundle,, KYOKAI
Housing, AIT, FEMFAM
Human femur bone, FIESTA, MODULEF
Hydraulic engine, STRUGEN

Imperfect cylinder, CASTEM

Jacket, FENRIS

Landing gear, FIESTA

Mast antenna, REST
Mining excavator, NE-XX
Missile impact, CASTEM
Mixing drum, HYBRID
Motorway bridge, MICRO-STRESS, NE-XX

Notched parts, CASTOR
Nozzle, BEWAVE, CASTEM, FIESTA,
    PDA/PATRAN, TITUS
Nuclear reactor, PANDA
  housing, ZERO-4

Offshore, AQUADYN, STDYNL
  flexible arch, FLEXAN
Outlet nozzle, CASTEM

Parabolic dome, STDYNL
Pipe, ADINA, PAID
  impact, CASTEM
Pipework system, THERMAL PACKAGE
Piston, FIESTA, PDA/PATRAN
Plane frame, DEFOR, S AND CM
Plate, BEWAVE, FEMFAM, HYBRID, SAMKE
  with circular hole, NE-XX
  with variable section, ESA
Portal frame, AFAG, RCAFAG
Pressure vessel, CASTOR, MEF/MOSAIC
  head, BOSOR 5
Pylon, MICRO-STRESS

Railway wagon, FEMFAM
Reactor vessel, RAPS
  shroud, BOSOR 4

Rigid-jointed frame, S AND CM
Ring stiffener, AXISYMMETRIC PACKAGE
Ring-stiffened cylinder, BOSOR 5
Rockets, BOSOR 4, BOSOR 5, PDA/PATRAN
Rotor, MODULEF
  disk, MEF/MOSAIC

Satellite structure, SIMP
Shearer arm, PAFEC
Shell structure, CASTEM
Shock absorber, MEF/MOSAIC
Silo construction, FEMPAC
Skew grid, S AND CM
Soil structure, OSTIN
Solar arrays, SIMP
  cell, SIMP
Space frame, DEFOR, THERMAL PACKAGE
Spherical dome, STDYNL
Spray nozzle, TITUS
Stator, MODULEF
Statue of Liberty, CASTOR
Steam generator, BEWAVE, TITUS
Steel structure, ESA
Stiffened torus, CASTEM
Structural steel work, STAR 2
Submarine finder, CASTEM
Support rail, ALSA
Suspended bridge, TITUS
Syphon tank, ROBOT

Tank, ESA
Tapered disc, AXISYMMETRIC PACKAGE
Tennis racket, DAPST
Tension tower, DEFOR
Thin-walled cylinder, AXISYMMETRIC
    PACKAGE
Three dimensional frame, IBA
Torus, CASTEM
Trolley reinforcement, SIMP
Truss, S AND CM
Tube, MSRC-RB
Tubular joint, SESAM-80
Tunnel, ADINA,
    ELASTODYNAMICS (2D), OSTIN
Turbines, ANSYS, CASTOR, FLASH,
    PDA/PATRAN
Turbo alternator, FIESTA
Turbo-jet, TITUS

Valves, CASTOR, FEMFAM, FIESTA,
    PDA/PATRAN

Wall, ESA
Water
  injection platform, SESAM-80
  pipe, ADINA
Watertank, BOSOR 5, CASTOR

X-braced frame, FENRIS
X-ray tube, ANSYS

RAYMOND H. FOGLER LIBRARY

DATE DUE

BOOKS ARE SUBJECT TO
RECALL AFTER TWO WEEKS

JUL 2 7 1987